Springer Series in Design and Innovation

Volume 2

Springer Series in Design and Innovation (SSDI) publishes books on innovation and the latest developments in the fields of Product Design, Interior Design and Communication Design, with particular emphasis on technological and formal innovation, and on the application of digital technologies and new materials. The series explores all aspects of design, e.g. Human-Centered Design/User Experience, Service Design, and Design Thinking, which provide transversal and innovative approaches oriented on the involvement of people throughout the design development process. In addition, it covers emerging areas of research that may represent essential opportunities for economic and social development.

In fields ranging from the humanities to engineering and architecture, design is increasingly being recognized as a key means of bringing ideas to the market by transforming them into user-friendly and appealing products or services. Moreover, it provides a variety of methodologies, tools and techniques that can be used at different stages of the innovation process to enhance the value of new products and services.

The series' scope includes monographs, professional books, advanced textbooks, selected contributions from specialized conferences and workshops, and outstanding Ph.D. theses.

Keywords: Product and System Innovation; Product design; Interior design; Communication Design; Human-Centered Design/User Experience; Service Design; Design Thinking; Digital Innovation; Innovation of Materials.

How to submit proposals

Proposals must include: title, keywords, presentation (max 10,000 characters), table of contents, chapter abstracts, editors'/authors' CV.

In case of proceedings, chairmen/editors are requested to submit the link to conference website (incl. relevant information such as committee members, topics, key dates, keynote speakers, information about the reviewing process, etc.), and approx. number of papers.

Proposals must be sent to: series editor Prof. Francesca Tosi (francesca.tosi@unifi.it) and/or publishing editor Mr. Pierpaolo Riva (pierpaolo.riva@springer.com).

More information about this series at http://www.springer.com/series/16270

Francesca Tosi

Design for Ergonomics

 Springer

Francesca Tosi
Department of Architecture (DIDA)
University of Florence
Florence, Italy

ISSN 2661-8184 ISSN 2661-8192 (electronic)
Springer Series in Design and Innovation
ISBN 978-3-030-33564-9 ISBN 978-3-030-33562-5 (eBook)
https://doi.org/10.1007/978-3-030-33562-5

This Springer imprint is published by the registered company Springer Nature Switzerland AG
The registered company address is: Gewerbestrasse 11, 6330 Cham, Switzerland

Foreword: The Growth
of the Ergonomics-Design Relationship

More than ten years ago—fourteen to be precise—my friend, Francesca Tosi, asked me to write the introduction to the book she was editing for PoliDesign: Ergonomia & Design. I gladly accepted.[1]

A few days ago, Francesca asked me the same thing for this new book: *Design for Ergonomics*. Flattered, I once again accepted. Almost immediately, however, I asked myself, somewhat concerned, what I could possibly write about a very similar, if not almost identical, subject. One thing, banal though it may be, calmed me. Fourteen years is a very long time when you are young, but it is longer again if the events and shifts that occur happen very quickly, as has happened and still happens.

In 2004, just as Francesca's book was released, Facebook (what a happy coincidence!) was founded; it has, in the meantime, become a juggernaut, even if it is a source of much debate these days. Two new quasi-monopolies, Amazon and Google, have also taken shape, while Apple, another Bay Area giant, has taken on new life.

Indeed, we have been talking about the end of human labour for some time. It is the prediction of many. Jerry Kaplan,[2] for example, wrote a successful book called Humans Need Not Apply, in 2015, which was translated into Italian by LUISS University Press the following year with the vivid title, Le persone non servono. If true, Ergonomics would be made redundant by the extinction of its subject of study, while the condition of Design would not be much better.

Happily, Ergonomics and Design are not redundant: they have changed, however, both individually and in their relationships.

Therefore, I had—or rather I have—material for a short presentation!

Let's quickly see what has changed.

Ergonomics has changed its central focus: at the turn of the century, this was still the man–machine relationship, seen as a connection between two physical bodies that must be compatible with their setting, a place where both operate and

---

[1]Cfr. Bagnara (2004).

[2]Cfr. Kaplan (2015).

manufacture: the workplace. Today, it does not focus on correcting Design and planning errors; it is no longer the Ergonomics of correction. The Design of machines, because it also involves Ergonomics, is now ergonomic. The workplace, too, is no longer fixed; it is often mobile or, as they say, smart, though still varied. It is only in truly exceptional cases that work and the workplace are for life, as was typical last century.

Today, almost all machines, from washing machines to cars, incorporate a smart component, and their relationship with people is substantially communicative and, therefore, cognitive. Now the prevailing focus of Ergonomics is on the interaction between man and the smart element of the machine, that is, the IT component that controls it and communicates its status and progress—its interface—rather than the relationship.

Ergonomics has moved from the analysis of simple manual human actions, which are often very infrequent, required by tasks (task) to the planning of these activities, which requires many different actions aimed at one or more objectives. Thus, Design has become increasingly interested in the planning of processes and services, rather than individual subjects.

Machines, then, not only carry out operations/actions that were once performed by people, perhaps using their own force, but are becoming increasingly independent, not only in the force they employ but in the way they function, so much so that they are capable of self-diagnosis. They do not just proceed automatically, but also suggest possible options, or communicate, on occasion, the choices and decisions they have already made independently.

One of the basic principles of Ergonomics (the human control over objects and machines and their usability) has been lost. The machine no longer communicates its state to the person so that it can be quickly understood; instead, it is the person who, often without knowing it, communicates his state to the machine. The classic Ergonomic problem of making the machine and its functioning transparent, clear and easily understood has been flipped on its head: it is necessary for the person to understand what the machine knows about him, what information it is acquiring and for what purpose. This is a common objective of Design and Ergonomics. The latter has changed[3] from a study of human labour (Ergonomics), to a study of the Human Factor in labour systems, to a study of the person as an active player in systems, changes and innovation (Human Actor), to a study and plan of the digital environment (Digital Ecology) in which we increasingly find ourselves.

The conflict between ergonomists and designers—which was still in effect at the turn of the century—could be accurately described as a clash between the status quo and creativity, between freedom in shaping ideas and anchoring ideas to the limits of what the human body is capable of, and has failed to dissipate in a substantial way.

Both have widened their focus, first from the body to the mind and then from the manual to the cognitive, and are now tackle the individual, the diversity between the person and their social setting, their community, their ethics and their values.

---

[3]Cfr. Pozzi and Bagnara (2016, pp. 59–65).

Ergonomics and Design tackle and confront, in an interdisciplinary manner, the important questions that Neville Moray identified in his great article from 2000: Culture, Politics and Ergonomics.[4]

Moray saw and indicated different priorities for study and intervention in Ergonomics—though I believe they also apply to Design—such as the environmental challenge, health and safety, poverty, the differences in gender and mental and physical ability, social promotion and, finally, happiness, which are still topical and unresolved, as shown in his last article.[5] In truth, none of these global issues have been conclusively overcome: some have been at least partially dealt with, while others have just begun.

For example, as it relates to the environment, Design and Ergonomics now work together in planning the entire life cycle of a product, up to the planning of how it will be disposed of and/or recycled. Much has been done for safety, but perhaps not enough as of yet, if there are still accidental deaths in the workplace and on the streets, as well as, ironically, in hospitals. And little is known about safety in the new digital environment, where privacy can be invaded with ease.

Things are better in terms of health, because the correct functioning of machines requires environmental conditions that are compatible with people's physical well-being. Mental health in the digital environment, however, is still unfortunately a black hole, as seen by the symptoms of net addiction and social isolation. Gender differences are slightly improved, even if these are still present and there is much to do.

Significant advances have been made in dealing with differences in physical and cognitive abilities. We have discovered that, for in many ways and at various moments in our lives (with age, for example), we are all differently abled, not in the proper sense but as a description of a situation that is often characterised by suffering. In this case, a new, interdisciplinary approach has been developed by designers and ergonomists, one that allows us to tackle problems in an appropriate manner, without seeing them as issues removed from "normality": Design for All, which will be discussed in detail in a chapter of this book. This is perhaps the greatest achievement of Ergonomics and Design in these years.

Work is being done, though still not enough, to tackle the major issue of poverty. There are interesting developments in Latin America, India and Africa, both in terms of Design and Ergonomics. But this is still too little! That is because today this also concerns countries who considered themselves to be free from this plight.

The goal of social promotion has been lost over long years of unreliable work and the dramatic increase in inequalities.[6] As far as happiness is concerned, perhaps Neville Moray, a visionary in his day, asked too much. Emotions have been discovered in both Ergonomics and Design.[7] Happiness is certainly an objective, but it

---

[4]Cfr. Moray (2000, pp. 858–868).

[5]Neville Moray died December of last year. His last collaborative work was Thatcher (2018, pp. 197–213).

[6]Cfr. Franzini (2013).

[7]Cfr. Norman (2004).

is still limited to brief periods and is usually linked to relationships with people and less so to relationships with objects and machines. Happiness remains, in a long-term state, as a powerful utopia. Let us content ourselves with the state of well-being that Ergonomics and Design doggedly pursue.

In summary, Ergonomics and Design now understand how to work together, but they still have not grasped thinking big. They know how to plan to make complicated—perhaps unnecessarily so—processes and objects simple. But they find it difficult to tackle that complexity,[8] even when it is in large infrastructures, such as cities, public administrations and networks, that are invisible[9] but altogether real (to the extent that, when they do not function, we clash and become angry) and govern our day-to-day lives, such as water, rail and air networks.

However, they have started, for some time, to be more truthful—Design more than Ergonomics—in tackling issues such as social innovation,[10] collective living and day-to-day matters. More importantly, they have realised that the designer must learn from those who will be living with the innovations they have implemented. The Human-Centred Design approach, in which Ergonomics plays a significant role, as well as the Participatory Design model, which constitutes a central axis of this book, represents this new dimension of Design and Ergonomics.

The designer and the ergonomist cannot design in isolation; they must collaborate. They also require the contribution of other sciences and professions and, above all, flesh and blood people. People with different individual and collective desires and needs, who will experience the designs once complete.

This book provides a complete overview of the current state of Design and Ergonomics and their position with respect to the objectives and priorities outlined by Neville Moray.

The message of the text can thus be summarised in an invitation: don't be afraid to think big. To dare confronting the real problems, which are global and political, without forgetting the small issues we face every day, which are often rooted in, and inspired by, the bigger questions. To spot, solve and plan for the micro-problems, aware that they are always a part of a macro-issue. Macro-issues are not solved once and for all in a single project, but should always be considered when dealing with a problem that seems outwardly banal.

There are no banal problems in the complex and hyper-connected society we live in. No man is an island in the "liquid" society.

Florence, Italy   Sebastiano Bagnara

---

[8]Cfr.Norman (2010).

[9]Cfr. Pozzi and Bagnara (2015, p. 5).

[10]Cfr. Manzini (2018).

References

Bagnara S (2004) Il rapporto fra ergonomia e design. In: Tosi F (ed) (2004) Ergonomia & Design, Polidesign, Milano

Franzini M (2013) Disuguaglianze inaccettabili. Laterza, Bari

Kaplan J (2016) Le persone non servono, LUISS University Press, Roma (1st ed.: Kaplan J (2015) Humans need not apply: a guide to wealth and work in the age of artificial intelligence. Yale University Press, New Haven)

Manzini E (2018) Politiche del quotidiano. Edizioni Comunità, Milano

Moray N (2000) Culture, politics and ergonomics. Ergonomics 43(7):858–868

Norman DA (2004) Emotional design. Basic Books, Cambridge

Norman DA (2010) Living with complexity. MIT Press, Cambridge

Pozzi S, Bagnara S (2015) Designing the future cities: trends and issues from the interaction design perspective. City Territory Archit 2(1):5

Pozzi S Bagnara (2016) The third wave of human computer interaction: from interfaces to digital ecologies. DigitCult 1(1):59–65

Thatcher A (2018) State of science: ergonomics and global issues. Ergonomics 61(2):197–213

Preface

This book, like many books stemming from the field of university research, represents the result of research and experimentation that has been carried out over a significant number of years in the field of Ergonomics and Design.

This book, which is being published today as *Design for Ergonomics*, was conceived, in its initial stages, as an updated edition of "Ergonomics and projects",[11] which was published by Franco Angeli in 2006.

The new edition had been requested by the publisher many times; after more than ten years, in fact, an update and new graphics were appropriate. Many books about this subject, which were published from the mid-90s in Italy but, above all, in other countries, particularly Anglo-Saxon ones, have been republished many times with revisions and updates. Releasing a new edition, therefore, is not an out-of-the-ordinary endeavour.

The work of updating, though, turned out to be tiring from the off. I would go so far as to say dull and increasingly difficult.

In fact, over the course of a few years, the frame of reference has changed drastically. The quantity and quality of the research and field experimentation carried out in the field known today as "Ergonomics and Design" has also shifted in a truly radical manner, broadening its fields of interest and progressively building an extremely diverse theoretical, methodological and applicative framework.

---

[11]"Ergonomics and projects" was the first of three volumes that summarised the research and experimentation carried out over approximately 10 years regarding the relationship between the intervention contents, methods and philosophy of Ergonomics and User-Centred Design (now Human-Centred Design and User Experience) and the Project, in particular, Design. This was concluded with the book "Ergonomics, projects, product", which outlined the theoretical and methodological base of the UNI11377 "usability of industrial products" regulation, which was published in 2010.

The three books, in order of their publication, are: Ergonomics and projects; Ergonomics and sensory-quality designs (with Gussoni M. and Parlangeli O.); Ergonomics, projects, product.

A huge number of books have been published on this subject and brought significant scientific contributions to this field, although the non-Italian publications are configured with an engineering matrix that is notably different from that of Italian design.

As previously mentioned, a huge number of new editions have already been published and partly updated in terms of their contents, particularly as it relates to the research experiments and case studies included.

As far as I'm concerned, the research that has been carried out in recent years and the various publications produced around Ergonomics and Design make it difficult to maintain the original layout of the text, or indeed its contents.

Thus, the result is a new summary of what has been done to date.

This book deals with the themes of Ergonomics as it relates to Design—and vice versa—and thus its reading is strongly focused on the project itself, as well as being limited with respect to the varied state of ergonomic research today.

The approach is focused on the specificity of Design, as it is understood in the Italian sense. This refers to the different variations of product design (physical and virtual) for interiors, communications and fashion, as it relates to the design synthesis of humanistic, artistic, technological and social skills and know-how.

This is a particularly topical subject today, given the growing role of Ergonomics in manufacturing, particularly for products and services, and the approach, including the addition and integration, to the evaluation and design methods for Human-Centred Design (HCD), User Experience (UX) and Design Thinking.

In addition to the role of Ergonomics in Design, it now makes sense to discuss the research areas in which both approaches meld, as well as the evaluation and design methods in which the central figure of the "person" represents the project's starting point and end goal.

Design, which refers to a capacity for design synthesis, is now a key factor in the approach to Ergonomics.

Design is expected to respond to people's needs and expectations—socially and commercially—and is wholly recognised as a strategic factor in development and a harbinger of innovation—both productive and social—when it merges with the fundamental principles of HCD, providing a capability to interpret and translate the needs that arise from the community into the project itself.

Thus, one must speak not only of Ergonomics and Design but also Design for Ergonomics, that is, the opportunity for innovation that the synthesis of these two approaches offers for the culture and practical side of design, as well as the strategic role Design plays in synthesising the skills and approaches of various professionals (and professions) involved in the design and production processes in the project.

The innovative contribution this book has in mind is a reading of Ergonomics/ Human-Centred Design (E/HCD) that specifically focuses on the project and, in particular, Design.

E/HCD, in fact, represents a vital strategic factor for the innovation of the product/system and the innovation of the process, as well as a core component of both design culture and its practice.

This assumes a particular and specific value for Design, particularly Italian Design, which has always been focused on innovation—both technological and formal—and closely linked to the reality of industrial production.

Today, it is Design that offers radically new tools and areas for experimentation in Ergonomics when compared to the past, a situation in which the contents and methodological tools #of E/HCD become concrete factors for innovation and Design offers an overview and the capacity for design synthesis—and, of course, intervention solutions—that can respond to people's needs, expectations and desires.

This book is composed partially of reworked and updated texts from the first edition and partially of new texts. These are both mine and those of my colleagues, and deal with some of the most topical elements of the relationship between Ergonomics and Design. The second part contains factual elements regarding both physical and cognitive Ergonomics. Some of the chief "Design issues and intervention criteria" and some case studies about design intervention are handled at the end of the book.

Though I am, of course, responsible for what is written here, and for any inaccuracies or errors that may be contained in this book, I have a number of people to thanks for making this work possible in a variety of ways.

Firstly, thanks to Adriana Baglioni, who, only a few years ago, supported and oversaw the development of the first edition of this book.

A heartfelt thanks to Luigi Bandini Buti, Silvia Piardi and Riccardo Tartaglia for their help and advice on the first edition, and to my colleagues at the Milan Polytechnic, who were with me when I started this research: Laura Anselmi, Fiammetta Costa, Sabrina Muschiato, Michele Ottomanelli, Giuseppe Andreoni and Lina Bonapace, among many others with whom these topics have been shared in various ways.

Thank you as well to my colleagues at the Laboratory for Ergonomics and Design at the University of Florence, for our great collaboration in recent years: Alessandra Rinaldi, Alessia Brischetto, Irene Bruni, Daniele Busciantella, Mattia Pistolesi, Ester Iacono and Claudia Becchimanzi.

Finally, thank you to the many people who I have listened to at meetings, round tables, conferences and informal discussions, who are too numerous to name.

Florence, Italy Francesca Tosi

Introduction: Ergonomics and Design—Design for Ergonomics

Ergonomics—with its components of Human Factors, Human-Centred Design, User Experience—is the term used to define all the interdisciplinary knowledge of Human Factors (from psychology, medicine, social sciences, engineering and design) as well as the set of methods and procedures aimed at evaluating the needs and expectations of people in their interaction with products, environments, services and systems in general, which people interact with in a work setting or as part of their daily routine.

The term Ergonomics is currently used as a term synonymous with Human Factors, although in the USA and in Anglo-Saxon countries, the latter term is more widespread. To clarify, "Ergonomics and Human Factors are now terms, synonymous and accepted throughout the world, that describe the theory and practice of studying human characteristics and abilities, and then using that knowledge to improve people's interactions with the things they use and the environments in which they do it" (Wilson and Sharples 2015, p. 5).

The objective of Ergonomics is therefore the evaluation and the design of the interaction between people and the systems which people interact with during their work and daily life activities.

The methodological and operational tools made available by ergonomic research, and its philosophy centred on people's needs and expectations and the evaluation of Human Factors, represent an essential contribution to research and design practice and a concrete factor in the innovation of product and system development.

The peculiarity of the ergonomic approach, as well as its innovative value, resides, first of all, in the ability to synthesise knowledge from disciplinary sectors which are traditionally distant and to then use methods of collecting and assessing people's needs and expectations, with the aim of creating environments, products and services (physical or virtual) that can guarantee conditions of safety, usability and well-being to people when they engage in work and daily life activities.

This approach posits the evaluation and interpretation of the multiplicity of variables that define the interaction between people and the products/systems they use (i.e. people's characteristics and capabilities, the characteristics and objectives of the product/system and of the activities for which it is—or can be—used, the characteristics of the physical and social context) by identifying and interpreting, from time to time, the needs that users express or can express with respect to that interaction, their priorities, their mutual conditioning, their modification over time.

In other words, Ergonomics uses knowledge and methodological tools from different disciplines (in particular medicine, psychology, human sciences, polytechnic disciplines), and a philosophy of intervention that places the user and his needs at the centre of every evaluation and/or design process.

The two assumptions on which Ergonomics, by definition, bases its theoretical and methodological approach, are multidisciplinarity and centring people. The former involves the integration of knowledge and methods of evaluation of people's physical, sensory and psycho-perceptive characteristics and abilities, whereas the latter focuses on the role of any individual user as the centre of interest and specific goal of ergonomic intervention.

Based on these assumptions, the relationship between Ergonomics and Design, together with the development of industrial products, has been developing and consolidating over time.

In Ergonomics, the knowledge of human characteristics and abilities, as well as the effects on individuals of the product/system features and environmental variables, constitute the necessary scientific basis which serves as a point of reference for the development and testing of design and/or intervention choices. Similarly, understanding and assessing people's needs, attitudes and expectations is the starting point for any evaluation and/or planning process.

Today multidisciplinarity and centring people represent the conceptual and methodological basis of Ergonomics and Design, a design approach that places the human–system interaction as the starting point and the central objective of the design process.

In Ergonomics and Design, its theoretical content and its evaluation and intervention methods are based on a double perspective change: from the evaluation and design of products, environments, systems, to the evaluation and design of the interaction between people and those products, environments or systems; from the evaluation of the objective reality people interact with, to the projection in the future and the possibility of the reality of these interactions, and to the design of the most suitable solutions to respond to the possible needs and desires of the people.

In its almost 70 years of history, Ergonomics, its content and fields of intervention, has progressively expanded from the safety of workstations and their compatibility with human physical and cognitive characteristics, to the study of organisations, and to the usability of physical or virtual environments, products and services, addressing the modern user experience as a global quality of the physical, sensory, cognitive and emotional relationship experienced by people when they interact with products/systems.

The ergonomic approach has also progressively been extended from the objective evaluation of safety and compatibility conditions of environments, equipment and products, to a design action based on the knowledge and interpretation of people's needs and expectations, and of all aspects—objective and subjective—of the relationship that people establish with products and environments.

This path of progressive extension of Ergonomics content and fields of intervention is of specific interest for the relationship between Ergonomics and Design.

For a long time, Ergonomics has represented a Design tool, understood as an overview of useful design knowledge and intervention methods, which while useful are also completely external, if not extraneous, to the design culture.

The knowledge of human physical and cognitive characteristics and abilities and, since the 90s, usability evaluation methods, represented valuable operative tools for designers. Professional Ergonomists, who rarely have a design background, also collaborated in the design process as external consultants, called to provide a conclusive verification, usually required only at the end of the design process.

Since the 90s, in the Ergonomics field, and obviously in planning and the Design field, the research and professional specialisation area now called "Ergonomics and Design/Ergonomics in Design" has been growing. Within this field, theoretical content and intervention methods specifically aimed at the process of formation and development of products, environments and services have been developed.

In recent years, with the establishment of the Human-Centred Design and User Experience approach, Ergonomics has opened up to the study of all human activities—work and daily life—and of the complexity and globality of the experience—physical, sensory, cognitive and emotional—of the person who interacts and works with physical and virtual systems.

The relationship between Ergonomics and Design is therefore particularly relevant now, and for those working in the design field, the relationship between Design and Ergonomics, its evaluation and intervention methods and, in general, its approach centred on the needs, expectations and desires of people, both as individuals and as social groups, is also particularly relevant.

Design and the synergy between the intervention methods of the User/Human-Centred Design and Design Thinking are today described as one of the possible strategies for the innovation of European production systems and, in particular, for small and medium companies.

In the most recent European Union documents dedicated to the possible innovation strategies of European production and social systems, Design is described as a key activity which can bring new ideas to companies and to the market, based on the integration of the User-Centred approach and Design Thinking strategies.

> A more systematic use of design as a tool for user-centred and market-driven innovation in all sectors of the economy, complementary to R&D, would improve European competitiveness. Analyses of the contribution of design show that companies that strategically invest in design tend to be more profitable and grow faster[12].

The aim of Ergonomics for Design is to evaluate the global quality of the design of systems which people enter into a relationship with during their work activities and daily life. Quality that involves the globality of people's experience—physical,

---

[12] European Commission, *Implementing an Action Plan for Design-Driven Innovation*, EU Commission Staff Working document, Brussels 2013.

sensory, cognitive and emotional. To do that, all the variables that define the context in which that relationship is implemented, their reciprocal conditioning and their variability over time should be accurately evaluated and designed.

The content of Ergonomics for Design is in fact on two non-separable levels, constituted first by usability and user experience evaluation and interpretation methods, which can identify and evaluate people needs and expectations and secondly, by the basic interdisciplinary knowledge concerning people's characteristics and capabilities.

Its innovative value lies in the organic nature of the two cognitive levels—assessment methods and basic interdisciplinary knowledge—which constitute both the methodological and operational tools able to guide the entire product development, and an essential component of the training required for the designer operating within company structures and complex production processes.

Objectives and Volume Structure

This volume is aimed at designers and anyone working in the design field who is interested in the methods, tools and opportunities for in-depth analysis and development that Ergonomics can offer to the conception, production and testing process of products, environments and services, either physical or virtual. Obviously, it is also addressed to people studying Industrial Design or planning courses.

Its objective is to present a concise and structured picture of the contents of the ergonomic approach to planning, and of the methodological and operational tools developed by ergonomic research that can offer more directly a concrete contribution to the design process.

The point of reference of the volume is the role of Design as a strategic factor of productive and social innovation, its relationship with Ergonomics and Human-Centred Design and their theoretical and methodological contribution to the development of "design for the people".

The first part of the volume aims to provide an overview of the contents and evolution of the ergonomic approach to design, the theoretical and applied meaning of ergonomic requirements, intervention methods and tools of Ergonomics for Design. The three concluding chapters of the first part are dedicated to some of the research and experimentation areas in which the relationship between Ergonomics/HCD/UX and planning has been achieved and how it has evolved in recent years: Co-design and Innovation, Design Thinking and creativity, Design for All and planning for social inclusion.

The second part deals with the elements of physical ergonomics and cognitive ergonomics, essential components of design culture.

The third part of the volume contains the two chapters dedicated to some of the main "Design problems and intervention criteria" related to the design of environments, products and equipment, and to the design of communication, training and learning interface systems based on digital technologies.

The volume concludes with an appendix dedicated to the application of the "design path" to the contents and methods covered in the volume, and to some cases of design intervention carried out in recent years by the Laboratory of Ergonomics & Design—LED of the University of Florence.

References

European Commission (2013) implementing an action plan for design-driven innovation. EU Commission Staff Working document, Brussels

ISO 9241-210:(2010) Ergonomics of human-system interaction—Part 210: Human-centred design for interactive systems

Wilson JR, Sharples S (eds) (2015) Evaluation of human work. CRC Press, Boca Raton (1st ed.: Wilson JR, Corlett EN (1990), Evaluation of human work. Taylor & Francis, Londra e New York)

Contents

Part II The Components of Interaction Between People and Systems

Part I
Ergonomics and Design—Design for Ergonomics

Chapter 1
Ergonomics and Design

1.1 Introduction

The IEA (International Ergonomics Association) defines Ergonomics, or Human Factors: "Ergonomics (or human factors) is the scientific discipline concerned with the understanding of interactions among humans and other elements of a system, and the profession that applies theory, principles, data and methods to design in order to optimize human well-being and overall system performance.

Practitioners of ergonomics and ergonomists contribute to the design and evaluation of tasks, jobs, products, environments and systems in order to make them compatible with the needs, abilities and limitations of people".[1]

The central focus, therefore, is the **interaction**[2] that people establish or can establish with other elements of the **system** in which, and with which, they work and carry out daily activities. Interactions, therefore, take place within a complex system, in

---

[1]Cfr. IEA, International Ergonomics Association, www.iea.cc/whats/index.html (consulted in March 2018).

The Italian Society of Ergonomics and Human Factors provides a definition of Ergonomics that partially resembles that of the IEA, International Ergonomics Association: "Ergonomics is a corpus of interdisciplinary knowledge capable of analysing, designing and evaluating simple or complex systems, in which the person appears as an operator or as a user. It pursues competence and compatibility between the world around us—objects, services, living and working environments— and psycho-social and social human needs, while also aiming to improve the efficiency and reliability of systems. Its objective is to adapt the environmental, instrumental and organisational conditions in which human activities take place to the needs of the individual, defined on the basis of his physiological, psychological and socio-cultural needs, and those of the tasks he is asked to perform". Cfr. SIE-Italian Society of Ergonomics and Human Factors, www.societadiergonomia.it (consulted in March 2018).

[2]Interaction is the process by which two or more elements act with one another with consequent reciprocal modifications.

A system is defined as any object of study which, despite being made up of different elements that are mutually interconnected and interact with each other or with the external environment, reacts or evolves as a whole.

Cfr. Treccani dictionary online, www.treccani.it/vocabolario (consulted in March 2018).

© Springer Nature Switzerland AG 2020

F. Tosi, *Design for Ergonomics*, Springer Series in Design and Innovation 2,
https://doi.org/10.1007/978-3-030-33562-5_1

which each element conditions and modifies the others and which people, with their characteristics, abilities, needs and expectations, form an integral part of.

The goal of Ergonomics is to optimise, that is, improve to the highest possible degree, both the well-being of people and the overall performance of the system, through evaluation and design activities that aim to make systems and environments compatible with people's needs, abilities and limitations.

Thus, Ergonomics is based on a complex approach to the evaluation and design of the interaction between people and the systems they come into contact with, one that does not focus on the quality of the system itself but on the quality that is actually experienced by the specific group of people who have contact with it, depending on their characteristics, abilities, needs and expectations, the activities they perform and the collection of variables (physical, technological, environmental, organisational, cultural) that may affect that interaction on a case-by-case basis.[3]

The meaning of the terms used is essential for understanding the definitions.

The term **product** refers to its literal meaning of "a result of human activity" and, in the case of industrial products, the result of a design and production process that aims to respond to a specific need. Products, therefore, are the objects to be used, the environments, the services—and, more generally, the systems—whether physical or virtual, which will all be referred to in this book using the term *product*.[4]

The term needs refers to something that is "necessary for performing something"[5] and, more generally, *to the collection of needs, requirements, expectations and desires* that people express—knowingly or unknowingly—with regards to their use of, or relationship with, a product or a system.[6]

For brevity, the term *needs* will be used in the text to refer to this meaning.

---

[3]The meaning that the terms used to define the "relationship between the user and the product in a given context" assume derive from the meaning of interaction and its role as a subject of significant interest throughout every area of ergonomics.

In ergonomics, the term <u>user</u> refers to the person—or the group of people—who come into contact with the product with a specific usage context. The use, therefore, is not an indeterminate or generic figure, but a person with specific needs, whose features and skills and the activities he must and can perform with the product in question and, finally, the needs, expectations and desires that they express or can express towards their relationship with said product and all other variable in the usage context must be identified and understood.

The <u>context of use</u>, is defined by regulations as "users, tasks, equipment (hardware, software and materials), and the physical and social environments in which a product is used" (ISO 9241-210:2010 standard), user, that is the collection of conditions, limitations and possibilities within which the interaction between the user and the product take place, along with all other variables for a system in which the user and the product are integrated.

[4]The term <u>product</u> aligns, in this sense, with "artefact", in its currently consolidated meaning as resulting from human creation, that is, the creative capacity of human labour.

[5]Cfr. Treccani dictionary online, www.treccani.it/vocabolario (consulted in March 2018).

[6]See Sect. 4.2.3, "The needs of people".

1.1.1 The Ergonomics-Design Relationship

The basis for the relationship between Ergonomics and Design, and its growth over time, is the definition of Human-Centred Design. Today, this is largely comparable to the definition of Ergonomics; it represents its most recent component and is the closest to the culture and practice of design.

As we will see in later chapters, User/Human-Centred Design (HCD) describes a design approach that is aimed at the quality of the interaction between people and systems they come into contact with, one based on collecting and processing essential data so as to understand people's needs through structured and verifiable methods for study and evaluation.

According to ISO standards, "Human-centred design is an approach to interactive systems development that aims to make systems usable and useful by focusing on the users, their needs and requirements, and by applying human factors/ergonomics, and usability knowledge and techniques. This approach enhances effectiveness and efficiency, improves human well-being, user satisfaction, accessibility and sustainability; and counteracts possible adverse effects of use on human health, safety and performance".[7]

HCD, therefore, is an approach to design that uses the person as a point of departure and goal for any intervention and, at the same time, is a methodological approach to the study, evaluation and interpretation of the people's needs and expectations—both conscious and unexpressed—and their translation into the design process.

The HCD approach, together with the very broad overview of evaluation methods and techniques that have become available through research and experimentation in the field of safety, usability and user experience and the ergonomic awareness of features and human capabilities, represents an essential asset both for the culture and practice of design. Ergonomics operates in many sectors of intervention, into which its various components and skills are integrated based on common goals and working methods. To paraphrase Rubin (1994, p. 10),[8] Ergonomics and, in particular, the Human-Centred Design approach can be defined as the collection of methods and procedures that make it possible to conduct each evaluation and design intervention, starting from an awareness of people's needs and expectations, but also, above all, an intervention philosophy that places the person at the centre of the design and production process for products, environments and systems.

Ergonomics, therefore, is a **methodological approach**, that is, a research tool and professional activity in Ergonomics and, furthermore, an **intervention philosophy**, one in which the different areas and skills recognise and identify common goals for the intervention.

---

[7]Cfr. ISO 9241-210:2010 standard, Ergonomics of human-system interaction—Part 210: Human-Centred Design for interactive systems. The chief definitions contained in ISO 9241-210:2010 standard are outlined in Sect. 5.3, "Usability".

[8]Rubin writes: "UCD represents not only the techniques, processes, methods, and procedures for designing usable products and system, but just as important, the philosophy that places the user at the centre of the process" (1994, p. 10).

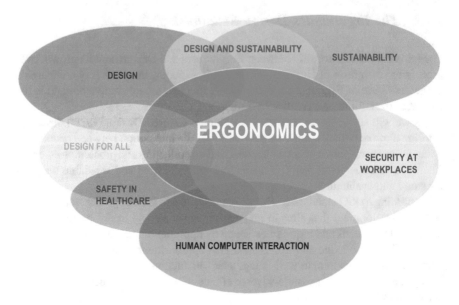

Fig. 1.1 The areas of ergonomics. Elaboration of the author

There are also many fields of application for Ergonomics and specialisations for individual researchers and/or professionals who work in its various environments.

The contents, methods and research results offered by the scientific literature of the sector, and presented at the international Ergonomics conferences,[9] offer a vast panorama: from workplace safety to Human-Computer Interaction, from clinical risk management to the User Experience, from the study of organisations to the design of products for use. The picture that emerges is one of a very vast puzzle of research areas, professional skills and specialisations—each characterised by its own scientific references—investigation and intervention methods, often from its own languages, whose contents wholly belong to both specific disciplinary areas and/or professional environments in which each ergonomic researcher or professional works, and the various specialisations of Ergonomics (see Fig. 1.1).

The IEA identifies three specialist areas—Physical ergonomics, Cognitive ergonomics, Organisational ergonomics—which are articulated and in close collaboration with the various application and intervention sectors.

---

[9]To consult the framework of international contributions in the field of ergonomics, see the main conference sites:

- IEA, International Ergonomics Association triennial Conference, www.iea.cc;
- AHFE International, Applied Human Factors and Ergonomics Conference, http://ahfe.org; Please refer to the numerous conferences of the FEES, Federation of European Ergonomics Societies, which can be consulted at the website, ergonomics-fees.eu, and the conferences of the National Ergonomics Society, which can be consulted at the websites of each individual scientific Society. For Italy: SIE Italian Society of Ergonomics and Human Factors www.societadiergonomia.it.

Physical Ergonomics

is concerned with human anatomical, anthropometric, physiological and bio- mechanical characteristics as they relate to physical activity. (Relevant topics include working postures, materials handling, repetitive movements, work related musculoskeletal disorders, workplace layout, safety and health).

Cognitive ergonomics

is concerned with mental processes, such as perception, memory, reasoning, and motor response, as they affect interactions among humans and other elements of a system. (Relevant topics include mental workload, decision-ma- king, skilled performance, human-computer interaction, human reliability, work stress and training as these may relate to human-system design.)

Organizational ergonomics

is concerned with the optimization of sociotechnical systems, including their organizational structures, policies, and processes. (Relevant topics include communication, crew resource management, work design, design of working times, teamwork, participatory design, community ergonomics, cooperative work, new work paradigms, virtual organizations, telework, and quality management.)

Fig. 1.2 Areas of ergonomic specialisation. *Source* IEA, International Ergonomics Association, www.iea.cc, 2018

As Shorrock and Williams (2017, p. 3) write: "Human Factors and Ergonomics (HF/E), as a professional activity, has now been introduced to almost all economic sectors. In the primary sector, HF/E helps to improve human involvement in mining, oil and gas extraction, agriculture, and forestry. In secondary sector, HF/E is embedded in manufacturing and construction to produce finished products. In tertiary (service) sector, hospitals and health-care organization, telecommunications, wholesale and distribution organizations, and governments all employ or contract HF/E services. In the quaternary (knowledge based) sector, HF/E practitioners are employed in information and technology, media, education, research and development (R&D), and consultancy organizations" (Figs. 1.2 and 1.3).

1.2 The Evolution of the Ergonomic Approach

The definitions[10] that were provided for Ergonomics in the near 70 years of its history reflect, on one hand, the progressive evolution of its contents and fields of

---

[10]The analysis of the definitions of Ergonomics and their evolution is extensively treated in the works of Wilson (1995), Re (1995), Attaianese (1997), Pheasant and Haslegrave (2006), Bandini Buti (2008), Shorrock and Williams (2017), Wilson and Sharples (2015).

Chief sectors of Ergonomic intervention:

- Health and safety of work stations and environments
- Musculoskeletal disorders
- Human error and reliability of systems
- Work organisation and psycho-social factors
- Usability
- User Experience
- Systems design (HMI, HCI)
- Human performance
- Clinical risk management
- Design For All
- Product design and interior design
- Communication design
- ....

Fig. 1.3 Ergonomic intervention sectors. *Source* SIE, Società Italiana di Ergonomia e fattori umani, www.societadiergonomia.it

intervention over time and, on the other hand, the more or less extended interpretation that Ergonomics has given to researchers and professionals.

As highlighted by Wilson (1995, p. 3), the differences refer solely to the boundaries of ergonomics and do not relate to fundamental disagreements regarding its fundamental principles and goals. Furthermore, the content, which is still highly innovative in both its aims and the perspective from which ergonomics analyses and designs the human-environment-product interaction, remains unchanged.

As Ivaldi (1999, p. 11) writes, Ergonomics is seen as a revolutionary approach since it was first defined as *"the adaptation of work to the person (… which) is presented as a wholly original concept in terms of the design methods for technology and work that were established at the turn of the century (ed. 1900s). This is a revolutionary concept that goes beyond mere innovation, because it significantly inverts the relationship between people and machines and, as seen in the decades that followed (and the emergence of ergonomics), does not limit its field of application to the more technical aspects of work."*

The contents and development of the ergonomic approach can be read through both certain keywords found in its different definitions and through the progressive extension of their meaning following the extension of the capacity for ergonomic intervention in new sectors that have been opened up by technological and social innovation (Fig. 1.4).

Interdisciplinarity has represented the distinct character of Ergonomics since its formation. In fact, Ergonomics was born not as a *discipline*, that is, as a research and teaching sector defined by the specificity of the scientific approach and the identifiable nature of the object being studied, but as a *corpus of knowledge*, an interdisciplinary approach, an osmosis of knowledge, that is, a field of study and intervention into

SOME OF THE CHIEF DEFINITIONS OF ERGONOMICS

- Adapting work to people. (Murrell H., 1949)
- The study of human behaviour during work. (Grandjean E., 1983)
- The subject of ergonomics is all of the relationships that humans established and experience in the course of their activities, through interactions with the environment, objects, work tools and other items. (Cribini G., 1991)
- Ergonomics is the science of work, of the people who do it and the ways in which it is done, of the tools and the equipment they use, of the places they work in and the psycho-social aspects of working situations. (Pheasant S., 1996)
- Ergonomics: the study of human factors. The scientific discipline concerned with the understanding of interactions among human and other elements of a system, and the profession that applies theory, principles, data and methods to design in order to optimize human well-being and overall system performance (ISO 9241-210:2010 and ISO 6385:2004).
- The methods of Ergonomics are designed to improve product by understanding or predicting human interaction with those devices. (Stanton N., 2014)
- Today, Ergonomics and Human Factors are the terms – which are synonymous and used the world over – that describe the theory and practice of human features and abilities, and then using this knowledge to improve people's interactions with the items they use the environments they operate in. (Wilson J.R., Sharples S., 2015)
- Ergonomics (or human factors) is the scientific discipline concerned with the understanding of interactions among humans and other elements of a system, and the profession that applies theory, principles, data and methods to design in order to optimize human well-being and overall system performance. Practitioners of ergonomics and ergonomists contribute to the design and evaluation of tasks, jobs, products, environments and systems in order to make them compatible with the needs, abilities and limitations of people. Ergonomics helps harmonize things that interact with people in terms of people's needs, abilities and limitations. (IEA, International Ergonomics Association, 2018)

Fig. 1.4 Some definitions of ergonomics

which knowledge and methodological tools from the different disciplinary sectors are integrated.[11]

---

[11] As noted by Wilson (1995), the ability to open Ergonomics up to the new fields of intervention that are continually opened up by innovation overcomes the questions posed by the definition of Ergonomics as a "discipline" or as a "corpus of interdisciplinary knowledge", as well as its role of "basic research" or "applied research".

Wilson and Sharples also note that "*Although we looked at definitions of ergonomics and human factors earlier in this chapter, it is more important that E/HF should be seen as an approach (or as*

As Mantovani (2000, p. 20) states, Ergonomics deals with the problems of the individual in his relationship with the environment and with objects and *"the reality of scientific research, and of ergonomic research in particular, is made up of problems, not disciplines. The history of science is a story of problems that have been constructed, explored and solved, not of disciplines that grow on their own, each in its own garden. (...) If Ergonomics occupies a special position in the overview of contemporary knowledge, this is due to the singular social relevance and mobility of its theme, which has prevented it from lying idly in a neatly defined disciplinary area. Ergonomics studies human work, which changes constantly and can do this only by developing period-appropriate skills, environments and different circumstances in which human labour is expressed."*

1.2.1   The Adapting of Work—And Its Activities—To People

Work, which is today understood as *a set of human activities*, represents the second keyword of ergonomics. The Greek term *ergon* has a series of meanings, from the more limited "work", that is, what we do for a living, to the broader "activity", which introduces a field of research and intervention aimed at the study and design of the interaction between people and the physical and virtual systems they interact with during their work activities and daily lives.[12]

The adaptation of work—and activities—to the person introduces a radical change in perspective to the study of the relationship between people and systems, which presupposes moving the focus from the features and performance of the system (product, environment or service) to the effect these products for the person.

Today, the original goal of *adapting work to the person* has developed with the aim of ensuring usability and the quality of the experience in this relationship with the system, through study and intervention methods that can evaluate the design the efficiency, reliability and physical, psychological and emotional conditions of well-being offered by both simple and complex systems, in and with which people carry out their activities.

The complexity of the ergonomic approach, in particular the User/Human-Centred Design approach, is based on the awareness and interpretation of the "context of use",[13] that is, all of the variables that contribute to defining the interaction between people and the system in which they operate, their reciprocal conditioning, their variability over time and the overall experience that was enjoyed.

---

a philosophy) of taking account of people in the way we design and organise; in other words, as designing for people. In this view, E/HF itself is primarily a process, to an extent a meta-method, which makes the clear understanding and correct utilisation of individual methods and techniques even more important". Cfr. Wilson and Sharples (2015, pp. 15).

[12]Cfr. Pheasant and Haslegrave (2006).

[13]For the definition of context of use, see the list of definitions contained in the ISO 2041-210:2010 standard reported in par. 5.3.

A global approach, therefore, to the evaluation and design of the interaction, which first considers the different ways—physical, perceptive, cognitive and emotional—in which people interact with a system, the collection of needs and expectations people have, based on their characteristics and capabilities, the activities they perform, the environmental, technological and organisational variables and, finally, the complexity of the design and production process and the collection of limits—time, costs, materials, organisation, etc.—that influence the development of design.

As Lupacchini (2008, p. 69) writes, the ergonomic approach to design starts *"from the principle of the whole, that is, that people do not perceive themselves to be a collection of distinct features (body, mind, abilities, etc.) but rather as a collection of a series of features that make them unique. The same is true for objects and environments; they must be evaluated as a whole, even if they are characterised by other components."*

Starting from this general principle, the ergonomic methodology explores the design problem from top to bottom, managing to control the parameters that form the specific system, but, at the same time, studies the relationships by using a global investigative approach that aims not to neglect any particular detail (Fig. 1.5).

As it evolved, Ergonomics has followed two lines of development: the first marked by the constant expansion of its fields of interest and the second by its progressive focus on the *user*, then the *person*, and the specificity of his relationship with the system in which he operates, as a point of departure and final goal of its intervention.

The *individual* is no longer generically held to be the subject of the study, but rather the *user/person* and the context he operates in. The individual becomes an integral part of this context, with the objects and systems with which he interacts, his own actions and the physical, organisational and social environment he acts in.

Awareness of the abilities and the needs of the *individual* (or group of individuals), checks on the safety and usability of products and systems and, finally, the user experience, which now represents the privileged areas for ergonomic research and experimentation, are added to knowledge about the anthropometric, physiological and psycho-perceptive qualities of the *individual*.

Work efficiency and the adaptation of working conditions and tools to the needs of individuals have obviously not been discovered in Ergonomics, but have been a part of human history since its origins. The goal of Ergonomics is to design and construct objects and environments that are suitable for human use and appropriate for their intended use. In this sense, most traditional work tools and items that we use can be defined as ergonomic. Chairs, tables, crockery, hammers, etc. are now made with materials and manufacturing processes that were once unknown, but their shape, their size and their basic technical features have been formed and refined over the centuries as a result of a non-scientific knowledge of people's features and need.[14]

The artisan who designed and manufactured objects and artefacts in pre-industrial times, worked within a deposited and shared technical culture, knew the characteristics of the materials and the available technologies, which generally remained the same throughout his life, and, finally, shared his direct knowledge of their use and

---

[14]See also: Bandini Buti (2008).

Fig. 1.5 Made to be used: work tools and objects of everyday traditional life have always been made to best perform their function and be suitable for human use

the needs to which the object must respond with users. It is the industrialisation of work, and, with it, the fragmentation of the phases of production and the tasks that the worker is required to perform, that requires a control action and the verification of the design and manufacturing process for products, as well as the systemisation of the knowledge regarding the abilities and needs of the individual.

The coining of the term Ergonomics—from two Greek nouns, *ergon* and *nomos*, or work and natural law (or control) respectively, literally meaning "natural law of work" or "control of work"—is attributed to Hywel Murrel and coincides with the foundation of the Ergonomics Research Society and the date of its first conference. This was held in Oxford in July 1949. However, the term had already been used by the Polish naturalist Wojciech Jastrzebwski, who introduced the term Ergonomics to literature in 1871.

The problem of the efficiency of human activities within the production process was widely addressed at the end of the 1800s with "the scientific study of people at work" by F. W. Taylor, who devised the study of "work movements and times" to increase the efficiency and productivity of human work. The collection of data regarding human characteristics, the means of performing the required operations by individual industrial process and, finally, the "choice" of the most suitable workers to carry out individual tasks, aimed "to adapt people to their work".

References to the History of Ergonomics[15]

The coining of the term Ergonomics—from two Greek nouns, *ergon* and *nomos*, or work and natural law (or control) respectively, literally meaning "natural law of work" or "control of work"—is attributed to Hywel Murrel and coincides with the foundation of the Ergonomics Research Society and the date of its first conference. This was held in Oxford in July 1949. However, the term had already been used by the Polish naturalist Wojciech Jastrzebwski, who introduced the term Ergonomics to literature in 1871.

The problem of the efficiency of human activities within the production process was widely addressed at the end of the 1800s with "the scientific study of people at work" by F. W. Taylor, who devised the study of "work movements and times" to increase the efficiency and productivity of human work. The collection of data regarding human characteristics, the means of performing the required operations by individual industrial process and, finally, the "choice" of the most suitable workers to carry out individual tasks, aimed "to adapt people to their work".[16]

In the early 1900s in the USA, studies were developed around the theme of the "human factor" (human factors, human engineering), that is, the study of human relations within the industrial process.

The interdisciplinary study of the person at work, which focused on evaluating the "human factor" within production processes, in particular for heavy industry and the military, started to find a structure in the first half of the 1900s, becoming a field of theoretical and applied research in which knowledge and contributions from disciplines that had been historically separated by their scope and research methods converged.

The synthesis of knowledge and theoretical and applied tools from physiology, psychology, engineering and design allows us to study the man-machine systems, with the aim of improving both the efficiency and reliability of the system and the well-being of its operators.

Studies on the human factor developed during the First and Second World War, when the increasing complexity of the military equipment put the spotlight on the efficiency and reliability of systems, as well as the performance levels required of operators and the risks connected to human error.

The usability of industrial machines by female operators, and the performance and reliability of aeronautical pilots, particularly in reading and controlling the steering equipment, are the most noted examples of the studies carried out during the two World Wars.

These experiences came together in the founding of the first interdisciplinary work group, which formed in Oxford in July 1949 as the Ergonomics Research Society. The term "ergonomics" was coined at its institutional meeting, in order to define a new means of studying and resolving human problems in the workplace environment.

In 1957, the Human Factors and Ergonomics Society is founded in the United States, with the aim of adapting the design of systems and equipment to workers. At the same time, ergonomics associations are formed in many European countries. At the end of the 50s, the IEA (International Ergonomics Association) is formed, with its first official meeting held in Stockholm in 1961. The SIE (Italian Society of Ergonomics and Human Factors), a member of the IEA, was founded in Italy in 1961.

Theoretical and applied research departments are formed within the military and aerospace industries first and, subsequently, in the civil industry, within the large production groups that operate, in particular, in the steel and mining sectors and, then, in transport, telecommunications and, finally, the production of hardware and software. Research and professional groups develop in parallel, characterised by a strong sense of applied action; these also operate outside of the scientific community and national associations.

In Italy, ergonomics developed over a long period as a specialisation within specific disciplinary sectors—particularly occupational medicine, engineering and, to a much smaller extend, architectural technology and design—and, similarly, as a professional specialisation.

Starting from the 80s, ergonomics was introduced into Italian faculties (known today as departments), first with a few seminars and then with official courses characterised at first by application and, today, fully integrated into the educational offerings of Psychology, Engineering and Design courses.[17]

Starting from the 90s, "compliance with ergonomic requirements" is introduced into the Italian legislative system (the most famous being Legislative Decree 626/1994 and 626/1994, "Protection of health and safety in the workplace", 17/2010, "Machinery directive", 37/2010 "Medical devices") and the technical regulations that now incorporate the extensive international regulatory framework regarding ergonomics and machine safety.

In particular, the ISO international standards,[18] and UNI national standards, provide the definitions, methods and parameters required for the evaluation and planning of health and safety, comfort (visual and acoustic) and dimensional suitability of environments and products. Finally, the regulations regarding usability and the contents and methods of Human-Centred Design[19] are very expansive.

Some authors have proposed a reading of the later evolutionary phases of ergonomics based on the relationship between ergonomics and technological evolution.[20]

An initial period, which came partially before the official birth of ergonomics, is placed between the '40s and the first half of the '50s. In this phase, the ergonomic approach is identified as *the adaptation of work to the person* which is intended as a reversal of the typical Taylorism analytical perspective.

The focus of the first ergonomic studies is aimed, in particular, at the adaptation of work equipment and the environmental conditions that define the workplace. The central focus is the worker in relation to the machine, within the microcosm of the work station, analysed outside the organisational and social context in which he is situated. In this phase, Ergonomics is the *science of work*, with its intervention required by the problems posed by industrial development and the need to make human performance reliable and safe within production processes. The subject of the study is the psycho-physical conditions of the individual and the means of carrying out the operations required by his job; therefore, in this first phase the anthropometric, biomechanical and physiological aspects assume a prominent role in the evaluation

---

[15]For further information about the history of ergonomics, see works by Eastman Kodak Company (1983), Pheasant (1997), Stanton (1998), Chengalur et al (2003), IEA, International Ergonomics Association (2006), Pheasant and Haslegrave (2006), and to the Italian texts of: Re (1995), Attaianese (1997), Bandini Buti (2008).

For the evolution of Ergonomics in Italy, see SIE, Italian Society of Ergonomics and Human Factors, *The history of ergonomics*, available on www.societadiergonomia.it.

[16]Human resources are studied and used by F. W. Taylor for machines in operations. "A trailblazer of ergonomics for his experimental studies on tools and working methods (his studies on the optimal size of the blades for the loading and unloading of cast iron are a classic example of adapting a tool), Taylor is, at the same time, the negation of ergonomics when taken as the adaptation of the machine to the person and to the real conditions in which the work takes place. (…) On the contrary, with Taylor, work becomes very similar to an 'agonistic' activity, where time is of little importance and the isolation of the productive moment is theorised from the existential dimension. This dimension is so far removed from ergonomics that Taylor himself underlines that only one worker in seven can be considered skilled and included in the scientific organisation". See Re (1995, pp. 4–7).

[17]On this matter, see Sect. 2.3 "Design for Ergonomics".

[18]See the ICS (International Classification for Standard) classification of ISO, the International Organisation for Standardisation used by international regulatory bodies and also adopted by the UNI, the Italian Standardisation Body. In particular the ICS codes 13.180 "Ergonomics" and 13.110 "Security of machinery". See note 4, Chap. 5.

[19]Cfr. ISO 9241-11:1998 standard, Ergonomic requirements for office work with visual display terminals (VDTs)—Part 11: Guidance on usability.

[20]Cfr. Bandini Buti (2008), Green and Jordan (1999), Mantovani (2000).

of the man-machine relationship. A second period is placed between the end of the '50s and the mid '70s, when the cost of labour and the growing attention paid to ensuring workplace safety forces industries to guarantee the safety on the conditions under which industrial processes# are carried out.[21] The subject of study is no longer just the performance of the individual in his workplace, but the relationship between individuals and the context defined by the work equipment, the surrounding environment and the organizational characteristics of the work. In this period, Ergonomics' field of intervention is mainly the high-risk industrial-processing sector. *"In the 70s, the context (in particular in Italy) limited the intervention on the conditions in which workers' tasks were carried out. The methodologies were analysing elementary work operations and reconstructing them in a more complete sense, from the analysis of work movements to adapting the technology for the comfort of its user"* (De Gregorio 1999, p. 82). This is particularly true in the steel, mining and automotive industries, where the most urgent problems are the safety of the workers' tasks and their health.

Starting from the 80s, and more markedly in the two decades that followed, the focus of Ergonomics extends to the environmental and organisational aspects of the workplace, which now includes all production areas and office jobs.

The dissemination of IT technologies in the primary and tertiary sectors and, subsequently, in services, equipment and, finally, everyday products leads Ergonomics to new fields of interest and, in particular, towards the study of human behaviour within both simple and complex systems, in which the individual is no longer considered an external user but an integral part.

The elimination of a large part of the strictly manual work, which is replaced by automated control tasks in the industrial production lines, the now widespread computerisation of office work and, finally, the appearance of new means for organising and managing industrial production and relationships within companies moves Ergonomics' focus to the study of these interfaces, the usability of the man-machine systems in which the "machine" is no longer understood to be simply a mechanical machine, but a collection of devices, hardware and software that is controlled by the operator.

At the same time, Ergonomics extended its fields of research and intervention to the study of human needs in their interaction with environments and objects during daily activities.

Ergonomic interventions, which were initially corrective in nature, tended to move to the forefront of design and to find a place in each phase of the design, production and evaluation process for a product.

The belief that true prevention is achieved by intervening in the design before the harmful event occurs emerges, leading to the development of suitable forecasting methods. In fact, between the end of the 80s and the 90s, we see an increasing

---

[21] In Italy the approval of the Workers' Statute in 1970 establishes the right for workers to control the application of the regulations for the prevention of accidents and occupational diseases and to promote the research, drafting and implementation of all suitable measures to protect their health and their physical integrity. See Bandini Buti (2008, pp. 18–19).

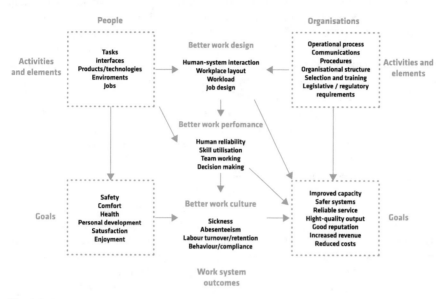

Fig. 1.6 The goals of ergonomics. Reworked by Wilson and Sharples (2015, p. 8)

involvement by the ergonomist, not only in evaluations but also in the design of buildings, machines, products and IT systems.[22]

1.3 Ergonomics Today: Intervention Sectors

Today, the goals of Ergonomics are the usability and safety of systems that people interact with during their activities (as operators, users and, finally, as purchasers of products, equipment, environments and services) and the quality—physical, cognitive and sensory and emotional—of the experience for the person as they interact with the system (Fig. 1.6).

In this context, the usability, safety and user experience of the system represents the essential conditions to ensure both the well-being of the individual and the functioning of said system, not only in the workplace but also in domestic and private settings.

The user/person, therefore, is understood to be the real user that the design is addressed to, and the user must be guaranteed well-being, safety and easy of use, conditions that take into account his specific characteristics and capabilities, needs and the tasks that he is asked to perform within the context he operates in.

---

[22]Cfr. Grieco and Bertazzi (1997).

Today, Ergonomics is established as the study and design of complex systems, the efficiency—and the existence—of which is determined not only by the functioning of the system itself, but by the broader technological, economic and social context in which it is placed. Its activities are no longer limited to the study and chiefly preventative intervention of the man-machine interaction within the productive cycle, but extends to the product (that is, the result of the productive cycle) and intervenes in each phase of development, with an expressly design-based focus.

It is precisely this progressive expansion of its investigative fields and "*the mobility of its subject that allows the history of ergonomics to show us situations in which previously established approaches disappear, as new themes and approaches emerge. (…) The necessarily contextual nature of ergonomics forces us to continuously revise theories and methods, in order to keep up-to-date with the changing nature of work in industrial and post-industrial societies*" (Mantovani 2000, p. 17).

As noted by Re (1995, p. 16), the birth of Ergonomics coincides with the explosion of problems related to the performance and reliability of military technology and is the direct result of the scope of these problems during the Second World War. The emergence of new fields of interest, in parallel with the technological and social evolution in which Ergonomics operates, and the progressive expansion of the disciplinary knowledge and skills needed to deal with these innovation-based issues, do not, however, involve distorting the initial approach or eliminating references to the initial corpus of knowledge on which ergonomics is founded.

The emergence of new fields of interest, in parallel with the technological and social evolution in which Ergonomics operates, and the progressive expansion of the disciplinary knowledge and skills needed to deal with these innovation-based issues, do not, however, involve distorting the initial approach or eliminating references to the initial corpus of knowledge on which ergonomics is founded.

The ergonomic approach can be interpreted in a theoretical and applied manner, particularly, according to the subject of the intervention. According to Hendrick (1999, p. 5), Ergonomics can be considered to be the *technology of the human-system interface*. As a *science*, it refers to the development of awareness regarding people's abilities and limitations and other features related to the design of the interface between the individual and the other components of the system. As a *professional practice*, Ergonomics concerns the application of this knowledge and these analysis intervention methods to the design and evaluation of the systems, with the aim of increasing the safety, well-being and quality of life of the users.

1.4 Disciplinary Matrices

Now, the panorama offered by these lines of research, methodological tools and the experiments performed in the field of ergonomics appears to be very vast and difficult to track within a single organic framework.

There is *"a feeling of fleeting 'breadth' that, in various ways, assails those who hope to understand the disciplinary development according to a broad and continuous vision; the breadth of the different knowledge that comes together to form its multidisciplinary dimension (…); the breadth of social and economic interests that ergonomics invests (…); the breadth of the contents and areas of interest, which take on various connotations in relation to the priorities and objectives that are attributed to ergonomic action, and which contribute to delineating (…) various conceptions of ergonomics"* (Attaianese 1997, p. 5).

An initial element of the problem concerns the relationship between the different matrices on which the theoretical and methodological system of Ergonomics is based. The research and experiments conducted in the fields of psychology, medicine and engineering coexist today with a markedly design-based approach to ergonomics, while the professional practice is closely linked to the constraints and resources of the production processes.

If there is substantial agreement on what the basic contents and objectives are in the field of ergonomics, both the basic formation from which we "arrive" at Ergonomics and the research fields in which ergonomists operate are very different. As privileged ground for its own professional and/or research activity, we have arrived to Ergonomics not through a specific course of studies, but starting from a previous university education. Ergonomists' varying initial training and the marked specialisation that characterises research and professional practice in the field of ergonomics make it difficult to disseminate the knowledge and methods of ergonomics outside the sectors in which they were developed in many cases and, as we shall see in the next section, for ergonomists to communicate with the other players in the product-formation and development process.

In terms of the relationship between Ergonomics and Design, a second aspect relates to the availability and usability of the data and tools developed in the various research areas that come together in ergonomics (in particular, anthropometry, biomechanics, the physiology and cognitive ergonomics) in the field of design.

In the fields of anthropometry and the biomechanics of posture and movement, today there is plentiful, structured data. The methods for analysing the variability of data, the criteria for its correct use and its adaptability based on the design problem have been extensively discussed by some authors,[23] but, in many cases, the "useful design data" is contained in specialised publications which are often difficult to find.

The conditions for environmental well-being represent a slightly different case, for which there is a wide availability of data referring to the behaviour of the environment (and of objects), in particular as it relates to lighting and acoustic and thermo-hygrometric aspects. The scientific knowledge in these sectors is obviously structured, as are the methods for measuring performance, the tools and test methods and the acceptability thresholds, the framework of which is completely defined by both legislation and national and international technical regulations. However, the

---

[23]Cfr. among others: Panero and Zelnik (1979), Grandjean (1986), Grandjean and Kroemer (1997), Pheasant (1995, 1997), Pheasant and Haslegrave (2006).

evaluation methods refer to the *performance of the object of analysis*, that is, the in-use behaviour of an environment, a material or a component.

These problematic aspects are matched by the progressive formation of a regulatory corpus on the subject of "Ergonomics" and "Safety" of machinery, which today offers a sufficiently organic and structured framework, not only of definitions and terms but also of procedures and methods for verifying ergonomic requirements.

In recent years, national and international regulatory bodies have produced a considerable number of guidelines, for both ergonomic-specific matters and those in related fields, in which ergonomics is, in any case, referred to as an essential requirement for the design and construction of environments and products.

The contents of the Italian and international standards on Ergonomics, Machinery safety and Human-Centred Design allow us to rely on definitions, intervention and evaluation criteria and test methods that are recognised and shared internationally. The framework for the technical regulations relating to the characteristics and physical capabilities of people, which concern both the definitions and the general principles of ergonomic intervention, and the criteria necessary for the correct sizing of some products and equipment (in particular, furnishings and industrial machinery) and some workstations, and the limits of acceptability for the movements and strength required by physical tasks are also very well established.

Finally, numerous technical standards refer to the visual, acoustic and thermo-hygrometric qualities of the environments, materials and products. Even if these norms fall under the classifications relating to ergonomics and safety of machinery to a limited extent, and refer mainly to the physical and technical behaviour of the environment and objects, they represent, in any case, an essential point of reference for ergonomic design and the evaluation of the conditions for environmental well-being.

On the one hand, then, there is a field of research—or rather many fields of research—that is continuously evolving and expanding, in which highly specialised skills and experiments are placed, and the results of which are difficult to find and scarcely disseminated in the design field. On the other hand, there is an increasingly broad framework of legislative requirements and regulatory recommendations that require the compliance of products and workplaces with ergonomic principles and parameters and, at the same time, impose knowledge of the principles and tools of ergonomic design upon an ever-growing number of designers and operators.

It should also be emphasised that the focus of companies on intervention and the contribution of ergonomics is closely linked to the regulation level of products and production systems by national and international legislation and technical regulations. In fact, compliance with ergonomic requirements is, naturally, considered necessary by the company only if prescribed or recommended by regulations and/or required by the market, that is, by the purchasers of environments and products.

In connection with the mandatory nature of legislative and regulatory constraints, there is, however, a growing focus from companies towards the levels of safety and usability guaranteed by their products, and towards the user experience that they are able to offer. This focus originates not only from the action of regulatory limitations, but also from the willingness of companies to promote and safeguard their image in the market, from the growing awareness of potential buyers and from the safeguarding action exercised by consumer associations.

The possibility that a product causes an accident is, in fact, an unsustainable risk for the company, not only because of the possible legal and economic consequences of an accident, but also because of the reputational damage that this can cause. Similar reputational damage can also be caused by products that seem potentially dangerous or too difficult to use.

1.5 Ergonomics in the Design Process

The role of ergonomics in the design process refers to both the placement of ergonomic intervention, and the professional ergonomist, in the product-formation process and the relationship between the ergonomist and the other players in said process.

The skills of the professional ergonomist are generally required within complex production processes, in which the economic and operational limitations determine not only the resources for the ergonomic intervention, but also the room for dialogue between the different professional figures and the ways in which the skills and the methodological tools provided by Ergonomics can be placed in both the design phases and the engineering and production phases.

As highlighted by Norman (1988, p. 176), an initial consideration refers to the often totally different relationship that producers, designers and users have with the object or environment that they produce, design or use respectively.

- The *designers* have detailed knowledge of the item, its means of functioning and, during the design phase, acquire a significant confidence with the item or the environment made, to the point of taking some of the usage procedures or part/space articulations, etc. for granted.
- The *ergonomists*, in particular professional ergonomists, have highly specialised skills and, often language, and, as already mentioned, do not necessarily come from a design training background.
- The *producer* has a precise overview of the costs, the resources invested in design and construction and the assembly times, in the labour required, etc.
- The *user* typically has full confidence in the task he must perform (which he has, to date, done with different items and/or environments) or very consolidated ideas about the activity he wants to perform with a specific item and/or in a specific environment.
- Lastly, the *purchasers*, when they are not said users, are experts in the management aspects related to that task, but rarely perform the same task.[24] In terms of production, it can be said that awareness of the role played by the safety and usability of a product in responding to the needs and expectations of the users to whom it

---

[24]Cfr. Norman (1988, p. 176).

THE FIVE FUNDAMENTAL FALLACIES

- this design is satisfactory for me — it will, therefore, be satisfactory for everybody else;
- this design is satisfactory for the average person — it will, therefore, be satisfactory for everybody else;
- the variability of human beings is so great that it cannot possibly be catered for in any design — but since people are wonderfully adaptable it doesn't matter anyway;
- Ergonomics is expensive and since products are actually purchased On appearance and styling, ergonomic considerations may conveniently be ignored;
- Ergonomics is an excellent idea. I always design things with ergonomics in mind but I do it intuitively and rely on my common sense so I don't need tables of data or empirical studies.

Fig. 1.7 The five fundamental fallacies. Reworked by Pheasant and Haslegrave (2006, p. 11)

is addressed is widely disseminated within companies, as are the economic and reputational risks the producer and/or designer faces in the event of accidents caused by their products. On the contrary, awareness of the usability and security verification procedures made possible by Ergonomics is much less widespread, as is the contribution they can offer to the product design, production and verification processes. The main obstacles to the integration of Ergonomics into the product-formation process may derive both from the marked specialisation of the tools and information made available by the ergonomic intervention, and by the ways in which they are presented (Fig. 1.7).

It is these "errors" that lead the designer or the company to ignore ergonomic principles and to remain fixed to a sort of blissful ignorance about its contents.

However, in this context, which snapshots some aspects of the relationship between Ergonomics and the design process as they currently appear, it is possible to note some signs of change today. Even if awareness of the contents, methods and strategic value of Ergonomics is consolidated only within advanced production and design sectors, the results of the experiments, conducted within the most advanced companies and/or research structures, are beginning to be disseminated in non-specialized publications too, making their contributions known even to those who are not strictly involved in this work.

At the same time, the growing number of guidelines regarding ergonomics, usability and the safety of work systems and equipment is today leading to an equally growing interest from companies and designers in the skills and procedures necessary to guarantee the compliance of their products with ergonomic requirements.

It is also important to emphasise that the theoretical and methodological tools produced by Ergonomics, and many of the topics currently addressed by ergonomic research, are in line with the most recent orientations of research carried out in the design field. In particular, the issues of building quality and quality control in production processes (of buildings and products strictly understood) and the methods for assessing and interpreting user requirements are objects of research and debate that directly involve the whole area of the design.

In this case, the obvious problem is the difficulty with which the theoretical and operational tools, which developed in highly specialised sectors, manage to expand and find comparative plans with other research areas and other professional figures.

1.6 From Design for End Users to Design for People: The Ergonomic Quality

The analysis of people's needs as a basic reference for design, and the compliance of products and systems with the requirements deriving from these needs, characterise not only the ergonomic approach but, on the contrary, constitute the assumption that unite all the sectors of research on quality (for systems, products, building artefacts, etc.).

The definition of the product requirements based on the *end user needs* also constitutes the conceptual basis for a large part of the national and international regulations, which are structured, to put it briefly, around the definition of the system needs-requirements, with reference to a specific object of analysis (product, component, environment, system), of the performances that this object must guarantee to meet the requirements, and finally the acceptability thresholds and the measurement methods necessary to verify these performances. Within this approach, the correspondence to the *end user needs* refers to the generally understood needs of users, as analysed outside the specific context of use and "measured" based on the physical and psycho-perceptive characteristics of the average user.

The technical regulation also defines quality as the capability of a product, a service or an organisation to meet the needs and expectations of customers and interested parties. In particular, "*the quality of products and services includes not only their intended functions and services, but also their perceived value and benefit for the customer*".[25]

Both in terms of the concept of the user/person and that of quality, the innovative contribution of the ergonomic approach to design lies in focusing attention not on the end user, but on people who really relate to the product, the environment or the service, and in assessing the quality of the interaction that they establish, placing it within the specific context in which it is implemented.

The concept of quality for the product/system in Ergonomics, therefore, coincides (and extends) with the quality of the interaction that people establish or can establish with that product, and can be defined in terms of *compatibility* between the

---

[25]Cfr. UNI EN ISO 9000:2015 standard, Quality-management systems. Foundations and terminology.

characteristics and the abilities of the individual and the characteristics of the product, based on the function of the different conditions of employment, of *compliance with the needs of people* who actually use or will be able to use that product (needs that must be identified and evaluated by considering all the elements that define the interaction), and, finally, the *value* that can be attributed to the relationship with the product and to the experience that derives from it, that is, the judgment that each of us expresses, knowingly or not, in terms of annoyance, appreciation or familiarity.

The specificity and innovative value of the ergonomic approach to design consist, in fact, of their ability to evaluate the variety of variables that define the interaction between people and what they interact with (that is, the characteristics and capabilities of the users, the characteristics of the product and the activities for which it is designed or the characteristics of the physical, social and organisational context can be used), by evaluating their mutual relationships and their possible modification over time, and identifying and interpreting, from time to time, the needs and expectations that people express or can express with respect to these interactions.

Design intervention is based on the ability to understand, interpret—and imagine—the different realities and the different possibilities with which people can, or could, interact with the system, by identifying the variety of variables involved and the complexity with which their reciprocal relationships are determined, defining the system of requirements for the product and, finally, the parameters and criteria necessary to evaluate and design the quality of the product.

Regardless of the characteristics and complexity of the system, Ergonomics operates using the most appropriate theoretical and methodological tools, in order to analyse user needs and to design the interaction within the system, of which the person is an integral part and focus of design. Ergonomics also operates on the applied level, as well as the theoretical level, and within the constraints of the specific design problem. The goal to be achieved is not, in fact, the optimal solution, but the *best balance possible* between individuals and what they use with respect to operating conditions and available resources.

1.7 User/Human-Centred Design—People-Centric Design

User/Human-Centred Design (HCD) describes an approach to design aimed at the quality of the interaction between people and physical or virtual systems (products, equipment, environments and services) with which they interact. It is based on the ability to gather and process the information needed to understand people's needs and expectations in the relationships they establish, or can establish, with the system, through structured and verifiable methods of investigation and evaluation (Fig. 1.8).

The literal meaning of Human-Centred Design—a person-centred design—clarifies that the conception of design is aimed at directing all of the design choices and the entire process of developing and producing the finished product, based on user data and the types of use for which that product is intended.

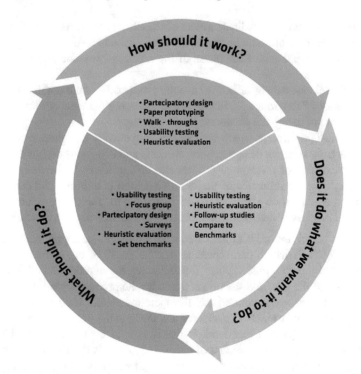

Fig. 1.8 Human-Centred Design approach: questions and methods for answering them. Reworked by Rubin and Chisnell (2011, p. 13)

As we saw at the beginning of this chapter, HCD represents not only an intervention philosophy that places the user at the centre of the process of designing and manufacturing products, but also, and most importantly, the techniques, processes, methods and procedures necessary to verify and design the usability of products and systems. However, this definition must not be considered only in relation to the use for which the product is designed, but must extend to the totality of the relationships that people can establish with the product, and to its entire life cycle, which goes from the design phase, to the production, sales, use, maintenance and, finally, disposal phases.

Woodson[26] defines HCD as "The practice of designing products so that users can perform required use, operation, service, and supporting tasks either a minimum of stress and maximum of efficiency" and, as Rubin writes (1994, p. 12), its objectives can be summarised in three fundamental principles:

- the ability to immediately focus attention on the user and the task. This means identifying and categorising the user (real or potential). Therefore, systematic and structured approach to gathering information from and about users is necessary. The user-selection criteria and the data collection and processing methods for users

---

[26]Cfr. Woodson (1981). Cited by Rubin (1994).

are the central and most delicate aspect of usability checks. As we will see in the next chapter, there are now numerous investigative and testing techniques aimed at providing user data at each phase of the product-formation and development process;

- the empirical assessment of how to use the product. Here the emphasis is on the assessment of behaviour and, in particular, assessing the ease of learning about, and the ease of using, the products;
- iterative design, through which the product is cyclically designed, modified and tested. A truly iterative design favours accurate and conscious design through the early assessment of conceptual models and design ideas. If the designer is not available, then the influence of the iterative design is minimal and cosmetic. In summary, a real iterative process makes it possible to "form the product" through a cyclical process of design, evaluation, new design and new verification.

The three principles of HCD indicated by J. Rubin can be supplemented by others[27] (see Fig. 1.9) which can represent the "slogans" of ergonomic design and underline its highly pragmatic nature, thus allowing the introduction of all the variables relating to the user and the interaction with the product within the design, by analysing, in a direct and structured way, the operating methods and the various objectives and contexts of use. HCD evaluates and designs the interaction starting from the analysis of the elements that define it, that is, people, the system with which they interact, the activity and the context of use. Its objective is the creation of products that are adapted to the needs of its users, the use of which is satisfying and pleasant for people and ensures the possibility of carrying out the required actions effectively and efficiently. Its objective is also to eliminate or reduce conditions in which the methods for using an object, the purpose and use for which it was designed and, finally, the consequences of human actions, are incomprehensible or inappropriate for human use.

The study of human error introduces the evaluation of the behaviour of individuals in relation to the information that the product is able to provide about its function, its functioning and the effect produced by the actions that the user can perform (or has already done) with the product to HCD.[28]

The theoretical and methodological bases of HCD consist of the procedures for verifying usability and safety in use, conducted through observation and evaluation of the interaction and/or tests with users who, as we will see in the next chapters, can be used in every phase of the product life cycle, from the moment of its conception and design to the production, sales, usage, maintenance and, finally, disposal phases. The type of tests to be used, their objectives and their criteria must be discussed as specific contents of design. Product tests can directly contribute to the development cycle, by focusing on user needs or behaviours that can be critical for product effectiveness.

---

[27]Cfr. Pheasant (1997), Porter (1999), Pheasant and Haslegrave (2006).

[28]These aspects are discussed in Chap. 13 "Cognitive aspects in user experience design: from perception to action" by O. Parlangeli and M. C. Caratozzolo, and in Chap. 5 "The design of ergonomic requirements" regarding the safety in use and the usability of the products, limited to their design usability.

USER/HUMAN CENTRED DESIGN

1. **User-centred design is empirical.** It seeks to base the decisions of the design process upon hard data concerning the physical and mental characteristics of human beings, their observed behaviour and their reported experiences. It is distrustful both of grand theories and intuitive judgements — except insomuch as these may be used as the starting points for empirical studies.

2. **User-centred design is iterative.** It is a cyclic process in which a research phase of empirical studies is followed by a design phase, in which solutions are generated which can in turn be evaluated empirically.

3. **User-centred design is participative.** It seeks to enrol the end-user of the product as an active participant in the design process. It is non-Procrustean.

4. **User-centred design is non-Procrustean.** It deals with people as they are rather than as they might be; it aims to fit the product to the user rather than viceversa.

5. **User-centred design takes due account of human diversity.** It aims to achieve the best possible match for the greatest possible number of people.

6. **User-centred design takes due account of the user's task.** It recognizes that the match between product and user is commonly task-specific.

7. **User-centred design is systems-orientated.** It recognizes that the interaction between product and user takes place in the context of a bigger socio-technical system, which in turn operates within the context of economic and political systems, environmental ecosystems, and so on.

8. **User-centred design is pragmatic.** It recognizes that there may be limits to what is reasonably practicable in any particular case and seeks to reach the best possible outcome within the constraints imposed by these limits.

Fig. 1.9 The principles of User/Human-Centred Design. Reworked by Pheasant and Haslegrave

Each test can provide useful information for a specific phase of the process and allow the requirements and objectives of the next phase to be redefined. It is not just the so-called "end user" who uses the product; the interaction with the product concerns those who produce it, sell it, produce it, repair it, use it and, finally, those who dispose of it. These people enter into a relationship with the product in a way that is often radically different, have different needs and, finally, a different skill level when using the product and its parts. An example may be a dishwasher: the seller and user only know the control and adjustment parts, while, on the contrary, fitters and maintenance technicians will understand all of its internal components. The user's needs refer to the comprehensibility of the controls and the accessibility and usability of the basket to be used for washing; the latter, on the other hand, will require the accessibility of the parts and internal components and their ability to be rapidly taken apart and replaced.

The HCD approach is, in fact, essentially methodological in nature and consists of the possibility of acquiring and assessing the needs of users through structured and verifiable methods, and of being able to translate them into equally structured and verifiable design tools.

HCD is configured as a cyclical system, based on the continuous verification of hypotheses and design solutions, and on the possibility of acquiring and translating targeted and usable information at each stage of the product-formation and development process.

The theoretical and methodological principles of Human-Centred Design and User Experience, and their operational application, are outlined in length in Chap. 3 "From User-Centred Design to Human-Centred Design and User Experience", and Chap. 6 "Human-Centred Design—User Experience: tools intervention and methods".

References

Attaianese E (1997) La città malata. Liguori, Napoli
Bandini Buti L (2008) Ergonomia olistica, il progetto per la variabilità umana. FrancoAngeli, Milano
Chengalur SN et al (2003) Kodak's ergonomic design for people at work. Wiley, New York (1st ed.: Rodgers SH (1983) Ergonomic design for people at work. Wiley, New York)
De Gregorio R (1999) La formazione-intervento come metodologia ergonomica. In: Ivaldi I (ed) Ergonomia e lavoro. Liguori, Napoli
Eastman Kodak Company (1983) Ergonomics design for human at work. Van Nostrand Reinhold, New York
Enciclopedia Treccani. Online: www.treccani.it/enciclopedia
Grandjean E (1986) Il lavoro a misura d'uomo: trattato di ergonomia. Edizioni Comunità, Milano (1st ed.: Grandjean E (1979) Physiologische Arbeitsgestaltung: Leitfaden der Ergonomie. Ott, Thun)
Grandjean E, Kroemer KHE (1997) Fitting the task to the human: a text book of occupational ergonomics. Taylor & Francis, London and Philadelphia (1st ed.: Grandjean E (1963) Fitting the task to the man. Taylor & Francis, London)
Green WS, Jordan PW (eds) (1999) Human factors in product design. Taylor & Francis, London and Philadelphia
Grieco A, Bertazzi PA (eds) (1997) Per una storiografia della prevenzione occupazionale e ambientale. FrancoAngeli, Milano
Hendrick HW (1999) Ergonomics: an international perspective. In: Karwowski W, Marras WS (eds) The occupational ergonomics handbook. CRC Press, Boca Raton
IEA—International Ergonomics Association (2006) 50th anniversary booklet
IEA—International Ergonomics Association (www.iea.cc)
ISO 9241-210:2010. Ergonomics of human-system interaction—part 210: human-centred design for interactive systems
ISO 9241-11:1998. Ergonomic requirements for office work with visual display terminals (VDTs)—part 11: guidance on usability
Ivaldi I (ed) (1999) Ergonomia e lavoro. Liguori, Napoli
Lupacchini A (2008) Ergonomia e Design. Carocci, Roma
Mantovani G (2000) Ergonomia, lavoro, sicurezza e nuove tecnologie. Il Mulino, Bologna

Norman DA (1988) The design of everyday things. Doubleday, New York (1st ed.: Norman DA (1988) The psychology of everyday things. Basic Books, New York)

Oxford Dictionary. Online: www.oxforddictionaries.com

Panero J, Zelnik M (1979) Human dimension & interior-space. Whitney Library of Design, New York

Pheasant S (1995) Anthropometry and the design of workspaces. In: Wilson JR, Corlett EN (eds) Evaluation of human work. Taylor & Francis, London

Pheasant S (1997) Bodyspace: anthropometry, ergonomics and the design of work. Taylor & Francis, London (1st ed.: Pheasant S (1986) Bodyspace: anthropometry, ergonomics and design. Taylor & Francis, London and Philadelphia)

Pheasant S, Haslegrave CM (2006) Bodyspace: anthropometry, ergonomics and the design of work. CRC Press, Boca Raton (1st ed.: Pheasant S (1986) Bodyspace: anthropometry, ergonomics and design. Taylor & Francis, London and Philadelphia)

Porter S (1999) Designing for usability. In: Green WS, Jordan PW (eds) Human factors in product design. Taylor & Francis, London and Philadelphia

Re A (1995) Ergonomia per psicologi. Cortina, Milano

Rubin J (1994) Handbook of usability testing: how to plan, design and conduct effective tests. Wiley, New York

Rubin J, Chisnell D (2011) Handbook of usability testing: how to plan, design, and conduct effective tests. Wiley, Indianapolis (1st ed.: Rubin J (1994) Handbook of usability testing: how to plan, design, and conduct effective tests. Wiley, New York)

Shorrock S, Williams C (ed) (2017) Human factors and ergonomics in practice: improving system performance and human well-being in the real world. CRC Press, Boca Raton

Società Italiana di Ergonomia e Fattori umani. Società Italiana di Ergonomia e Fattori umani, La storia dell'ergonomia. SIE—Società di Ergonomia e Fattori umani.

Stanton NA (1998) Human factors in consumer products. Taylor & Francis, London

Stanton NA et al (2014) Guide to methodology in ergonomics: designing for human use. CRC Press, Boca Raton (1st ed.: Stanton NA, Young MS (1999) A guide to methodology in ergonomics: designing for human use. Taylor & Francis, London and New York)

UNI EN ISO 9000:2015. Sistemi di gestione della qualità. Fondamenti e vocabolario. Vocabolario Treccani della lingua italiana. www.treccani.it/vocabolarioonline

Wilson JR (1995) A framework and a contest for ergonomics methodology. In: Wilson JR, Corlett EN (eds) Evaluation of human work. Taylor & Francis, London

Wilson JR, Sharples S (ed) (2015) Evaluation of human work. CRC Press, Boca Raton (1st ed.: Wilson JR, Corlett EN (1990) Evaluation of human work. Taylor & Francis, London and New York)

Woodson WE (1981) Human factors design handbook: information and guidelines for the design of systems, facilities, equipment, and product for human use. McGraw-Hill, New York

Chapter 2
Design for Ergonomics

2.1 Introduction

There are many points of contact—and, today, of integration—between the theoretical and methodological approach to Design and the approach to Ergonomics, particularly to Human-Centred Design and User Experience.

Firstly, the centrality of the user—or rather the centrality of people—in the design and development process of the product or the system,[1] which constitutes the main point of contact, both from an ethical point of view, starting from the definitions of Ergonomics and Design themselves, and a methodological and operational point of view.

Furthermore, Design and Ergonomics operate chiefly at the problem-solving level, rather than the abstract theoretical level, and are characterised primarily by the intervention portion rather than the disciplinary field.

In fact, this capacity for design intervention, and the intervention methods that aim to synthesise innovative solutions to design, are the specific contributions that Design bring to Ergonomics and its multi-disciplinary set-up.

Today, it is in this context that we see the progressive approach and, often times, the overlap between the approaches of Human-Centred Design and those of Design-Driven Innovation and Design Thinking. Today, in particular, the synthesis of the methods for evaluating usability and the user experience, which are typical of the Human-Centred ergonomics approach and the Design-Driven Innovation and Design Thinking approach, represents one of the most up-to-date areas for research and intervention, one that can offer innovative solutions for both training processes and the production of products and services, along with social innovation.

The central aspect of the progressive integration of Ergonomics and Design is the capacity for innovation that the contents and methods of Ergonomics/Human-Centred Design (E/HCD) bring to the process of designing and producing system

---

This chapter's texts are partially taken from chapters by the same author, Tosi (2016c).

[1] See Chap. 1, notes 2, 3 and 4.

© Springer Nature Switzerland AG 2020
F. Tosi, *Design for Ergonomics*, Springer Series in Design and Innovation 2,
https://doi.org/10.1007/978-3-030-33562-5_2

products and services today, directing it towards the actual needs of the people these things are intended for, as well as the new areas for development and experimentation offered by the project and, in particular, Design.

As previously outlined in this book's presentation, Design provides E/HCD not only with the capacity to synthesise knowledge and research results from different disciplines and professions—and, obviously, concrete solutions for intervention— but also, most importantly, a design-based overview of the analysis and problem-solving process.

2.1.1 The Role of Design

Design, and its design methods, are now recognised on a global level as strategic factors for innovation and social and economic growth. In the European Union documents regarding strategies for innovation, Design is defined as *"a key discipline and activity to bring ideas to the market, transforming them into user-friendly and appealing products or services (…) Design provides a series of methodologies, tools and techniques that can be used at different stages of the innovation process to boost the value of new products and services. When applied to services, systems and organisations, user-centred design thinking drives business model innovation, organisational innovation and other forms of non-technological innovation. These methodologies may also be instrumental when addressing complex and systemic challenges, for example in redesigning public services and in strategic decision-making processes"*.[2]

The role of Design is defined chiefly as *design-intervention activities* based on the capacity for innovation and, moreover, *the capacity for synthesis and connection* between the different professional skills involved in the training and development processes for new products and services and, more generally, the decision-making processes in production and social areas.

In fact, Design is—by definition—a capacity for creative synthesis, based on the ability to imagine and make innovative design solutions achievable, and develop them into a finished product.[3]

The role of the designer is linked to his design ability, that is, his ability to intervene on what already exists, based on his capacity to interpret the complexity of the innovative and changing factors that surround us and to develop intervention solutions that can respond to people's needs, expectations and desires, while also proposing new behaviours and suggesting new lifestyles.

This innovation can be configured in many ways: from the application of a new material to the original application of traditional materials, from the ability to provide new answers to consolidated needs and expectations to the ability to interpret as-yet unexpressed needs. More generally, the innovations can be configured as *"advances*

---

[2]Cfr. European Commission (2013). On this matter, see European Commission (2011).

[3]For the meaning of "product" used in the book, see note 1.1 and Chap. 4.

in the performance of products that are made possible by frontier technologies and product improvements suggested by a more efficient analysis of client needs. The former is the domain of radical technology-driven innovation (technology push); the latter is the domain of incremental market-based innovation (market pull)" (Verganti 2009, p. 4).

A second, no less important, aspect to be highlighted is the role of Design as a connective factor between different disciplinary and professional skills and specificities. The professional designer and, more generally, Design as a field of research and intervention, operates within a system of skills that are typically well articulated, in which each intervention is tackled using operational tools and perspectives that are very distant.

As Celaschi (2008, p. 21) writes: *"The Design we study today appears as a relations-based knowledge among other skills. As a discipline that seems to consolidate around the awareness of not producing autonomous knowledge (or, in any case, of not having yet done so) with the possibility for analysis and with a knowledge of other diverse and historic disciplines in the modern sciences; if anything, by respecting the statutes and analytical knowledge gathered from the other disciplines, it uses this as a design input, as a base on which to develop transformative actions that are organised for the world of goods that surround us"*. Design as a discipline *"rests in the middle of four systems of knowledge (inputs) that traditionally do not communicate well with each other: the "humanities" and technology/engineering on one axis, and art/creativity and economics and management on another axis, perpendicular to the first"*.

Both the articulation of the meanings attributed to Design and the intervention methods that the different versions of Design can now offer to the complexity of the processes for training and developing products, goods and services are essential.

"There has been a shift in understanding during the last 10–15 years towards a more strategic view of design in business, and towards design as an essential activity for user-centred innovation in business, academia and (although to a lesser extent) in policy making. This has resulted in a number of schools of thought about the contribution of design, and new terminology including labels such as 'strategic design', 'design management', 'concept design' and 'design thinking'. The schools of thought may all have their own particularities, but they also have a number of points in common, namely:

- *Focus on user-centred problem solving: Design is seen as a way of identifying and solving user problems by for example studying users and/or by involving them through visualisation and participatory design techniques such as co-creation. User-centred design innovation stresses human needs, aspirations and abilities, and strives for holistic and visionary solutions.*

- *Design as a multidisciplinary and cross-functional innovation activity: The designer facilitates cross-disciplinary innovation processes and interactions by bringing together individuals from different corporate functions within a company, such as management, engineering and marketing, but may also bring in expertise from disciplines such as psychology, sociology, anthropology and arts.*
- Design as a holistic and strategic activity: Design considerations—i.e. putting the user at the centre—permeate the innovation process, from product development, customer service and management up to the highest levels of hierarchy. Rather than 'design as styling' added on towards the end of the product development process, the user is the focus in earlier (more strategic) phases. Design is a core element of company strategy and helps visualise possible scenarios to support strategic decision making".[4]

Thus, Design is described and required as a tool for innovation, both as it relates to the capacity for design intervention—that is, the ability to provide a design response to complex requests and requests coming from the company and the production system, from the social system and the reference user (which are not always the same and can not always be reconciled with each other)—and as it relates to the intervention methods which Design can act within and use to interact with this system.

Even when designing products that are not very complex, such as furnishings or individual everyday products, Design operators within design and production processes that involve production technicians, IT engineers, marketing managers, etc. and, naturally, acts within and/or in close collaboration with the corporate structure and its organisational features.

The complexity of the production system requires the ability to synthesise different problems and requests and, when necessary, to be able to communicate and compare very different skills, languages and work and intervention tools.

The designer is required to find the most suitable solution based on the instructions and the limitations which concern the availability and processing methods for materials, product times and costs, organisation—and geographic location—of the production line(s), the market expectations as identified by marketing and, of course, the procurement methods, the supplier network for materials and semi-finished goods, the distribution network for products, etc.

In this context, it is essential to refer to a precise definition of the term *Design* and the term *Project*.

---

[4]European Commission (2011).

The term "Design" in English[5] coincides in a broad sense with the Italian term "Progetto", which can refer to an object or a building, but also an organisation, a study or research programme, a travel programme or a life programme.

The meaning attributed to the term "Design" in Italian is different, and is globally attributed to "Italian design".

Design is the act of giving shape to objects by (creatively) integrating the complexity of the factors involved, whether functional, cultural, technological or economic.[6]

Leaving aside the merits of Italian Design's history—which has always been closely tied to the production sector and, particularly, the industrial products sector, and a factor for innovation that is closely integrated into the logistics and limitations of industrial production—it is worth highlighting the specificity of Italian Design, which is placed precisely within design culture—specifically in the area of Architecture and Design—and born from the synthesis of humanistic and technological culture.

As Trabucco (2015, p. 53) writes: "*Rooted in the architectural tradition, Italian Design—in both its radical manifestations and those that are more consistent with industrial logistics—is refined and cultured, and finds its foundations in history for the construction of an independent and, at times, eclectic modern aesthetic. Perhaps this is why it does not simply represent a production-sales cycle, but is instead able to translate the potential of production and the technological and linguistic innovations that are being defined in the explosive acceleration and shift (...) in*

---

[5]Oxford Dictionary defines "design" as:

- As noun: "A plan or drawing produced to show the look and function or workings of a building, garment, or other object before it is made. The art or action of conceiving of and producing a plan or drawing of something before it is made. The arrangement of the features of an artefact, as produced from following a plan or drawing".
- As verb (with object): "Decide upon the look and functioning of (a building, garment, or other object), by making a detailed drawing of it. Do or plan (something) with a specific purpose in mind". Cfr. "Design" on Oxford Dictionary online: www.oxforddictionaries.com (consulted March 2018). The definition of Industrial Design in the Treccani Encyclopedia is interesting in this regard:
- *industrial design*: design of items that are intended to be produced industrially, that is, with machines and in lines. This meaning of design is best expressed with the Anglo-Saxon term industrial design, thanks to the English distinction between *design* and drawing.

Cfr. Treccani Encyclopedia online, www.treccani.it/enciclopedia (consulted March 2018).

[6] "*Industrial design is a design activity consisting of determining the formal properties of industrially produced items. Formal properties refers not only to features but, above all, to functional and structural relationships that make an item a coherent unit, from both the producer's and user's point of view.*

While the exclusive concern of the external features on an item often hides the desire to make it more attractive or even to hide construction weaknesses, the formal properties of an item—at least as I understand it—are always the result of integrating many factors, be they functional, cultural, technological or economic.

In other words, while the external features treat this as a foreign entity, one that is not linked to the item and was not developed with it, the formal properties, on the contrary, represent a situation that corresponds to its internal organisation, which is linked to it and has been developed with it." (Maldonado 1961).

the communicative, aesthetic, performative and ergonomic quality of new industrial products, thus becoming a powerful indicator of societal modernisation".

In the field of academia,[7] Design is defined in Italy as design capacity that is based on theories and intervention methods specifically aimed at operating within complex production and social systems and declines on the basis of varied sectors of intervention: Product design, Interior design, Communicative design, Fashion design, along with the history of Design, which represents its essential cultural and cognitive foundation.

Over the years, the four basic sectors of Design have been complemented by transversal specialisations that can be clearly identified today by the specificity of the fields of intervention, objectives, methods and languages. These include, chiefly, Sustainable design, Service design, Design for all, Ergonomics and Design, Interactive design, etc. These Design specialisations are defined based on the theoretical contents and structured and congruent methodological tools—which are, in turn, applied or applicable to specific intervention areas—which complement and integrate into the four traditional sectors.

These include: Mobility design, Health design, Equipment design, Digital interface design, etc.

2.2 The Ergonomics-Design Relationship, from Dialogue from Afar to Progressive Integration

For a long time, the relationship between Ergonomics and Design and between Ergonomics and design has been on the fringe of the respective intervention fields. Until the 90 s, the relationship between Ergonomics and Design was regarded as one of two distant worlds: the former, one of rules, evaluation methods and technical regulations, the latter, of design ability and creative freedom.

In the field of ergonomics, the intervention of the designer—architect or designer—was mainly required at the end of the evaluation process conduct by professional ergonomists, who were specialists in the field of physical or cognitive Ergonomics. The designer was asked to provide a design response for what emerged

---

[7]"In the field of Design, the scientific disciplinary contents refer to theories, methods, techniques and tools for the design of items—material and virtual—with reference to their morphological features in their relationships with: the needs and behaviour of users; productive, constructive, performance, security and quality features of the industrial systems; functional, ergonomic and economic, social and environmental sustainability requirements; visual languages, artistic practices, aesthetic and cultural meanings. The scientific contents refer to the theoretical and historic-critical status of the items and the specific forms of design thought as an interdisciplinary practice and a moment of synthesis for the various knowledge involved in the design of items in their life cycle, as well as the strategic preconfiguration of socio-technical scenarios and the configuration of new solutions through the application and transfer of technological innovations. The fields of research and application refer to product, communicative, interior and fashion design and their integrated systems." Cfr. Italian Ministry of University Declaration DM 30-10-2015 n. 855 "Reassessment of macrosectors and insolvent sectors."

during the evaluation phase, and to use his skills to find the solution—design, to be exact—during the final phase of each intervention.

In the field of design, the architect's training did not include—nor does it include now—any teaching or awareness of the field of ergonomics. The designers—researchers and/or professionals—turned to the skills of ergonomists for instructions and advice outside of, or parallel to, the formation and development of design.

Obviously, there are some exceptions, many of them well known, chiefly represented by large companies that have always been mindful of the safety and well-being requirements of their products and aware of the essential contribution made by Ergonomics.

The role of Ergonomics is precisely located in research centres with specific skills dedicated to evaluation safety, the anthropometric and postural aspects and/or cognitive aspects of the finished products and, in parallel, the health and safety of work stations and locations within the company.

One of the best known is the FIAT Research centre, in which the sectors dedicated to Ergonomics have been traditionally divided into physical and cognitive Ergonomics and the Whirlpool Ergonomics and Usability (today, User Experience) Centre, the activity of which is named after the evolution of the User/Human-Centred approach.

In terms of the world of design, the role of certain designers and architects who have always been closely linked to Ergonomics and active in the field of ergonomics since the '70s is very important. Two of these are Luigi Bandini Buti and Isao Hosoe.

Bandini Buti, the first and most famous Italian designer to call himself an "ergonomist", is responsible for design and research explicitly placed in an ergonomic context that represented the point of departure and reference for all those who developed their own design businesses—researchers and/or professionals—in the field of Ergonomics and Human-Centred Design. His academic contribution at Milan Polytechnic and the many other Italian colleges where he taught and continues to teach is of great importance; it has allowed generations of Design students to interact with Ergonomics in such a way as to create designs focused on the needs and desires of people.

The role of Isao Hosoe is of equal importance. His products, which are well known nationally and internationally, have represented perfect examples of Design and Human-Centred design for many decades.

Until the '90s, however, these were unique cases in the academic and professional image of Design; their design intelligence and ability is responsible for constructing the foundations of Ergonomics and Design (Figs. 2.1, 2.2, 2.3, 2.4, 2.5 and 2.6).

2.3 Design for Ergonomics

Even if it is not possible to identify a specific date or event that officially marks the beginning of a concrete integration between Ergonomics and Design, certain phenomena have marked their gradual progress and, in the last two decades, the

Fig. 2.1 Radio cube Siemens Alpha 2 RK-501. Luigi Bandini Buti, 1971

Fig. 2.2 Floor lamp for Kartell (left). Luigi Bandini Buti, 1965. Disco telephone Italtel Rialto (right). Luigi Bandini Buti, 1970s

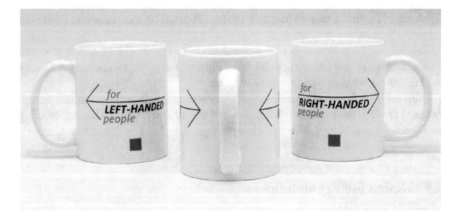

Fig. 2.3 Left-handed mugs. "Not everything is equal for left- and right-handed people". Luigi Bandini Buti, 2015

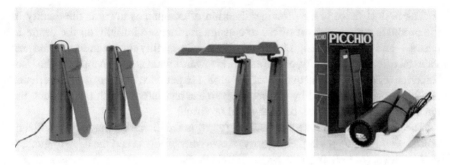

Fig. 2.4 Picchio, table lamp for Luxo. Isao Hosoe, 1980s

Fig. 2.5 Tama, table lamp for Valenti. Isao Hosoe, 1975

Fig. 2.6 Hebi, table lamp for Valenti. Isao Hosoe, 1970

growth of an area of Ergonomics that is specifically directed towards design and, in parallel, an area of Design that takes its theoretical and methodological contents from Ergonomics, in particular, the Human-Centred Design approach.

The first of these is the growing attention of companies towards the quality of the products and services it offers to consumers, focused initially on the desire to ensure—and publicise—high levels of safety and usability for the products and services on offer and, secondly, on the attention towards what is known today as the User Experience, that is, on the quality of the global experience—in terms of experience and impact—experienced by the people who use and interact with the product, the environment or the service, be it physical or virtual.

The second, though it seems obvious today, is the dissemination of digital technologies and languages in the majority of everyday products and the need to guarantee high levels of user-friendliness in digital interfaces and their infinite applications. The dissemination of dialogue screens and means of digital interaction in household appliances, cars, communications devices, ticket dispensers, vending machines and, finally, care devices (from medical equipment to supports for movement that can also be used in homes), etc. has created new levels and means for market competition for companies, which is often based on the ease and instant understanding of how to use the dialogue interface and the aesthetic and sensory appeal of the shapes, materials and surface treatments used on the products themselves.

Making products safe, easy to use, pleasant and instantly understandable, and making these qualities immediately apparent at the moment of purchase, has become one of the chief factors in success on the market in recent years.

Design is called upon to create solutions capable of guaranteeing—and publicising—the quality of the product in terms of the user's safety and well-being, its ease of use and, finally, the user experience, focusing on the needs, expectations and desires of the people who will choose and/or use the product and favour it over the myriad others that are available on the market.

In fact, as Norman writes (1998, p. 48), "*Ease of use has many benefits for a company. Not only are customers more likely to be satisfied with the product, but the need for service desks should decrease. (…) In the mature marketplace, where there is fierce competition, relatively low prices, and low profit margins, a single call from a customer can often cost the company enough to wipe out any profit from the sale of the item. Here is where one of the largest economic impacts of good product development can be measured. (…) In addition, satisfied users become repeat purchasers, likely to recommend both the product and the company to friends and colleagues, enhancing the overall reputation*".

Therefore, it is up to companies to request a *Design centred on the user* then *son the person*, in a series of requests that begins with a focus on the safety levels when using the product and then on the comfort levels and well-being of household and office furnishings, on the simplicity of using the physical and digital interfaces (referred to and publicised as ease of use, user friendliness, up to the definition recognised as "usability", particularly for digital interfaces), up to the aesthetic and sensory appeal and, finally, the user experience.

This course, which also explains the strong development of Ergonomics and Design/Ergonomics in Design, both in Italy and in other countries that are dealing with a much less marked, or even absent, development in other areas of Design, particularly as it relates to Architecture, has naturally followed subsequent phases in both the professional and academic fields. Design is progressively approaching Ergonomics, which provides precisely this knowledge and these intervention methods to guide the entire training and product development process, starting from the needs and expectations of people and, in parallel, Ergonomics is approaching Design as a repository of design-intervention skills, capable, by definition, of "giving shape to the product"[8] by synthesising the contributions and knowledge of people's current and potential expectations in the design. An essential transitional moment in this integration process is the 1988 publication of the ISO 9241/11, Ergonomic requirements for office work with visual displays terminals (VDTs) Parte 11: guidance on usability regulation,[9] which has become an essential point of references for all designers involved in ergonomics.

As has been written by numerous authors on a number of occasions, the ISO 9241-11:1998 regulation (which has now been replaced by 9241-210:2010) contains definitions and methodological guidelines aimed at evaluating and designing information systems, but is general enough to be of interest in every field of design.

The publication of many of Donald Norman's books in Italy is also of great importance. The most famous of these is *The psychology of everyday things*,[10] which is written in a language that is accessible to those who are not experts in cognitive psychology. It has been adopted for many university design courses and, from the late 90 s, in the first courses for "Ergonomics for Design",[11] subsequently becoming a reference text for courses in Industrial Design and Design.

---

[8]Cfr. Maldonado (1961); see note 7.

[9]The ISO 9241-11:1998 regulation was implemented by European regulation UNI/EN 29241, Ergonomic requirements for office work with video display terminals (VDT)—Usability guide, part 1. ISO 9241 has now reached its 210th part, as already been cited for ISO 9241-210:2010, Ergonomics of human-system interaction. Human-Centred design for interactive systems, which majorly revises the contents of the earlier part 11, extending them to the themes of User Experience and broadening the concept of the user—understood as the person who uses a given product—to people who interact with the product, regardless of its specific use. The definitions contained in ISO 9241-210:2010 are outlined in Sect. 5.3, "Usability".

[10]Norman, *The Psychology of everyday things*, 1988, was publishes in Italy as: *La caffettiera del masochista, psico-patologia degli oggetti quotidiani.*

The new edition of the book *La caffettiera del masochista, il design degli oggetti quotidiani*, was published in Italy in 2014 (original: *The design of everyday things*, 2013).

[11]Ergonomics made its official entry into Italian degree courses in the mid '90s, in the Milan Polytechnic and Turin Polytechnic, and in the Faculties of Psychology in Siena and Turin.

In fact, despite being present in many official Architecture and Psychology courses, Ergonomics was mainly taught in themed seminars or lessons until the 1990s, without being recognised as an independent subject.

The first Ergonomics courses were introduced in the late 90s with various titles. Courses in the "Ergonomic requirements of industrial products" started at Milan Polytechnic, while specific Ergonomics subjects were taught in other settings through classes and workshops in "Environmental design" and "Component morphology".

In the field of Product design, the theme of usability is of strategic interest, both because the means of interaction between the user and the product pose very similar problems to those that characterise Human Computer Interaction (possibility of error, correct understanding of feedback, analogies between the arrangement of the commands and the parts being commanded, etc.) and because the digital language has become an integral part of everyday products.

From household appliances to vending machines, home phones to mobiles and train, bus and tube ticket dispensers, a huge number of everyday objects now boast digital displays which require, and demand, the use of a language that is based on the interactive methods and conventions for digital interaction.

This phenomenon, which has been pervasive for more than a decade, has been joined in recent years by touch-screen interaction methods, which have also been disseminated very rapidly and become a common means of communication, not just in the widely used digital devices of today (smartphones, tablets, etc.), but also commonly in other widely used devices, from ATMs to systems for booking and paying for medical appointments, transport systems, etc.

Design's interest in the contents of User-Centred Design and the means for evaluating Usability was initially born from the direct application of industrial products and evaluation methods born in the field of cognitive psychology to the sector, in order to evaluate the human-computer interaction (or, better yet, the interaction between the person and the digital interface of the ICT systems).

Their application in the field of design clashes, however, with objective difficulties, due to the evident diversity of the object of analysis, as well as the diverse disciplinary origins in which they were conceived, which comprise, in the case of designing products and physical environments, elements characterised by their physical, morphological, sensory, cognitive and emotional features, which require the global evaluation of the different dimensions in which people interact with them.[12]

In other words, the relationship with an environment or an object is based on the possibility of seeing, touching and handling shapes and materials, of perceiving their smell, temperature, consistency and, at the same time, on our ability to understand the ways they work and to appreciate (or not) their sensory and aesthetic qualities.

At the same time, the specificity of the design sector, and the complexity of the industrial production processes, require the definition of equally specific methods

---

In subsequent decades, Ergonomics courses were inaugurated with increasingly more specific names and contents, leading to the current titles of courses such as "Ergonomics", "Cognitive ergonomics" and "Ergonomics and Design" and the design workshops in "Ergonomics and design" found in almost all Italian degree courses in Industrial design, Design and Psychology. Courses at the first and second level of training (undergraduate and postgraduate degrees) were joined by the Masters in Ergonomics in the mid 90s, inaugurated in 1996 at Milan Polytechnic (director Adriana Baglioni) and Turin Polytechnic (director Alessandra Re), based on the Eur.Erg Ergonomics training programme. Today Masters in Ergonomics with different specialisations (Workplace safety, Product and environmental design, Organisational ergonomics, etc.) are present in Italy and other European countries, based on the new Eur.Erg training model. This was approved in 2011 by the Centre for Registration of European Ergonomists (CREE). On this matter, see Tosi (2016b, pp. 23–31).

[12]On this matter, see: Tosi (2006), Tosi and Anselmi (2006).

for evaluating the needs[13] of people, as well as the performance of the product, its ability to intervene on the quality of the process and the design by aiding with the design process and with the procedures and goals of industrial production.

The process of designing and producing industrial products—and, obviously, the reasoning can also be valid for the process of designing and constructing buildings, albeit with the appropriate differences—is, in fact, radically difference from what characterises the process of designing information and communication systems, which is strongly characterised by the constraints and priorities imposed by industrial production, by the need to evaluate the feasibility and cost of morphological choices, as well as the technological solutions being adopted, the materials that can be used and the methods and timelines for production.

The development of an area of Ergonomics specifically aimed at design, which is not intended as an application of the contents elaborated in other fields of research, but as an autonomous research sector in its own right, one characterised by the specificity of its theoretical and methodological contents, presupposes the development of a collection of definitions, criteria for intervention and evaluation methods specifically aimed at the design sector. This is capable of constituting a collection of shared, operational tools and references and a common basis for dialogue and comparison for the different professional figures that operate within this sector.

The evaluation and the design *of the Usability of the product*—physical and/or virtual—has required, over the years, the development of intervention methods and tools that are specifically aimed at this sector[14] which, based on the close collaboration between the different skills needed for the development of a process for evaluating User-Centred Design, have led to the integration of ergonomic skills into the training and development process for industrial products, which is the goal to be achieved for the design to represent the same definition of Ergonomics. As previously mentioned in the introduction, Ergonomics and Design, their theoretical contents and evaluation and intervention methods, are based on a dual, radical shift of attention towards both the object being interacted with and the evaluation and interpretation of the intervention's context.

In fact, Ergonomics and Design shift the attention from the evaluation and design of the product, environment or system itself to the evaluation and design of the interaction between people and the product, environment or system, within a specific context, taking into account all of the variables that could impact said interaction.

Attention has also shifted from the evaluation of the objective reality—that is, the current situation in which to intervene with corrective or adjustment actions—to the projection into the future, where possible, of the reality of the interaction, and to the definition of the most suitable design solutions to respond to the needs and desires

---

[13]For the definition of "needs" used in this book, see Sect. 1.1.

[14]The UNI 11377 "Usability of industrial products. Part 1: terms and definitions. Part 2: Intervention methods and tools" standard was published in 2010.

The Standard was drafted by the homonymous working group within the "Ergonomics" Commission of the UNI, "General principles" GL, which was composed of: Francesca Tosi (coordinator), Lucio Armagni, Paola Cenni, Luigi Bandini Buti, Laura Anselmi. Barbara Simionato, Fiammetta Costa and Sabrina Muschiato were also collaborators.

of the person. The ergonomic approach uses the description and understanding of all of the variables that define the "usage context" as a point of departure, that is, all of the factors that determine the conditions and means in which certain people interact (or are able to interact) with a given product or system.

To do this, one must therefore answer some questions to start: *what is the product? Who uses it? Why and for what? Where? How? For how long?*

In other words, one must clarify and explain the reference framework according to the 5W's of the Anglo-Saxon journalistic tradition (What? Who? Why? Where? When?), adding "How?".

Translating these questions into a design key means adding, in a simple manner, their declination to the conditional and the future, expanding the attention beyond the current, objective situation (what is happening here and now; what do users expect in the known situation) to its possible evaluation.

Thus, the questions become: who can or could use the product? When and how can it be used? Why? Where?

Moreover: in what way could it be used (in the correct way or the wrong way? In the expected way or for other activities or goals?)? In what context (for example, in different settings, with different technologies, etc.)?

And, as a result: *what could a new product be?*

Today, Ergonomics and Design is configured as Design for people, that is, an approach to the design that centres—by definition—on the person, with its theoretical and methodological references based on the contents and methods of Human-Centred Design, closely integrated with the interpretive and propositional tools used for the culture and practice of the design.

References

AA.VV (ed) (2014) Ergonomia, valore sociale e sostenibilità. In: Atti del IX Congresso nazionale SIE, Società Italiana di Ergonomia, Roma 27–29 ottobre 2010. Nuova Cultura, Roma

Anselmi L, Tosi F (eds) (2004) L'usabilità dei prodotti industriali. Moretti & Vitali, Bergamo

Bagnara S et al (ed) (2006) L'Ergonomia tra innovazione e progetto, sistemi di lavoro e stili di vita. In: Atti dell'VIII Congresso nazionale della SIE, Società Italiana di Ergonomia, Miano 9–11 febbraio 2006. Moretti & Vitali, Bergamo

Celaschi F (2008) Il design come mediatore tra saperi. In: Germak C (ed) Uomo al centro del progetto. Design per un nuovo umanesimo. Allemandi, Torino

Enciclopedia Treccani. Online: www.treccani.it/enciclopedia

European Commission (2011) Design as a driver of user-centred innovation. EU staff working document. European Commission, Brussels

European Commission (2013) Implementing an action plan for design-driven innovation. EU Commission staff working document. European Commission, Brussels

Germak C (ed) (2008) Uomo al centro del progetto. Design per un nuovo umanesimo. Allemandi, Torino

Karwowski W (ed) (2006) International encyclopedia of ergonomics and human factors. CRC Press, London and Philadelphia (1st ed.: Karwowski W (2001) International encyclopedia of ergonomics and human factors. Taylor & Francis, London and New York)

Maldonado T (1961) La formazione del Disegnatore Industriale, relazione al Congresso ICSID a Venezia

Micheletti Cremasco M et al (eds) (2013) L'Ergonomia verso un modello di città sostenibile. In: Atti del X Congresso nazionale della SIE, Società Italiana di Ergonomia, Torino, 18–20 novembre 2013. Società Italiana di Ergonomia, Milano

Norman DA (1988) The design of everyday things. Doubleday, New York (1st ed.: Norman DA (1988) The psychology of everyday things. Basic Books, New York)

Norman DA (1992) Turn signal are the facial expressions of automobiles. Norton, New York

Norman DA (1993) Things that make us smart. Addison-Wesley, New York

Norman DA (1998) The invisible computer. MIT Press, Cambridge

Norman DA (2004) Emotional design. Basic Books, Cambridge

Norman DA (2010) Living with complexity. MIT Press, Cambridge

Norman DA (2013) The design of everyday things. Basic Books, New York

Preece J et al (2004) Interaction design. Apogeo, Milano (1st ed.: Preece J et al (2002) Interaction design, beyond human-computer interaction. Wiley, New York)

Rubin J, Chisnell D (2011) Handbook of usability testing: how to plan, design, and conduct effective tests. Wiley, Indianapolis (1st ed.: Rubin J (1994) Handbook of usability testing: how to plan, design, and conduct effective tests. Wiley, New York)

Stanton NA et al (2014) Guide to methodology in ergonomics: designing for human use. CRC Press, Boca Raton (1st ed.: Stanton NA, Young MS (1999) A guide to methodology in ergonomics: designing for human use. Taylor & Francis, London and New York)

Tosi F (2006) Ergonomia per il progetto, progettare e valutare la qualità ergonomica dei prodotti industriali. In: Ergonomia, 5

Tosi F (ed) (2016a) La professione dell'Ergonomo nella progettazione dello ambiente, dei prodotti e dell'organizzazione. FrancoAngeli, Milano

Tosi F (2016b) La formazione in Ergonomia. In: Tosi F (ed) La professione dell'ergonomo. Nella progettazione dell'ambiente, dei prodotti e della organizzazione. FrancoAngeli, Milano, pp 23–31

Tosi F (2016c) Ergonomia e Design, Design per l'Ergonomia. In: Tosi F (ed) La professione dell'ergonomo. Nella progettazione dell'ambiente, dei prodotti e dell'organizzazione. FrancoAngeli, Milano, pp 141–151

Tosi F, Anselmi L (2006) Product usability evaluation. In: Karwowski W (ed) International encyclopedia of ergonomics and human factors. CRC Press, London and Philadelphia

Trabucco F (2015) Design. Bollati Boringhieri, Torino

Verganti R (2009) Design driven innovation. Etas, Milano

Wilson JR, Sharples S (ed) (2015) Evaluation of human work. CRC Press, Boca Raton (1st ed.: Wilson JR, Corlett EN (1990) Evaluation of human work. Taylor & Francis, London and New York)

Chapter 3
From User-Centred Design to Human-Centred Design and the User Experience

3.1 The Evolution of User-Centred Design

User-Centred Design (UCD) is a design approach that arose in the IT field in the 1970s–80s and was subsequently adopted and applied in the field of Design. The goal of this approach has always been to focus the development process of the product/service in such a way as to ensure a high degree of usability.

In the 90s, however, UCD underwent an evolution, thanks to many studies and experiments relating to Design; in particular, these defined investigative methodologies for defining users' needs. The most significant shift involved the role of the user in the design process (Fig. 3.1).

PRODUCTS Evaluation research	EXPERIENCES Experimental research	IDEAS Generative research
Requirement mapping Usability evaluation metrics Task Analysis Application of standards	Field observation Storytelling Analyses of activities Research with quick and dirty mock-ups and prototypes	Creative techniques Probes design Co-design
1970 - 1990 I	1990 - 2000 I	2009 >
Quality and performance	Relational needs	Envisioning for innovation

Fig. 3.1 Evolution of user-centred methods. Revised by Rizzo (2009, p. 12)

This chapter was authored by Alessia Brischetto.

© Springer Nature Switzerland AG 2020
F. Tosi, *Design for Ergonomics*, Springer Series in Design and Innovation 2,
https://doi.org/10.1007/978-3-030-33562-5_3

The evolution of the UCD approach was dictated by the increased complexity of daily-use products and technological innovation, which significantly changed the nature of the user-product-system or service interaction.

In 1990, the ISO defined Human-Centred Design as: "An iterative process that consists of studies with users and specific design and evaluation solutions, which ultimately aims to design products and services based on the needs of the final users."

In the space of twenty years, we have moved from predominantly tangible interactions to an increasing number of intangible ones. In many cases, the interaction between a user and a product is composed of physical interactions that alternate or equate to virtual ones.

Social experiences and relationships have been strengthened and, in some cases, taken to the extreme, creating a continuous exchange between the private and public sphere. In turn, the need for a design approach that can respond to increasingly more complex and ephemeral needs and experiences has grown too.

The Design sector has been forced to change its scope and expand its boundaries. Focusing the design activities on aspects like appeal, usability and safety is no longer sufficient. Conversely, as Rizzo (2009, p. 11) states, Design today "is required to act as a pragmatic approach and way of thinking (design thinking), for the design of real, daily-use contexts in people's lives, that is, because design is able to act in accordance with Human-Centred ethics, unlike other conceptual disciplines."

The UCD approaches of the 1970s, 80s and 90s developed within the Human-Computer Interaction sector and aimed to improve the usability of interactive system interfaces.

The beyond usability and user experience movement represented the driving force in the redevelopment of the UCD approach and methods.

The evolution of the approach within the Design sector chiefly related to ways of involving the end users in the design process. Users are playing an increasingly active role; they are no longer seen as simply sources of information but as holders of experience, who can be involved actively and iteratively during each phase of design development.

A new dimension was further defined in the 2000s: the widespread co-creation design between the designer and the user—Co-Design (see Chap. 7 by A. Rinaldi).

The need to measure and design parameters and aspects relating to the user experience have also taken over the Human-Centred design process, especially in recent years, thanks to the dissemination of interactive systems in our daily lives. In this area, a significant contribution has been made by the studies of Jordan (2000) and Norman (2004), who highlighted how the emotional elements of the user experience are strategic aspects of design activity.

The study of how easy it is to use daily-use or common items has also shown how important it is to consider the complexity of the interaction.

The definition of usability itself has undergone many changes. Queensbury (2003) is one of the most interesting; it identifies the following elements as fundamental goals of usability: efficiency, effectiveness, error tolerance, how easy it is to learn and engagement.

In particular, engagement focuses on the importance of emotional involvement when functioning and during interactions between items and systems. The evolution of the definition of usability and the shift from User-Centred Design to Human-Centred Design was caused by the established theory that a positive or negative emotion can considerably transform the user experience, regardless of the product's other usability indicators.

Norman himself tackled this issue, changing his thinking on the subject of Usability. If he identifies Usability as a key indicator of quality in the Design of daily-use products in his book, The Design of Everyday Things, his thought process evolves in his subsequent book, Emotional Design. Norman states that, even if an item is difficult to use, it will still be used and favoured over items that may be more reliable from an operational or ease-of-use point of view, as long as it is able to elicit emotion and a positive experience that lasts.

The rise of low-cost technologies has led to a shift in attention from physical to metaphysical implications for many of the most successful products and brands. Some examples include well-known brands like Alessi, Armani, Apple, Facebook, Ferrari, Google, IKEA and Philips.

For example, Alessi's products have concentrated on emotional involvement, while defining new meanings has been essential for companies like Apple (Giacomin 2014, p. 608).[1] This does not mean that usability and the quality of the interaction should not continue to be one of the key goals of design activities, but rather that it is necessary to expand and consider emotional and relational aspects during design activities.

In conclusion, if the initial goal of UCD was evaluating usability, this has evolved over time towards design activities that consider the entire user experience from the offset and which the user plays a strategic role in at every step of the design process.

The ISO 9241-210:2010 standard also supports this intervention philosophy. It supplies a comprehensive view of the strategic activities of the HCD approach.

Figure 3.2 illustrates the interdependence of the activities in the HCD approach. The diagram describes the non-linear process and shows how each Human-Centred design activity uses the results of the other activities. The processes are carried out iteratively; the cycle is repeated until the particular usability goals have been achieved.

The operational mode in Fig. 3.2 shows how the HCD process involves users from the first phases of the design intervention (the drafting of design requirements or specifications), but also, and most importantly, how it allows us to experiment with the early versions of the system and to correct issues based on the reactions and comments of users, in a process of testing and subsequent adjustments.

Once the need to develop a system, product or service has been identified, there are four essential steps that must be followed to integrate the usability requirements into the development process for the product/system: understand and define the usage context; define the requirements of the user or group of users; produce design solutions; verify the design solutions.

---

[1]For further information, see: www.ideo.com.

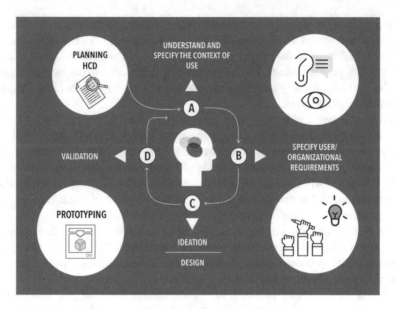

Fig. 3.2 Interdependence of the activities in the human-centred design process

The standard in point 4.1 describes six fundamental principles for the Human-Centred approach:

- the design is based on an explicit understanding of users, activities and environments;
- the users are involved in the design and development phases;
- the design is led and refined by human-centred evaluation;
- the process is iterative;
- the design addresses the entire user experience;
- the design team includes multi-disciplinary skills and perspectives. The activities in the HCD process are carried out iteratively; the cycle is repeated until the particular usability goals have been achieved. The iteration leads to a progressive improvement and perfecting of the product.

Evaluating the compliance of the requirements with the original goals, and the resulting ability to recognise that an error was made during the design phase, is an essential function to subvert traditional production policies and can lead to innovation.

Norman himself states that "the most difficult part of Design is discovering the right requirements", that is, the right solution to the problem (Norman 2013, p. 231).

David Kelly, the co-founder of IDEO, states that the purpose of design intervention must be to create prototypes and test them; his motto is well-known: "Make mistakes often, make them quickly".[2]

---

[2] Joseph Giacomin is director of the Human-Centred Design Institute (HCDI) at Brunel University London. See www.brunelac.uk.

Requirements identified in an abstract way are often the most incorrect, as well as those based on people's requests, since people do not always understand their own needs; the most correct way is to define the requirements of a product based on observations of how people act in their daily lives within their natural environments. This requires us to perform in-depth studies and evaluations in an iterative manner.

During the phase for defining requirements and the phase for validating proposals, the designers can use various surveying techniques or evaluation and design methods.

These operating techniques and methods are the result of research in the fields of Ergonomics, Cognitive Sciences, Human Factors, Human Computer Interaction (HCI), Sociology, etc. In recent years, we have also witnessed the definition of procedures aimed at evaluating the UX.

The latter discipline is a significant artery of the HCD approach and now supplied us with a system of techniques aimed at evaluating the experience and emotional dimension.

The theoretical framework is vast, the methodologies varied; there are no pre-defined strategies. Based on the identified problem, or the complexity of the product/system/service, we have a series of methods that we must choose and select on a case-by-case basis (see Chap. 6).

3.2 The User Experience

Recently, the term User Experience has spread throughout many different fields and contexts. This rapid uptake, however, was not accompanied by a precise definition of the concept, to which partially different meanings are attributed. These all share the ambition of overcoming the concept of usability and examination the interaction with the product/system/service (Fig. 3.3).

The User Experience (UX) is the sum of the emotions, perceptions and reactions that a person experiences when interacting with a product or service. In other words, it is equivalent to the level of subjective adherence between expectations and satisfaction when interacting with the system, be it physical (e.g. a ticket machine) or digital (e.g. online shopping). The User Experience, therefore, is a dimension of the design that revolves around the characteristics and needs of the users and focuses on the universality of the experiences within a specific usage context.

Dewey provides us with one of the most authoritative definitions of UX: "The experience is the irreducible totality of people who act, perceive, think, feel and make sense, including their perception and sensation of the good in context" (cited in McCarthy and Wright 2004).[3] The ISO 9241-210:2010 standard defines the User Experience as: "The collection of perceptions and reactions of a user, deriving from the use or expectation of a product, system or service." The usage experience, therefore, is subjective and focuses on the act of use. A series of additional notes (point 2.15) relate to this definition and explain that the usage experience includes all of the user's emotions, convictions, preferences, physical and psychological reactions, behaviours and actions, which occur before, during and after use (see Chap. 5, particularly Sect. 5.3) (Fig. 3.4).

---

[3]For further definitions, see: Law et al. (2009).

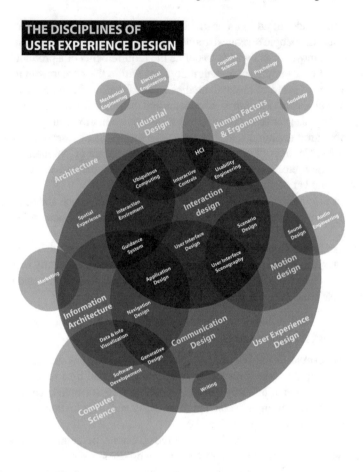

Fig. 3.3 The disciplines of user experience design. Reworked by the author from "The disciplines of user experience design" by Dan Saffer (2010)

The goals of usability are essential for designing the interaction and are implemented through specific criteria. The goals of the user experience are demonstrated in the outer circle and are less clearly defined.

In particular, the subjectivity of the experience produced through use of a product or service is highlighted, as well as the forecast for its use. The second aspect—the usage expectation—clearly separates the User Experience from the quality of use (defined in ISO/IEC 9126, which is now ISO 25010) and includes dimensions that are particularly relevant to the phase for selecting and purchasing a given product/system.

The temporal dimension of the interaction manifests when the item being evaluated moves from the usage phases (as an event) to the person. The evaluation of the User Experiences expands its scope from the usage phase and looks at aspects such

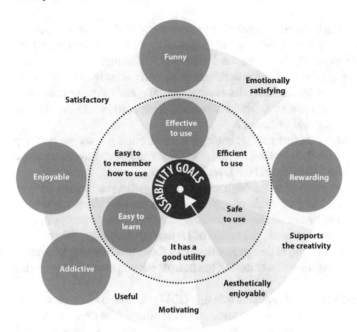

Fig. 3.4 Usability and user experience goals. Revised from Jennifer Preece's diagram. *Source* www.sharritt.com/CISHCIExam/preece.html

as, for example, anticipation, memory (the memory of said use) and attribution of meanings, desires and the construction of individual and collective identities.

The third note of the standard states that usability influences certain aspects of the user experience, for example, that the usability criteria can be used to evaluate aspects of the user experience.

Let's try to understand the difference between usability and User Experience. When we use the term "usability", we are typically referring to the pragmatic, non-emotional aspects of what the user experiences, including objective performance measures, subjective opinion measures and, obviously, qualitative data about usability issues. On the contrary, when we use the broader term, User Experience, we are typically referring to what the user feels, including the effects of usability, utility and emotional impact.

In short, usability, that is, the degree of effectiveness, efficiency and satisfaction in a person's interaction with the machine, is only one of the components of the User Experience: a web platform can be usable without necessarily ensuring a pleasant user experience.

It is important to recognise and understand the compromises between usability and the goals of the user experience. In particular, this allows designers to consider the consequences that stem from pursuing their different combinations in relation to the satisfaction of different users' needs. Naturally, not all usability goals and user experience goals apply to all interactive products that are developed. Some combinations can also be incompatible. For example, it may not be possible to design a control system (an emergency stop button, etc.) that is both safe and fun.

In summary, the user experience is the totality of the effects or perceived effects by a person as a result of their interaction with, and the usage context of, a system, device or product, including the influence of usability, utility and the emotional impact during the interaction and, finally, memory after interaction. "Interact with" is a broad concept that includes seeing, touching and thinking about the system or the product, including our admiration of the product and the effect of its presentation before any physical interaction.

The awareness that new technologies are offering greater support opportunities for people in their daily lives has led researchers and professionals to consider further goals around Usability and the User Experience.

The emergence of technologies like virtual reality, the Internet and mobile computing in a variety of fields of application has brought about a much wider series of considerations. In addition to focusing chiefly on the improvement of efficiency and productivity at work, the world of Design (in particular, Interaction Design) is increasingly looking at design practices that focus on designing the user experience.

The design dimension of the UX has been developed over many years, in particular, the UX models.

Garrett's model of experience design (2011) has found its place particularly in the Web Design sector. It is a pragmatic model for those who develop web technologies and platforms. It demonstrates how a technology product can be designed for the user experience and how the design development of such systems moves from developing abstract solutions to concrete development phases.

The goals of designing products that are interactive, fun, appealing and aesthetically pleasing chiefly revolve around the user experience, which involves making subjective terms explicit (Fig. 3.5).

Fig. 3.5 The Garrett model: the element of user experience. Revised from Garrett (2011, p. 29)

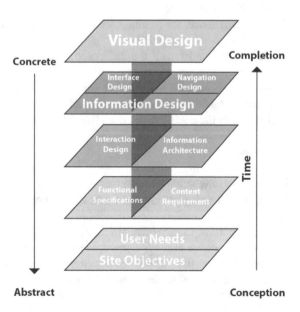

Surface

This concerns aesthetics and the instant sensory experiences of the user on a superficial level (when interacting with the product or interface).

In this phase, the designer must consider all of the cognitive and sensory aspects of the user. For example, factors like sight and cognitive load (the use of analysis tools like Eye tracking allows us to record aspects relating to attention, light, contrast, uniformity and consistency, etc.).

The Skeleton level

The Skeleton is on the model's penultimate level. It refers to design elements, including: information design, interface design, navigation design, etc. The design development of the information from the first three levels corresponds to this level. Specific aspects of the interface (spies, icons, buttons, etc.), the information structure (content grill and layout) and the means for navigating and interacting with the contents are developed during this phase.

The structural level

On an operational level, this level corresponds to the development of the information architecture, but also includes the specification of the interaction design.

The key goal of this phase is to establish a clear conceptual model and to simulate the levels of cognitive user-product interaction.

This level corresponds to the organisation of sequences and patterns regarding behaviour that the user will display in response to the system's content.

The purpose level

This phase involves identifying the technical operations (functional specifications) and the contents needed to create the product (content requirement—detailed description of the product's characteristics). It is essential that the purposes are expressed clearly and unambiguously. If the users have difficulties, we must be able to reformulate the purposes.

The strategic level

This concerns understanding the general goal of the design intervention. These strategies are chiefly focused on two aspects: real needs and the company's brand identity.

The operational goal of this phase is to understand what users want and what they expect to get from the product.

In this phase, techniques and methods aimed at understanding the needs of the target market should be used.

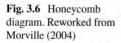

Fig. 3.6 Honeycomb
diagram. Reworked from
Morville (2004)

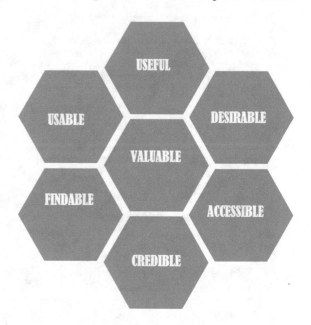

In order to understand the subjective dimension of the user experience and its operational dimension, we will now refer to Morville's model (2004), known as a honeycomb diagram. The positioning of the hexagons in Fig. 3.6 is completely subjective and demonstrates the factors that should stimulate the evaluation and design phases.

The Garrett and Morville models stem from the advent of new technologies and Web Design and demonstrate how the HCI sector has been forced to review its operational and theoretical apparatus.

The pioneer of this evolution was Donald Norman, who espoused the urgency of shifting the focus of HCI from practical aspects to emotional or subjective ones in his book Emotional Design, which was published in 2004. He distinguished between three processing levels for those who use an item: visceral, behavioural and reflexive. All three levels deal with aspects that have not traditionally been covered by classic usability. Let's look at them in detail:

- The visceral level is the automatic reaction to the item's appearance. It attracts or repels us before we can choose. It astounds or bores us. It is a largely automatic reaction, one that is emotional and visceral. A design must know how to consider the effect their product has at this level.
- The behavioural level has a more traditional effectiveness component, but also concerns the pleasure of using an item effectively.

- The reflexive level has implications that do not relate to effectiveness. It creates memories and involves self-image and a different concept of satisfaction, such as reward and personal fulfilment. Simple satisfaction was not sufficient and, if it was noticed, had not been well defined. Hassenzahl (2003) was one of the first to attempt to better define the aspects of the experience that are not strictly tied to productive environments and to hypothesise means of quantifying them. He particularly distinguished between pragmatic and hedonistic goals when using a product. The first concerns the desire "to do"; the second the need "to be". If the former can be evaluated using traditional metrics for effectiveness and efficiency, as well as evaluations of satisfaction and perceived ease, the latter required an entirely new theoretical definition. The author identified three hedonistic goals:

- stimulation: concerns the desire for personal growth and improving skills and knowledge.
- identification: concerns the desire for self-expression and interaction/relationships with significant others.
- evocation: concerns self-preservation and the creation of memories and feelings. It is clear to everyone that these three aspects are relevant for websites and apps and it would be sufficient to cite the dissemination and major success of social networks (which particularly meet the hedonistic goals of identification), news sites and blogs and online games and forums to prove it; these all concern hedonistic goals. Hassenzahl also designed and tested a tool that creates measurement scales for the pragmatic aspects of the experience, both hedonistic ones and those that make a site or app generally appealing. This is a psychometric tool based on the semantic differential. The semantic different was already used by some usability specialists to compensate for the paucity of subjective information obtained from the most used usability tests.

Meanwhile, the recently established ISO 252010 (a replacement for 9126) supports ISO 9241-210:2010 and defines satisfaction as being composed of the following features in terms of usage quality:

- usefulness (utility, understood as cognitive satisfaction): linked chiefly to the ease of use in achieving pragmatic goals;
- pleasure (emotional satisfaction): the level of user satisfaction from the perceived achievement of the hedonistic goals of stimulation, identification and evocation and the emotional response;
- comfort (physical satisfaction): the level of user satisfaction from physical well-being stemming from the interaction;
- trust (safety-related satisfaction): the level of user satisfaction and belief that the product behaves as expected and yields results that are perceived as acceptable and reliable.

Hassenzahl's work led to a greater understanding of how we can understand satisfaction when using a product. However, in his book Experience Design (2003), Nathan Shedroff proposes a "manifesto" for what he considers to be a new discipline. We can identify key elements in his work as:

- identity: a sense of authenticity is needed for identity and self-expression. The sense of authenticity is often noted only when it is gone. If you are involved in an experience and something reminds you that it is not real, the authenticity of the experience may be lost;

- adaptability: this concerns change and customisation, with changes in the difficult levels, rhythm and movement. Musical instruments are often cited as examples of interaction design. The task is not simplifying things; it concerns things that can be experienced with many levels of skill and fun;
- narrative: this concerns a good story, with convincing characters, a plot and suspense. Narrative is not only about fiction, though. A good narrative is equally important for a company's promotional video, a smartphone's menu structure or any other design issue;
- immersion: this is the feeling of being completely engaged in something, such as being completely immersed in a specific setting;

In conclusion, the User Experience and the UX designer (as a profession) arose when the methods and goals of Interaction Design and Ergonomics and Design needed to more closely consider the user's perspective and develop usable interactive systems, following the advent of the Internet. The UX Design sector has a number of methods,[4] which include: interviews, questionnaires, design methods like profiles, scenarios, wireframes and prototypes, which allow for the adoption of iterative testing and corrective procedures (based mostly on informal tests or reactions to design hypotheses found online). This allows us to gradually, in a structured manner, define a design that is better suited to the user's needs and those of the brand. Some interactive products designed to improve the interaction between the user and the product are given as an example:

- flow: this is the sense of fluid movement, a gradual shift from one state to another.
- Measuring Less to Feel More,[5] a domestic device for people with Type II diabetes. This product, which was created by Mickael Boulay, does not display numerical values, but the position of an LED light shows whether the blood-glucose level is high, low or balanced;
- BeoSound Orbit,[6] designed by Paulo Pannuzzo, Anastasia Ivanova and Tuomas Hämäläinen for Bang and Olufsen, offers a new way to listen to music at home, by bringing digital music to the physical world. In fact, BeoSound expands the graphic user interface beyond the device's limits. Song information is projected in orbit around the device, like a vinyl on a record player. It is also possible to play music by selecting a colour that matches your mood;
- Tio,[7] designed by Tim Holley, is a light switch that encourages kids to reduce energy consumption. It uses a shape that resembles a face and changes colours to resemble emotions, varying from relaxed to upset.

---

[4]For more details see Vermeeren et al (2010) and Wilson (2011).

[5]For further information, see: https://waag.org/sites/waag/files/media/publicaties/measuring-less-spreads.pdf.

[6]For further information, see: https://www.tuomashamalainen.com/project/beosound-orbit.

[7]For further information, see: http://timholley.de/2010/08/10/tio/.

References

Garrett JJ (2011) The elements of user experience—centered design for the web and beyond. New Riders, Berkeley

Giacomin J (2014) What is human centred design. Des J 17(4):606–623

Hassenzahl M (2003) The thing and I: understanding the relationship between user and product. Funology. Springer, Dordrecht, pp 31–42

ISO/IEC 25010:2011 (2011) Systems and software engineering—systems and software quality requirements and evaluation (SQuaRE)—system and software quality models

ISO 9241-210:2010 (2010) Ergonomics of human-system interaction—part 210: human-centred design for interactive systems

Jordan PW (2000) Designing pleasurable products: an Introduction to the New Human Factors. London: Taylor & Francis

Law E et al (2009) Understanding, scoping and defining user experience: a survey approach. In: Proceedings of human factors in computing systems conference, CHI'09, Boston, 4–9 Apr 2009

McCarthy J, Wright P (2004) Technology as experience. MIT Press, Cambridge

Morville P (2004) User experience design. Semantic Studios, Ann Arbor

Norman DA (2004) Emotional design. Apogeo, Milano (ed. originale: Norman DA (2004) Emotional design. Basic Books, Cambridge)

Norman DA (2013) The design of everyday things. Basic Books, New York

Queensbury W (2003) The five dimensions of usability. In: Albers MJ, Mazur B (eds) Content and complexity: information design in technical communication. Routledge, Mahwah

Rizzo F (2009) Strategie di co-design. Teorie, metodi e strumenti per progettare con gli utenti. FrancoAngeli, Milano, p 428

Saffer D (2010) designing for interaction: creating innovative applications and devices. New Riders, p 18

Vermeeren AP et al (2010) User experience evaluation methods: current state and development needs. In: Proceedings of the 6th Nordic conference on human-computer interaction: extending boundaries, ACM, pp 521–530

Wilson C (2011) Method 10 of 100: perspective-based inspection. In: 100 user experience (UX) design and evaluation methods for your toolkit. https://dux.typepad.com/dux/2011/03/method-10-of-100-perspective-based-inspection.html

Chapter 4
Designing for People: Design References

4.1 Introduction: Global Quality

As we mentioned in the first chapter, the concept of "quality" for a product[1] or a system is, in this book, dealt with starting from the assumptions and perspective of Ergonomics, in particular, the HCD approach. As we have already seen, this shifts the focus from the quality of the product or the system itself to the *quality of the interaction*[2] that a certain individual—or group of individuals—has or is able to have with that product or that system[3] within a specific context, one in which the individual plays a central role.

The aim of Ergonomics and Design, therefore, is the evaluation and design of the overall quality of the interaction between people and the products or systems they come into contact with—or can come into contact with—during their work or daily activities.

This quality covers every aspect of the experience—physical, sensory, cognitive and emotional—that the individual experiences and must be evaluated and designed by considering all of the variables that define the context—physical, technological, organisational and social—of the person/system interaction, their reciprocal conditioning and their variability over time. The individual and his/her features, abilities and attitudes form an integral part of these variables, along with the activities he/she performs.

---

[1] For the meaning of the term "product" used in this book, see Sect. 1.1.

[2] The assumptions of this analytical approach are contained in the main ergonomics texts regarding usability. These main texts include: Norman (1988), Wilson and Corlett (1990, 1995), Rubin (1994), McClelland (1995); Jordan et al. (1996), Jordan (1998), Stanton (1998), Green and Jordan (1999), Norman (2004), Rubin and Chisnell (2011), Wilson and Sharples (2015).

The idea of ergonomic quality as "quality of interaction" and "quality in terms of use" is also widely discussed in the texts of Buti (2001, 2008), Tosi (2001, 2005, 2008) and Anselmi (2003), which are cited in the bibliography.

[3] For the meaning of the terms interaction, system and product used in this volume, see Sects. 1.1 and notes 2, 3 and 4 in Chap. 1.

© Springer Nature Switzerland AG 2020

F. Tosi, *Design for Ergonomics*, Springer Series in Design and Innovation 2,

https://doi.org/10.1007/978-3-030-33562-5_4

As a result, the point of departure for any design (and/or evaluation) project is describing and interpreting the *complexity of the variables that define the context of the person/system interaction* and *complexity of needs*[4] expressed by the individual—knowingly or otherwise—in terms of their relationship with the system.

To this end, we must start by identifying:

- the group of people who come—or will be able to come—into contact with the product;
- the type(s) of use (professional or domestic, daily or sporadic, etc.) for which the product is intended;
- the context(s) of use in which the person comes into contact—or may come into contact—with the product.

On an operational level, it firstly involves responding to the questions previously stated in Chap. 2: *what is* the product or the system that must be designed? *Who are the people* who come into the contact with the product? *For what activities* and *objectives*? *Where* (in what physical environment, with what technology, systems and furnishings, etc.)? *How* (with what means and skills)? *When* (at what time, for how long, how frequently, etc.)? The questions that come immediately after are: *what are the characteristics of the person in question* (personal, physical, perceptive, cognitive, etc.)? *What is their skill level*, in their relationship with, and use of, the pro- duct? Finally, *who chose* the product? Or what role did or does the person in question have in choosing the product? Depending on the design problem, the reference users/individuals can be represented by a specific group or, on the contrary, by a wide range of users who may, in some cases, align with the overall population. The design of specialised products, which are aimed at professional use and/or used by select operators (for example, based on skill and professional training, or specific characteristics and/or physical capabilities) will need to be considered on the basis of said characteristics and capabilities.[5] Products that are intended for professional use are primarily used in organised contexts that require control over the physical environment and the usage procedures. On the other hand, the design of daily use products—which we can expect to be used by a very extensive group of people and in a variety of usage contexts—must take into account the groups of users, defined on the basis of more or less generic characteristics (for example, age and/or gender, spending capacity, etc.), and refer to a wide variety of physical and psycho-perceptive characteristics and differing levels of skill and usage experience.[6]

---

[4]For the meaning of the term "needs" used in this book, see Sect. 1.1.

[5]This is the case for industrial equipped or devices that are exclusively by specialised technicians, aircraft and racecar driving and steering equipment, or sports equipment for professional athletes and, finally, aids and equipment for the disabled. In these, and similar, cases, users are precisely pinpointed starting from their specific physical characteristics, skills or abilities.

[6]Daily use products are products (environments or equipment) that can be used by people who do not necessarily have specific skills or abilities. They include domestic appliances, furnishings and utensils for the home or office, but also complex products and environments, such as cars, computers or residential or public buildings that can potentially be used by the "entire population".

4.2 Who Enters into a Relationship with the Product—People

Identifying the needs and expectations of individuals is an essential element of the design brief, which is used to define the objectives and point of departure for the product-development phase. For companies, the user/individual identifies the potential buyer and/or target audience, or, better still, the individual who will buy the product on the market. For designers, the individuals targeted by the design are those who will enter into a relationship with the designed product (or be able to) and those people (or groups of people) who will find the solution proposed to be beautiful, useful, efficient and appealing.

The design brief may include a precise definition of the target audience. In the case of a mass market product (for example, a new type of utility vehicle or a domestic appliance), the reference will be the gender and/or age of potential buyers, their lifestyle and, naturally, their spending capacity. In other cases, the brief may be less explicit and the designer will be responsible for identifying the target audience by associating it with a segment of users (characterised by age, attitudes, spending capacity, etc.), or with "all users" when the characteristics of the intended buyers for the design can not be foreseen.

The opportunity to accurately identify the characteristics, needs and expectations of the potential audience/buyers for the product represents only the point of departure for the design process, of course. This aims to process the collected data, translate it into objectives and requirements and, in the end, into design solutions.

It is also obvious that the products can be marketed using strategies that highlight their utility and quality and create need but, once the product has been purchased, it will be the final users who judge it in use and ensure its success on the market. In this respect, the role played by individuals in choosing and purchasing the product is vital.

We can distinguish between cases in which the product user is also the one who chooses and purchases it and cases in which the user and the purchaser are two different people. As Buti (2008, p. 70) writes, "*We all make occasional use of items that we have personally chosen, while, at other times, we are forced to use items that are imposed on us by the choices of others. How the purchase occurred is a very influential factor in the relationship between people and items*" and on the criteria used by the company to produce and market its products.

In the case of transport, workplaces and public services in general, the people who use and come into daily contact with the product are, gene- rally speaking, completely removed from the decision-making processes that led to its purchase. The needs of people are, in these cases, mediated by various factors (designers, purchasing managers, administrators of the public or private company that needs the product, etc.), which will tend to privilege, according to their role and their skills, the products' compatibility with previous design choices, the quality/price ratio, compatibility with the work organisation, or simply the economic offer proposed by the manufacturer for the tender. In these cases, the stimulus to improve the product only partially follows the rules of the market and competition, but can, and must, be induced by the presence of legislative provisions and regulatory guidelines on quality.

The case of items for personal use (home furnishings, appliances, cars, clothing, communication products, etc.), which are generally purchased directly by those who will use them, is obviously different. These are chosen on the basis of one's needs, economic availability, desires, taste, etc. In this case, each of us focuses on the functional and aesthetic qualities of what is being chosen, the brand and the image that the object conveys with its appearance, its accessories and finishing, the quality of its components and materials (for example, whether it belongs to a low- cost or luxury category, whether it is fashionable or not, etc.), or, indirectly, the advertising campaign used to market it.

The aspects of this problem, naturally, are varied and widely considered in specialised studies. What is of interest to us here is that the interlocutors for the company and the designer are, in this case, those who will be using the product directly. They are targeted with advertising campaigns and marketing surveys, which evaluate the potential spending capacity as it relates to the quality and characteristics of the product, etc. In this situation, it is not just the regulatory obligations used to improve the products, but also the market laws and the direct evaluation of the purchasers that dictate their success or failure.

It should also be noted that the criteria used to choose and purchase the products partially derive from the growing awareness of potential purchasers, who are increasingly safety-conscious and aware of products' ease of use and aesthetic qualities, as well as their public image and the reputation of the manufacturer.[7]

4.2.1   The Characteristics and Capabilities of People

Designing "for a specific group of people" and designing "for the highest number of possible users", therefore, are not two contradictory principles, but two aspects

---

[7] Attention to safety and ease of use of everyday products is a concern for an increasing number of potential purchasers. The choice of a system or a piece of furniture, as well as a domestic appliance, is increasingly based on assessments that concern not only its aesthetic appeal and the reliability of the manufacturing company, but also its safety and reliability over time, and the immediate identification and understanding of the usage and command mechanisms.

A similar evaluation is performed for the usability of what purchasers consider to be technologically innovative products. In particular as it relates to products equipped with control or adjustment mechanisms that are not immediately understandable (think of a mobile phone or video recorder, but also washing machines and dishwashers with different settings for pre-washing, washing, energy saving, drying, etc.), the judgment made by the potential purchaser concerns the compliance of the product with their needs, and the simplicity and comprehensibility of its command and programming systems. Their judgement, therefore, takes into account, with varying degrees of awareness, the accuracy and completeness with which the product achieves the objectives (for example, mechanical functionality or programming flexibility), the time and physical and mental effort that use of the product requires, as well as, of course, its price and its usefulness.

of the same problem, which is linked to "designing for people who will come into actual contact with a specific product or system".

As we will see in the second part of this book, the ergonomic approach assumes that the aspects that characterise each individual (or group of individuals) can be studied and arranged by considering the fact that many of these are common to everyone, many are common to specific groups or segments of the population (defined based on their activity and/or specialisation, e.g., aeronautical pilots or professional soccer players; or based on age, e.g., children and the elderly). Others, however, are completely dependent on the individual.

In this paragraph, and those that follow, some necessarily rigid and non-exhaustive classifications and breakdowns are proposed. These are useful for analysis and, according to the writer, for understanding and dealing with these topics.

What follows is a breakdown of the characteristics of individuals[8]:

- *characteristics that are common to all humans*: these are obviously things that distinguish human beings from other living species and which are commonly referred to as "fundamental characteristics and abilities", even when considering all of the possible individual variations. For example, fundamental characteristics include the ability to visually perceive objects and, in particular, to perceive wavelengths between 0.380 and 0.780 microns or hear frequencies between 16 and 20,000 Hz;
- *characteristics that are shared by large segments of the population*: these are aspects related to age (for example, reduced visual acuity for people over 65), geographical provenance (the ability to read and write for educated populations in the West) or to very widespread abilities (driving a car, using a telephone, etc.);
- *characteristics that are shared by precise user segments*: for example, these include weight and height characteristics for particular categories of sports stars, the ability of employees in tertiary companies to use a word-processing program to an acceptable level, and more. In this case, we are referring to characteristics and abilities that are common to a select group of users, independent of considerations regarding the individual;
- *characteristics that are explicitly individual*: these can be physical, perceptive, cognitive, cultural and generational and characterise the individual in this way (or the group of individuals under consideration). These provide us with specific information about psycho-physical characteristics and abilities (for example, height, motion capabilities, visual acuity, etc.) and also about habits and expectations, cultural formation, etc. The individual characteristics are difficult to trace back to numerical values and structured evaluation parameters and, in parallel, the ability to precisely define the reference group of individuals/users does not conclude the analysis phase for the problem. For example: being Italian university students, i.e., young people between 18 and 25 years of age with a secondary school diploma and a sufficient knowledge of the English language, is not enough to define the physical characteristics, visual capacity, cognitive abilities and, finally, the habits, tastes and expectations of each individual student that is attending a specific course.

---

[8] See Buti (2008).

The individual characteristics can also be evaluated on the basis of some varying factors:

- age and gender;
- level of ability or disability;
- geographical provenance;
- cultural background.

The anthropometric characteristics, capacity for strength and movement, sensory capacities, and more naturally depend on the age and sex of the person. They vary with age and are different between men and women. The physical factors of human variability, which are discussed in the second part of the book, can be classified and measured on the basis of universally accepted parameters and measurement methods. Data regarding the anthropometric and bio-mechanical characteristics of the human body, along with data related to sensory and perceptive capacities, represents a fundamental design reference, which helps to guide the most appropriate choices for the design problem.

In addition to the strictly physical aspects, there are, of course, other factors that depend on age and sex, which can affect the availability and attitude of people regarding performing and learning new tasks and choosing and using new products. The training received, generation of the user, the period in which the user's studies and/or professional training occurred, the period when the user was young, etc. strongly impact expectations and availability regarding new products or activities. An obvious example is the exceptional ease with which young people who grow up using electronic devices and digital systems are able to use the buttons and functions of a new mobile phone or a new computer program, which may appear complex to those who only learned how to use computers as an adult, or even incomprehensible to those who were excluded due to age, training or financial availability.[9]

Further variables include the cultural provenance and instruction of the individuals, which is linked to their geographical provenance, the rules that govern behaviour in their native community, along with their habits and conventions. Driving on the left in Anglo-Saxon countries, horizontal white street signs in Europe and yellow ones in the United States, street signs, written and oral language, and more are conventions that must be understood in order to design products for foreign markets.

---

[9]The case of the Internet is of considerable interest in this respect, as its dissemination is causing a sort of cultural marginalisation which victimises anyone who does not have the ability to connect to the Internet. Internet access, and the potentially infinite amount of information and communication that it can provide, is a privilege reserved for those who, in addition to possessing or having the necessary equipment available, are also able to use it. The access and research procedures, and the logic with which one moves within the network, represent an incomprehensible and hostile universe to those who are unfamiliar with its language and its codes. The problem obviously concerns the social strata, who, for economic or cultural reasons, are unable to buy a PC and/or use one, but it also concerns, perhaps in a more serious and obvious way, the elderly, for whom learning new languages and dialogues (which are now used not only for computer systems, but also for the interfaces of mobile phones, appliances, public utility services, etc.) is objectively different from those that are learned and consolidated over time; this causes exclusion from social life and from the forms of communication that we consider "normal" today.

Similarly, the habits and rules of behaviour are variability factors that are common to large groups of users on the basis of their geographical area of origin. Within the same geographical area, the cultural background pertains to the community, social group, generational provenance, etc. These aspects can strongly impact the behaviour, tastes and expectations of individuals.

4.2.2 The Usage Skill of Individuals

The skill employed to use the product generally depends on the level of knowledge about the product (or the type of product) possessed by the individual and the type of use (professional or domestic, frequent or occasional) for which the product is intended.

Every day, each of us uses products that we understand perfectly and have used for a long time, with buttons, functions and usage methods that allow us to use them with ease or, as they say, "with our eyes closed". For example, your car, your TV, but also the spaces and furnishings in your home or office. Other products, environments or pieces of equipment require us to learn to use them, that is, to understand how they function, their buttons, etc. and, in many cases, to endure the stress and frustration of the often insurmountable difficulty in using and learning how to use them. These are products we use for the first time, products we use infrequently, unknown or changed environments, that is, products or environments with which we have little or no skill or regular use. The examples are obvious and numerous: the interface of the new TV, a ticket machine we have never used, routes and directions within an unknown public building, etc.

The usage skill refers to the type and level of awareness that the individual possesses (or is believed to possess) regarding the use of the product and may be defined thusly:

- *Regular*: this is the collection of awareness that is believed to be possessed by at least 90% of individuals, i.e., it is possessed by most people or by extensive portions of the population. Skills that are considered regular are: –characteristics shared by everyone, that is, those that are innate or developed during human development; for example, standing, walking, talking, etc.–characteristics shared by large segments of the population; for example, the ability to read in high—and medium—income countries, the ability to use word-processing programs for works in office settings, etc.
- *Specific*: these specifically refer to the use of the product. This knowledge is acquired through a short training session, which, upon its termination, allows 90% of people to use the product. It may be acquired through:—using the product;— reading the instructions;—quick training from another user.
- *Specialised*: this is the knowledge acquired through targeted training (driver's licence, a professional training course, etc.). Competence and experience in the use of a product translates into the ease or difficulty with which the person will

be able to complete a given task. Naturally, having already performed a given activity will make it easier to perform it a second time and, in the same way, having already used a product for a given purpose will allow the user to use it more easily for different purposes. Finally, the use of a product can be facilitated by prior experience with similar products. Well-designed products, in fact, allow people to perform general operations based on prior experience, adapting and exploiting this to achieve new goals. A second aspect, in fact, is knowledge of the task, which may be independent from knowledge or familiarity with the product needed to perform it, that is "knowing what has to be done makes it easier to do it".[10] As Norman writes, complexity is necessary; it is important only to distinguish between complexity and complication. Complexity is a state and a "is part of the world, but it shouldn't be puzzling: we can accept it if we believe that this is the way things must be. (...) But when that complexity is random and arbitrary, then we have reason to be annoyed" (Norman 2010, p. 4). Complication, conversely, is the source of confusion. The problem, therefore, is not the apparent complexity of things and situations, but the ability to understand them.

4.2.3 The Needs of People

Once we identify who the people that will come into contact with the product are (or could be), and what their physical, perceptive and cognitive characteristics are (or could be), we can identify their current or potential needs and what the requirements of the product need to be. This process consists of analysing the needs, expectations and desires (or needs, to be brief) of a specific group of people as they relate to the product and translating them into design requirements and specifications.

Certain requirements can be precisely identified on the basis of knowledge about the users' physical characteristics. For example, we can identify the size requirements of a table or chair once we know the anthropometric characteristics of the people who use, or will use, it. In this case, the product requirements are quantifiable and measurable and their verification can be based on precisely defined acceptability thresholds, such as, for example, the minimum movement space for the arms and legs, the height of the elbow when seated, etc.

Parameters that derive from the cultural or professional characteristics of the users (for example, the language and terminology to be used to describe the buttons, the skill level of potential users, etc.) can be similarly identified.

In other cases, identifying the needs/requirements system requires in-depth knowledge of the lifestyle and attitudes of users. Designing a backpack for children or young people requires us to know who our product is targeted at, what their habits are and the context in which they will use it: what will the backpack be used for? To go to school? Nursery, primary or secondary? For trips in the mountains, short outings? and more.

---

[10]On this subject, see: Jordan (1998).

Similarly, designing a smartphone application, for example, one for booking travel and hotels, requires us to define the skills and habits of the target audience: are they people with an extensive or limited experience with digital systems? Are they used to booking in advance or are they looking for last-minute offers? Does the application need to be designed exclusively for smartphones, or for PC and tablet as well? Should it allow the reservation to be printed or should this requirement be considered irrelevant?

Other factors are not strictly related to the functionality of the product or its suitability for use, but concern people's expectations and taste, and the reasons that lead them to purchase a certain product. Taking the backpack example, information about the expectations and tastes of people will allow us to identify the sense of pleasure from having a new model, or a particular brand, and the different judgments of the user experience.

In the case of the smartphone app, the quality of the user experience will depend on the simplicity and fluidity of the intended steps, on the clarity and aesthetic quality of the graphic interface and the user's immediate involvement in using the product.

Needs vary from person to person on the basis of their physical, sensory and cognitive characteristics, the characteristics of the environment in which people perform their activities (microclimate, humidity, lighting, noise, crowding, etc.) and vary for everyone as time goes by, based on age, health conditions and independence levels. In terms of a specific product, the needs of each individual also vary based on the reason for which the product is used, contributory factors (haste, fatigue, distraction etc.) and, finally, the expectations the user has for the product, which may relate to its functionality, price, appeal, aesthetics, simplicity and ease of use, etc.

People's needs, therefore, cannot be generalised, except where they pertain to the fundamental needs related to the basic characteristics of human beings (refer to the previous section). The needs analysis is only valid if it is conducted via structured and verifiable investigative methods, which consider the characteristics, attitudes and abilities of a specific group of people under consideration and the activities and context for which the product has been designed or is used.

Like those that came before, the following classification has a purely instrumental value and may be used only to facilitate analysis. Starting from the characteristics of the user, needs can be:

- *Common to all people,* or fundamental needs. For example, these could refer to walking in an upright position or the need to sleep for a minimum period of time every day;
- *Shared by large segments of the population,* which refers, for example, to the need to wear glasses for sufferers of myopia or the need to live in heated environments for people who live in countries with winter temperatures that drop below 15–18°;
- *Shared by a precisely defined segment of the population.* These are needs that derive from specific physical, cognitive or cultural characteristics of the user, e.g. the dimensions required for a wheelchair to pass, the use of the Cyrillic alphabet in Russian-speaking countries, etc.

- *Explicitly individual.* These derive from the characteristics of a specific individual or group of individuals, e.g. the need for movement and visibility relative to the driving position of race cars. For groups of people with identical characteristics and needs, the latter can be further divided into three fundamental levels:

  - *Primary*: these are needs that motivate people to use or purchase it. These typically refer to: eating, sleeping, staying cool or keeping warm, moving, sitting, communicating, staying informed, etc.
  - *Functional and safety*: these are the needs that are strictly related to use (avoiding risks to health and safety, being able to perform a given activity in a satisfactory fashion);
  - *Aesthetic and symbolic*: these are needs that relate to the aesthetic quality of the environment, product or system, and the symbolic value that can be attributed to them.

The three categories of needs and expectations obviously do not come up separately and, on the contrary, there are no primary needs that exclude the desire to also satisfy the functional need or aesthetic need. In fact, needs arise according to a hierarchy of priorities that will lead us to worry about the quality and image of our clothing or the furnishings in our home only when we have satisfied the primary need of clothing ourselves or shielding ourselves from the elements.[11]

Jordan (1999, p. 209) proposes a classification of user needs that is based on Maslow's hierarchy of needs. This model views the human being as an "animal that needs" and is rarely completely satisfied. The principle can be summarised with the idea that, when a person has satisfied their needs, which belong to a specific level of the hierarchy, he is driven by the desire to satisfy those of the next level up (See Chap. 14, "Cognitive aspects in user experience design: From perception to emotions" by O. Parlangeli and M. C. Caratozzolo, in particular, Sect. 14.2).

This does not mean that having not yet satisfied the primary needs will not lead to frustration at not satisfying the needs on the next level, but rather that the fundamental needs are essential for all human beings (Figs. 4.1, 4.2 and 4.3).

As has been previously stated, the need for safety and functionality pertains to the ability to use a suitable environment or product for the required activities and objectives, without creating risks to the user's health and safety.

The functionality and usefulness of a product can obviously become secondary when its aesthetic value is considered to be absolutely more important than any other consideration. Similarly, a fashionable object or dress is difficult to judge on the basis of its usefulness or comfort but, on the contrary, is purchased for its beauty and/or the value of its social representativeness. In other cases, the functionality and usefulness of a product can be ignored or forgotten, because the focus is completely absorbed by the activity that it performs. During an exam, while listening to a particularly admired piece of music, and, in general, when you are completely immersed in what you are doing, your needs can go completely unnoticed and we can sit for a long

---

[11] See Jordan (1999, p. 209).

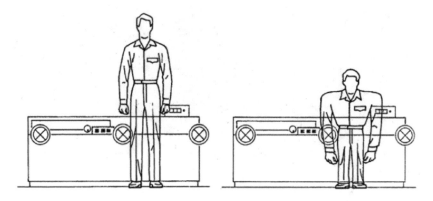

Fig. 4.1   The user-product relationship. *Source* Eastman Kodak Company (1983)

time in an uncomfortable chair, put up with an overly high temperature or extremely bright lighting, or we can forget about drinking or eating.

4.3   What Is the Product?

Of course, the fact that products and systems must be designed and manufactured to respond to the real needs of the people who use them, or will have to use them, and that they must be appropriate for their intended use, is not a discovery of Ergonomics.

Since the dawn of time, human beings have designed and created useful tools that were suited to their goals. The innovative role of Ergonomics does not lie in recognising the importance of products' usability, in fact, but in the ability to design and evaluate it, based on the interaction between the product and the people who will be able to use it and the specific contexts within which said interaction will occur.

The technological development of recent years has also created an urgency in evaluating the real needs of people starting from the preliminary phase of the project, and to guarantee the usability of products/systems as an essential condition for their safety in use and for market success. As a result, it is the increasing complexity of the products and the growing presence of technology—in particular, computer technology—that channels the attention of companies, designers and the target audience of products/systems towards the usability of products and, generally speaking, towards the quality of the interaction between them and their users.

Norman (1993, p. XI) has repeatedly highlighted how our «society has unwittingly fallen into a machine-centered orientation to life, one that emphasizes the needs of technology over those of people, thereby forcing people into a supporting role, one for which we are most unsuited. Worse, the machine-centered viewpoint compares people to machines and finds us wanting, incapable of precise, repetitive, accurate actions (…). The result is continuing estrangement between humans and machines,

D3 - Riding chair. Extend the pleasure of riding.

D4 - ballet chair by French can-can called "the Toulouse Lautrec". Fun in a Second Empire decor.

A6 - Bent hammer. Its particular shape allows it to easily hit the most inaccessible nails.

A7 - Poly-hammer. The rapid rotation of this eight-headed hammer saves a lot of time for professionals and do-it-yourselfers.

E6 - Fork for snails.

D8 - Bed for acrobats. An acrobat with this piece of furniture can keep training even during sleep.

E18 - Coffee maker for masochists. We believe that the drawing, sufficiently explicit, allows us not to insist on details that could be painful.

Fig. 4.2 Impossible products. Carelman (1978, pp. 13–43)

A product is ergonomically designed if:

Try using it. Think forward to all of the ways and circumstances in which you might use it.

- Does it fit your body size or could it be better?
- Can you see and hear all you need to see and hear?
- Is it hard to make it go wrong?
- Is it comfortable to use all the time (or only to start with)?
- Is it easy and convenient to use (or could it be improved)?
- Is it easy to learn to use?
- Are the instructions clear?
- Is it easy to clean and maintain?
- Do you feel relaxed after a period of use?

If the answer to all of these is 'yes' then the product has probably been thought about with the user in mind.

Fig. 4.3 Pheasant and Haslegrave provide a short check list of sorts, taken from a booklet from the Ergonomics Society, entitled "Ergonomics: fit for human use" which can be easily used to evaluate is a product has been produced in accordance with ergonomic criteria. *Source* Pheasant and Haslegrave (2006, p. 9)

continuing and growing frustration with technology and with the pace and stress of a technologically centered life».

The technological innovation of products and production processes, and, in particular, the dissemination of information technology, has had a profound impact on the organisation of work and everyday life, rapidly populating the world around us with products that require awareness of the logic and language of computer interfaces, and a familiarity with tools and command and regulation devices that no longer require—at least not exclusively—a "physical" action, such as rotating, pressing or pulling, but perform procedures based on an understanding of the possible options envisaged, of the hierarchy and sequence of the actions required and, finally, of the final outcome.

The dissemination of IT technology represents, what is now universally recognised as the most important technological event of recent decades, the pervasiveness of which has taken on all the characteristics of a "technological revolution".[12] The change that computerisation has produced in the organisation of work and services

---

[12] According to C. Freeman, a technological revolution is a profound transformation induced by far-reaching innovations (such as the introduction of electricity or the creation of railways) that not only favour the creation of a new range of products or services, but also involve innovation of processes (productive and economic) and affect every other sector of the economy (and social organisation), by modifying the structure of the production factor costs and the conditions of production and distribution for the whole system. The characteristics of a technological revolution can be summarised as:

- "A drastic reduction in the costs of many products and services and a widespread perception of new profitable investment opportunities;

has taken place in a radical way and in a very short time span, and has reach every, or almost every, aspect of daily life.

Furthermore, technological evolution poses the problem of controlling complex systems, whose command and programming procedures leave room for human error or difficulty in understanding the response provided by the system. This interaction with modern technology poses the problem of comprehensibility, not only of its operation and of its procedures for use, but also of its responses and, in particular, of the consequences deriving from the operations carried out. When technology is not designed to place man at the centre of all design considerations, it does not reduce the frequency of human error, nor does it minimise the impact of errors when they occur, but, rather, it tends to leave room for human error, in particular, errors caused by distraction or a lack of understanding of the information or procedures by the operator. The clarity and the ease of understanding of the feedback—return message—provided by the system represents the essential element on which the ability to correctly use a product, and the security with which the required operations can be performed, depend.[13]

Not only do people frequently make mistakes when using products, but they use them incorrectly or improperly even more frequently. Furthermore, errors do not occur only with new or particularly complex products, but also frequently occur with objects that we use on a daily basis or with extremely simple objects. Errors in the interpretation of the information provided by the product, errors in reading or understanding the instructions and, finally, errors due to habits or distraction, which can cause us to miss a usual passage, etc., are everyday problems that can be predicted and avoided with appropriate planning.

Human error is a research topic in psychology and cognitive ergonomics and is also one of the central themes of research and professional practice in Ergonomics, both in the field of risk management in organisations and work procedures (for example, in the aeronautical sector and, in general, transport, in the production sector, in the management of clinical risk and care safety) and in the field of Ergonomics for Design (from risk assessment in the use of products and physical environments to the study of error in the dialogue procedures of digital interfaces).

- a considerable improvement in the technical characteristics of many products and processes, in terms of their reliability, precision, speed and other performance characteristics;
- the political and social acceptability, particularly as it pertains to the spread of technological innovations outside the sector in which they arose and the legislative, educational and regulatory changes that may require;
- environmental acceptability, as it pertains to the environmental costs that the dissemination of technological innovation may entail;
- the general effects on the economic system, i.e. the ability of innovation to impact the conduct of the entire system and not only a specific social or economic sector. Based on these considerations, we can say that nuclear energy cannot be defined as a technological revolution, while this definition is entirely appropriate in the case of microelectronic and IT technologies". See Freeman (1985, pp. 193–195).

[13] See Norman (2013, p. 25).

The evaluation methods for human error and the majority of the studies regarding safety of use, which employ evaluation procedures for products and systems that are similar to those used to verify usability, belong to this field of research.

The theme of human error is covered in Chap. 13, "Cognitive aspects in user experience design: from perception to action" by O. Parlangeli and M. C. Caratozzolo.

According to Rubin and Chisnell (2011, pp. 6–12),[14] there are certain fundamental reasons that difficulty in using products derive from; the most widespread of these are those that stem from an incorrect focus on the design problem.

- During the development of the product, attention was focused on the product or system and not on the final user.
- There are three principle components for each type of human performance: the individual, the context, the activity. Since the development of a product or system aims to improve human performance, the designer must consider these three components during the entire process of developing the product. Unfortunately, designers, engineers and programmers have traditionally favoured problems associated with the activity, and the most suitable technology to solve them, neglecting the needs established by users and the usage context.
- Designing usable products is difficult. Designing usable systems is a difficult and unpredictable endeavour, so many organisations treat it as if it were only a matter of "common sense". As many authors note, much has been written about the concept of usability and very different definitions have been provided too. The problem is now partly solved by the definitions and measurement and assessment procedures introduced by the ISO 9241 standard and by the results of research and experiments conducted in this sector in recent years.
- When technology made its appearance on the "main road" of the market, the characteristics of end users changed radically, while the criteria for the development of IT products have remained partly the same. The initial users of IT products were a few experts who possessed a very specific skill set and were proud of their ability to understand and solve every problem. Those who designed and developed those products were very similar to users, and, in many cases, designers and users were ultimately the same people. The type of design was therefore a next-bench design (that is, a project for the colleague one desk over).
- Today, the situation has changed radically. We all use IT products for work and in our everyday life, and we generally use them while ignoring how they really function. What most of us know how to use is the dialogue interface of our word processor or mobile phone. The user that digital products are targeted to today is not a computer expert; he has little or no technical knowledge and wants the product to be understandable and easy to use.

---

[14]See Rubin and Chisnell (2011), Rubin (1994).

4.3.1 The Usage Skill Required by the Product

In this case, the skill level, which has already been analysed as it pertains to the user, is the skill level that the product requires from its user. In the regulatory documents dedicated to the "easy of use for everyday products",[15] the latter is analysed on the basis of the ease with which they can be correctly and comprehensively used, with or without learning specific instructions or skills. In this way, the products can be divided into:

- *easy to use*: these are daily-use products that can be used by those with skills that are considered to be regular. They are products that at least 90% of users (including the elderly, children, the disabled, etc.) are able to use for their primary function, without learning instructions (they are, or should be, "easy to use", such as small home appliances, ticket machines, etc.);
- *usable with instructions*: these are products that require the user to read and/or study a small number of usage instructions. Once studied, at least 90% of users will be able to safely use the product without further assistance (for example, the remote control of a TV or a microwave can generally be used after reading the instruction manual or after hearing an explanation from another user);
- *usable after a period of training*: these are products that require complex and more or less extensive training (for example, complex IT programs, cars, motorbikes, etc.).

From the point of view of design, the usage capacity and skills required by the product must be in line with the capacity and skill level that is actually possessed by the people targeted by the product. In other words, products that required special skills must be aimed at (and designed for) people with an appropriate skill level.

On the contrary, daily-use products, which are potentially intended for "everyone", must provide for every usage method and every possible skill and competence level.

Therefore, it must be taken into account that the skill level and competence in using the product can be very different depending on the population groups under consideration, particularly for commonly used products. As previously highlighted in Sect. 4.2, the different physical and sensory characteristics and skills, the generational and cultural differences, the different learning abilities, all of which can strongly condition the relationship with the product, must always be considered. The lack of attention to the skill level required by the people that the design targets leads to an immediate discomfort on the part of those who are forced to use products they cannot understand the functions or master the programming or operating procedures of every day: the result is a growing feeling of frustration and inadequacy of one's own

---

[15]This refers, in particular, to the ISO 20282: 200, "Ease of operation of everyday products requirements for user characteristics" standard, and to the ISO/IEC 71: 2001 guidelines, "Guidelines for developers to address the needs of older persons and persons with disabilities" (today known as ISO/IEC 71: 2014 guidelines, "Guide for addressing accessibility in standards", which adopts the usability definitions contained in ISO 9241/11, (transposed into ISO 9241/210: 2010) and interprets them within the framework of the design criteria for accessible and inclusive products. For this subject, see Chap. 9 "Ergonomics and Design for All: Design for inclusion".

abilities.[16] The result is that the user tends not to complete the operations requested, or to perform them incorrectly, or to use only part of the functions and/or equipment of the product.

Last but not least, the products can also be classified based on the physical, sensory and cognitive characteristics and capacities of the individuals, that is, based on their ability or disability, even where temporary in nature.

- *Barrier-free products.* These are products that consider the needs of users with restricted capacity for movement or impaired vision, and can therefore be used by everyone, including the elderly, children and the disabled.
- *Products for the average user (average products).* These can be used by people with "average" characteristics. These are products that do not take into account any limitations on the physical, sensory and cognitive abilities of potential users, and whose use may be impossible or risky for people with reduced mobility, strength, vision, cognitive skills, etc.
- *Special products.* These are aimed at people with particular abilities or disabilities, for example, aids for the disabled or, conversely, competitive sports equipment.

This classification, which relates particularly to daily-use products, partially reflects the more general interest of ergonomic research in the themes of Design for All. In this context, the objective of the ergonomic approach is the inclusion of the needs and expectations of people with skills and/or abilities—physical, sensory or cognitive—that are outside the norm, when defining the requirements of the product and, in particular, of daily-use products, which are used, by definition, by all users. Design for All, in particular, issues of social inclusion,[17] which is the focus of Chap. 9, "Ergonomics and Design for All: Design for inclusion", is also the target of many research projects at the European and international level, as well as numerous experiments and more recent design projects.

4.4 Why We Use or Relate to the Product—Activities

People come into contact with products, environments and systems for a specific use—that is, to carry out activities regulated by a work procedure or specifically directed at a goal—or for unforeseen or random reasons.

---

[16]By way of example: a smartphone's interface may be "easy to use" for a young person who is accustomed to using computer programs and digital interfaces, and absolutely incomprehensible and impossible to use for an elderly person or for those who do not understand the language and the dialogue logic; the correct and complete use of an electrical appliance notoriously assumes the user will read instruction booklets, the language of which often appears understandable only to specialised technicians and, similarly, numerous furnishing elements are often accompanied by equally complex instructions for assembly and use.

[17]"Social inclusion" is a specification of "design for inclusion" as it pertains to the use of products, services and systems.

As we will see in the next section, the type of "activity" carried out depends on the type of use (professional or daily), the role of the user (operator, technician, casual user, etc.) and his objectives.

An initial interpretation is the role within which the individual comes into contact with a product or system.

If it is true that *"machines, systems (and products) are designed and built for a single purpose during the final analyses: to make it possible for individuals— users and operators (including supervisors and managers)—to do something easily, simply, quickly and well"* (Chapanis 1996, p. 5), it is also true that the relationship with the product changes radically, based on the methods and the role with which the user employs it. A. Chapanis also notes that people use products in four main ways: as general users, operators, maintenance workers or salespeople.

- *General users*: these are users who interact with commonly used products (domestic appliances, dishes, personal computers, etc.) as part of their day-to-day activities (personal or work), without the need for specific preparation.
- *Operators:* these are users who use complex machines and devices that require specific professional preparation. The operator, therefore, is a user characterised by specific skills and/or physical characteristics, which allow him to perform a specific task by following structured procedures, in many cases, within a controlled context. The difference between the general user and the operator, therefore, lies firstly in their skill level and, furthermore, in the level of control they can exercise on the procedures used and on the physical and organisational environment in which they operate. The operator, naturally, refers to the pilot of an airplane, an athlete who uses competitive equipment or a technician in charge of overseeing a production line. The operator is the worker who uses a small circular saw to carry out pre-arranged machining procedures inside a factory. The general user is an individual who uses the same circular saw at home as a hobby.
- *Assembly and/or maintenance workers*: these are users who repair and maintain products, machines and devices in good working order. In some cases, the needs of maintenance workers are identical to those of users, because they are, above all, another category of user. In other cases, their needs are specific and must be considered during the design process. For example, the need for simple diagnostic procedures, easy access to the parts on which action is to be taken, the usability of suitable maintenance tools, etc.
- *General assembly or maintenance workers*: these are often the same users who assemble or repair a product without possessing any specific skills or sufficient instructions, e.g. assembling a piece of furniture, changing an ink cartridge, replacing a component in a domestic appliance. In this case, the difference from the previous category also lies in the skill level and the presence and/or availability of structured, controlled procedures.

Every role naturally corresponds to different activities and objectives, which can be further analysed on the basis of different contexts.

Further interpretations can stem from the different activities and objectives for which the product can be used and the skill level and experience needed to perform said activity.

Said product can be used by different people and for different objectives, of course, and be used by the same person for different purposes and using different methods, times and attention levels from time to time.

Finally, the same activity can be carried out by different people, with different levels of competence, experience and skills, even if the objectives are similar. One example of this is medical devices used in domestic settings. Thermometers for measuring temperature, glucose meters for measuring blood sugar, aerosols, blood pressure meters, pulse oximeters for measuring heart rate and many other similar devices used in the home—and, therefore, outside hospital or treatment and care facilities—can be used by people with different expertise and experience. In this case, the individuals involved are, or may be, the so-called "formal" care workers, that is, medical or nursing staff, but also the recipients of said care and the "informal" care workers, that is, the family members and the people being cared for, who generally do not have any medical skill.[18]

The analyses of the activities represent the basic evaluation method for the development of any evaluation and/or design intervention in the field of ergonomics, and is specifically discussed in Chap. 6, "Human-Centred Design—User Experience: tools and intervention methods".

4.5 *How, Where and When* We Come into Contact with the Product

Another criterion for classifying products concerns the usage method for which the product has been designed, which depends on the functions for which the product is intended and the frequency with which it is or can be used.

The chief distinction lies between products and systems for professional use and those for daily use.

The procedures for using professional products or systems, that is, those that are specifically intended for use by specialised workers, are, or should be, planned and regulated, supervised and appropriately explained to those who must perform them and carried out in the correct manner.

Products for professional use are used, or should be used:

- by people with suitable characteristics and skills for performing the required activities;
- by people who are specifically trained to perform the intended tasks;
- in organised contexts, such as companies, production plants, etc.;
- based on regulated and controlled procedures;

---

[18]On this theme, see the volume of: Tosi and Rinaldi (2015).

- with specific, controlled objectives.

For example, the use of industrial machinery requires the necessary procedures to be carried out by operators who understand the task required of them, have built up their skill through a period of training and followed regulated procedures. The tasks are also carried out within an environment that allows the operations being carried out to be overseen and, possibly, for corrective actions, programmed maintenance of the machinery, etc.

In the case of daily-use products, and, in general, of products, environments, equipment and services that are used outside of controlled procedures, it must be taken into account that the user can be "anyone" and can use the product "in any way" instead. In fact, daily-use products can be used[19]: by "anyone", that is, by people who may have unpredictable characteristics and abilities (different ages, physical and psychological characteristics, skill levels and, finally, with differing abilities to perceive/understand the risks associated with incorrect or improper use), and similarly; by people who have any level of training and/or skill in using the product; in contexts other than those intended; inappropriately and/or for unintended purposes; for objectives other than those stated; incorrectly, as a consequence of the three previous points, and, in particular, for reasons deriving from the user's restricted physical or perceptive ability or contributing factors (haste, carelessness, need to perform more actions simultaneously, etc.).

In this case, the procedures and methods for use are learned by simply observing the product, trying to understand "how to do it", or by reading the instructions or indications on the product, or explanations received informally from another person. As a result, their behaviour when using the product can be correct, that is, as foreseen by the designer, but it can also be anomalous or unexpected and become a "risk behaviour".[20]

As we will see in Chap. 5, intentional or unintentional risk behaviours, which can lead to errors or accidents, can in fact derive from:

- a lack of perception of the risk by the user;

---

[19]The distinction between specialised use and daily use may not be easy to define in some cases, and the same product can be used by users with different skill levels (and/or physical or psycho-perceptive ability) with different objectives and in different contexts.

A personal computer and its programs can be used at home or in an office with a medium level of skill (which today is considered normal), without there being structured procedures that regulate their use nor controls on the correct execution of such procedures.

Regardless of the environment in which the computer is used (home or office), in this case, the user is a "generic user" and it is not possible to determine a priori their characteristics (for example the age, the anthropometric characteristics or the visual capacity) or specific needs and expectations.

On the other hand, the same personal computer can be used for the management of very extensive databases and require specific professional preparation. In this case, the user is an "operator", who requires a high skill level in using the program, which follows procedures that are structured according to the objectives required by the management, and operates within an environmental context and controlled system.

[20]See Sect. 5.2 "Functionality, security, accessibility".

- failure to understand how the product/service functions, due to insufficient or inadequate training;
- physical, perceptive and/or cognitive limitations that prevent correct usage;
- emergencies;
- difficulty in performing the required operations;
- absence or inadequacy of information relating to the consequences of the actions.

The anomaly may pertain to *the way* in which the product is used for its primary function (opening a drawer on guides by forcing it beyond its range of motion to reach the bottom with your hand), or to *the purpose* f or which it is used (the same drawer used by a child as a ladder to reach the top of the wardrobe).

The correct use of the product obviously also depends on the person's knowledge of the product and the frequency with which it is used. In the case of both specialised and daily-use products, it must be considered that those who use a certain product (or certain environments or equipment) may do it frequently or only occasionally and that, in addition to learning to use it the first time, they may be forced to re-learn its operation after a period of non-use.

In fact, the frequency of use is one of the parameters that the "usability measures" identified by the regulation refers to and allows products to be classified based on:

- *frequent or continuous use*: this refers to a consolidated habit that requires no further learning of instructions and procedures. Frequent repetition of the required operations and knowledge of the product make it possible to consolidate and maintain the user's competence over time;
- *periodic use*: this refers to the way in which we use a product, an environment or a service only for certain periods of time or for certain purposes. Using skis during the winter holidays, entering a restaurant that we only visit a few times a year or using the appliances in a holiday home are examples of periodic or occasional use, which require us to "regain" knowledge and skills that we already possess every time;
- *occasional use*: this occurs when we use a product only a few times and for occasional reasons only. This can happen when we use an object that we encounter for the first and probably only time in our life, which we want to figure out to carry out the necessary activity at that moment, and which we will probably forget about quickly as well. For example, the ticket dispenser at a subway station in another city or the replacement car provided by the garage.

A different case occurs when we use a product in a particular condition due to an emergency or urgent situation. This case is particularly critical, because, under conditions of psychological pressure, our abilities to understand how an object functions and to fully control our actions are drastically reduced. In these cases, the ability of the product to clearly communicate its functions and how it works is essential, while its ability to prevent human error through the elimination of risk factors, alarm signals and self-stopping mechanisms for incorrect procedures is even more vital.

References

Anselmi L (2003) Quale qualità, cosa si intende per qualità d'uso e come è possibile verificarla. Poli, Design, Milano

Buti LB (2001) Ergonomia e prodotto. Design, qualità, usabilità e gradevolezza, Il Sole24ore, Milano

Buti LB (2008) Ergonomia olistica, il progetto per la variabilità umana. FrancoAngeli, Milano

Carelman J (1978) Catalogo d'oggetti introvabili. Mazzotta, Milano, pp 13–43

Chapanis A (1996) Human factors in system engineering. Wiley, New York

Eastman Kodak Company (1983) Ergonomics design for human at work. Van Nostrand Reinhold, New York

Freeman C (1985) Prometeo liberato. In: AA.VV., Paradigmi tecnologici, saggi sull'economia del progresso tecnico, FrancoAngeli, Milano

Grandjean E (1986) Il lavoro a misura d'uomo: trattato di ergonomia, Edizioni Comunità, Milano (1st ed.: Grandjean E (1979) Physiologische Arbeits- gestaltung: Leitfaden der Ergonomie. Ott, Thun)

Green WS, Jordan PW (1999) Human factors in product design. Taylor & Francis, Londra e Philadelphia

Jordan PW (1998) An introduction to usability. CRC Press, Londra

Jordan PW (1999) Pleasure with products: human factors for body, mind and soul. In: Jordan PW, Green WS (eds) Human factors in product design. Taylor & Francis, Londra e Philadelphia

Jordan PW et al (1996) Usability evaluation in industry. Taylor & Francis, Londra e New York

Karwowski W (2006) International encyclopedia of ergonomics and human factors. CRC Press, Londra e Philadelphia (1st ed.: Karwowski W (2001) International encyclopedia of ergonomics and human factors. Taylor & Francis, Londra e New York)

McClelland I (1995) Product assessment and user trials. In: Wilson JR, Corlett EN (1995) Evaluation of human work. Taylor & Francis, Londra e Philadelphia

Murphy's law and other reasons why things go wrong. Price Stern Sloan, Los Angeles, 1977

Norman DA (1988) The design of everyday things. Doubleday, New York. (1st ed.: Norman DA (1988) The psychology of everyday things. Basic Books, New York)

Norman DA (ed) (1993) Things that make us smart. Addison-Wesley, New York

Norman DA (2004) Emotional design. Basic Books. Cambridge

Norman DA (2010) Living with complexity. MIT Press, Cambridge

Norman DA (2013) The design of everyday things. Basic Books, New York

Nielsen J (2000) Web usability. Apogeo, Milano; (1st ed.: Designing web usability. New Riders, Thousand Oaks, 2000)

Pheasant S, Haslegrave CM (2006) Bodyspace: anthropometry, ergonomics and the design of work (3rd ed.) CRC Taylor & Francis group, Boca Raton

Preece J et al (2004) Interaction design, Apogeo, Milano (1st ed.: Preece J et al (2002) Interaction Design, beyond human-computer interaction. Wiley, New York)

Rubin J (1994) Handbook of usability testing: how to plan, design and conduct effective tests. Wiley, New York

Rubin J, Chisnell D (2011) Handbook of usability testing: how to plan, design, and conduct effective tests. Wiley, Indianapolis (1st ed.: Rubin J (1994) Handbook of usability testing: how to plan, design, and conduct effective tests. Wiley, New York)

Stanton NA (1998) Human factors in consumer products. Taylor & Francis, Londra

Tosi F (2001) Progettazione ergonomica: metodi, strumenti, riferimenti tecnico-normativi e criteri d'intervento, Il Sole24ore, Milano

Tosi F (2005) Ergonomia, progetto, prodotto. FrancoAngeli, Milano

Tosi F (2008) Ergonomia e progetto della qualità sensoriale. FrancoAngeli, Milano

Tosi F, Rinaldi A (2015) Il Design per l'Home Care. L'approccio Human-Centred Design nel progetto dei dispositivi medici. Didapress, Firenze

Wilson JR, Corlett EN (1995) Evaluation of human work. Taylor & Francis, Londra

Wilson JR, Corlett EN (eds) (1990) Evaluation of human work. Taylor & Francis, Londra e New York

Wilson JR, Sharples S (2015) Evaluation of human work. CRC Press, Boca Raton (1st ed.: Wilson JR, Corlett EN (1990) Evaluation of human work. Taylor & Francis, Londra e New York)

Chapter 5
The Design of Ergonomic Requirements

5.1 Introduction

The main design requirements for the product,[1] which are presented in this chapter, are understood in Ergonomics as the requirements the design must meet comply with the user's needs,[2] that is, as was explained in previous chapters, the set of needs, expectations and desires expressed explicitly or implicitly by the people who may come into contact with the system/product.

The term *requirement* is used as defined by technical standard,[3] meaning "transposition of (the user's) needs to a technical level" and as accepted by the scientific literature, although in many texts reference is made, with a largely similar meaning, to "product characteristics", "project objectives," "expected result" or similar definitions. This meaning can be summarised as "what the product must ensure to meet the needs of the people targeted by the design".

5.2 Functionality, Security, Accessibility

The functionality of a product or system literally means its "compliance with the function for which it is intended".[4]

To guarantee its functionality, the product must obviously be "suitable for use", that is, be compatible with the characteristics and capabilities of those who must, or can, use it, and with the characteristics of the context in which it can be used.

---

[1] For the meaning of the term "product" used in this book, see Sect. 1.1.

[2] For the meaning of the term "needs" used in this book, see Sect. 1.1.

[3] See the following standards: UNI 8289: 1981, Building. Functional requirements of final users. Classification; and UNI 8290: 1983, Residential building. Building elements. Analysis of requirements (consulted in March 2018).

[4] See Treccani online dictionary www.treccani.it/vocabolario (consultated in March 2018).

© Springer Nature Switzerland AG 2020
F. Tosi, *Design for Ergonomics*, Springer Series in Design and Innovation 2,
https://doi.org/10.1007/978-3-030-33562-5_5

The goal of Ergonomics is to craft a physical and organisational context—in which the postures, movements and efforts required of people (operators or users) are compatible with their characteristics and their physical, perceptive and cognitive abilities—and, of course, products, environments and systems that allow us to carry out the activities for which they were designed.

The aspects that pertain to the design of work equipment, and the definition of the physical and cognitive tasks necessary for their use, are covered comprehensively by the technical legislation regarding "Safety of machinery",[5] which provides the parameters we need to assess their acceptability according to their type of employment (professional or domestic) and the characteristics of the planned work operations (type and duration of the required postures, extension and frequency of movements, intensity of effort, mental workload, etc.).

The design criteria for work stations and the methods for assessing bio-mechanical loading risks, which are proposed by legislation and scientific literature, can be applied in the design context, in particular as regards the prevention of risks that may derive from postures, movements and efforts required for the performance the work tasks.[6]

Similarly, the analysis of safety risks related to human error and/or the failure of the product, environments and systems to be easily usable are essential references for design and a basic component of the Human-Centred Design approach.[7]

The requirement for safety is defined by the Italian legislation as "the set of conditions relating to the safety of the user, as well as the defence against, and prevention of, damage due to accidental factors".[8]

The safety conditions foresee, primarily, the absence of sources of danger and dangerous areas, such as elements that can cause injuries, bruises or abrasions in the event of a collision or be a hindrance or obstacle to the movement of the individual, and the elimination or containment of risk factors, for example, the risk of accidents relating to the safety of the person, such as slips, falls, collisions, burns, electrocutions, injuries due to explosions. Alternatively, consider the risk of accidents due to

---

[5]The international technical standards are classified according to the ICS (International Classification for Standard) adopted by the ISO, International Organization for Standardization. The "Ergonomics—Ergonomics" and "Safety of machinery—security of machinery" articles are classified based on codes 13.180 and 13.110 respectively.

The ICS classification is also adopted by the UNI, the Italian Standardisation Body. To consult its catalogues, see www.iso.org/standards-catalogue/browse-byics.htmlandstore.uni.com/catalogo/index.php/(consulted in March 2018).

[6]Requirements relating to the design of work stations and the acceptability of the tasks required by work activities are contained in the legislation on workplace safety and, in particular, Legislative Decree 81/2008 "Protection of health and safety in the workplace". For these subjects, see Chap. 12 "Elements of biomechanics of occupational interest" by Marco Petranelli, Nicola Mucci, Francesca Fazzini, Giulio Arcangeli.

[7]For these subjects, see Chaps. 13 and 14 by Parlangeli and Caratozzolo, which are dedicated to cognitive ergonomics, 5.3 and 5.4, and Chap. 2.

[8]See UNI 8289: 1981, Building. Functional requirements of final users. Classification. The definition refers to the building system and can be extended to any system used by people.

deficiencies or gaps in organisation and/or procedures envisaged for work activities (and, in general, in the procedures envisaged for any activity).[9]

The security conditions also provide for the preparation of "barriers" or protective constraints, which are both physical (for example, physical barriers to reaching dangerous areas or parts), or controls established at different stages of a work flow and/or the use of products and systems.[10]

One essential aspect concerns forecasting potential usage methods, that is, as we saw in Chap. 4, the ways in which the same product, environment or system (whether physical or virtual) can be used for different purposes and/or with different levels of skill and ability.

The identification and elimination of possible risk factors is the objective of some technical standards developed in the areas of Ergonomics and the "Safety of machinery".[11]

The recommendations are aimed at predicting potential *risk factors* connected to possible shortcomings or gaps in the work procedures and/or use of products, environments or systems, including potential *unintentional risk behaviours* and/or

---

[9]For these subjects, see Chaps. 13 and 14, which are dedicated to cognitive ergonomics, 5.3 and 5.4, and Chap. 2.

[10]In Annex 1 of Legislative Decree 17/2010 "Machinery Directive", the following are defined:

(a) "danger": a potential source of injury or damage to health;
(b) "dangerous area": any area inside and/or near a machine where the presence of a person constitutes a risk to the health and safety of that person;
(c) "exposed person": any person who is wholly or partially within a dangerous area;
(d) "operator": the person or persons in charge of installing, operating, regulating, cleaning, repairing and moving a machine or performing maintenance;
(e) "risk": a combination of the probability and severity of an injury or damage to health that may arise in a dangerous situation;
(f) "barrier": an element of the machine specifically to ensure protection through a material barrier;
(g) "protection device": a device (other than a guard) that reduces risk, alone or in conjunction with a barrier;
(h) "intended use": the use of the machine that complies with the information provided in the instructions for use;
(i) "reasonably foreseeable misuse": the use of the machine in a manner other than that indicated in the instructions for use, but which may derive from easily predictable human behaviour.

The Machinery Directive also indicates that: machines must be suitable for operation, to be operated, to be regulated and to undergo maintenance without these operations exposing users to risks, as long as they are carried out under the conditions provided and reasonably foreseeable incorrect use is taken into account.

The measures adopted must eliminate any risk during the foreseeable existence of the machine, including the transport, assembly, disassembly, dismantling (putting out of service) and scrapping phases.

Given their pertinence to any product or system, the definitions in the Machinery Directive can be applied to everyday products.

[11]See note 6 in this section.

incorrect conduct that may result from fatigue, distraction, neglect, instinctive reactions during use or reduced perception and/or control of risk.[12]

If the risk behaviour is not intentional, it can be caused by a poor understanding of the functions or the functioning of the product as a result of insufficient or inadequate instructions, the difficulty of performing the requested operations and, finally, the absence or inadequacy of the information relating to consequences of our actions.

Usability, which is covered in the next section, represents the essential requisite to ensure actually safe working conditions. In particular, the ability to understand the functions and interfaces (whether physical or virtual), as well as the operations to be performed and how to correct any errors, are the basis of the safety conditions offered by the product, whether it is a piece of industrial machinery or daily use product.

The possibility that a product or an environment may be used improperly and that users adopt, consciously or otherwise, risk behaviour or incorrect behaviour, must be considered as a basic input factor in the design.

The designer must start from the assumption that, according to Murphy's Law, "If anything can go wrong, it will", from which we may understand that, if an error is possible, someone will eventually make it, and that all possible errors will probably be made. The project must be set up in such a way as to minimise not only the possibility of errors, but also the possible consequences of the error once this has occurred, that is, the errors must be easy to identify, must have minimal consequences and, if possible, their effects must be reversible.

It is important to remember that, as we have seen in Chap. 4, in the case of daily-use products, and, in general, products, environments, equipment and services that are used outside controlled procedures, we must consider that the user can be "anyone" and can use the product "in any way".

When assessing risk, and, of course, designing, particular attention must be paid to[13]:

- the interaction between the person (s) and the product, including the need to correct any malfunctions;
- the interaction between people;
- stress-related elements;
- the ability of people to be attentive and/or fully perceive the risk factors, a skill that depends on training, experience, ability (physical and cognitive) and, finally, age;
- the conditions of fatigue;
- distraction due to external and/or personal factors;
- constraints to physical, perceptive and cognitive skills due to age and/or disability.

---

[12]See UNI EN ISO 12100 2010, Safety of machinery. General principles for design. Risk assessment and risk reduction.

[13]Reworked from UNI EN ISO 12100:2010, Safety of machinery. General principles for design. Risk assessment and risk reduction (p. 19).

The aspects considered may occur singularly or jointly and can, most importantly, occur unexpectedly. The perception of risk is very low in children, who lack the experience and awareness of their own actions that adults typically possess, but it can also be very low in an adult, during times of stress, fatigue, haste and, in general, distraction while performing duties. A reduced physical, perceptive and cognitive ability, which can obviously derive from disabilities, is common in elderly people and, in general, anyone who is temporarily incapacitated, e.g., in terms of movement, visual perception, attention span, etc.

If, however, the risk behaviour is intentional, it may stem from the need to perform the requested procedure in a shorter time, to respond to too many requests at the same time, from external pressure, and so on.

Finally, intentional risk behaviours originating from the deliberate desire to cause damage must be considered. This so-called vandalism must be taken into account, because it can occur in any case, and also because this behaviour can stem from a lack of awareness of the real consequences of one's actions, however improper or incorrect.

All equipment, components and objects that are intended for use in publicly accessible places must be designed so as to eliminate or minimise the possibility that they will be damaged, tampered with or removed by vandalism. The examples are well known in this case: taps installed in the toilets of a motorway service station, bus shelters, benches in a public park, etc. are objects that will undoubtedly be tampered with, damaged or used improperly. On these subjects, see Sects. 5.3 and 5.4 (Fig. 5.1).

The analysis of possible risk factors also refers to an "intended use" that includes use by elderly people, children and people with limited ability, but not the conscious and deliberate assumption of risks by people.

The first problem that arises, therefore, is the perception of risk by people and their awareness of the potential consequences of their actions.

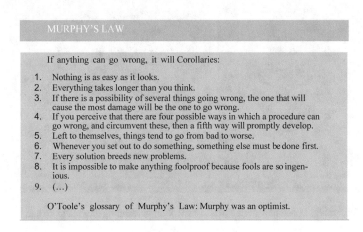

Fig. 5.1 The Murphy Law. Taken from Bloch (1977)

For example, periodic cleaning of doors and windows or the adjustment of their opening mechanisms, etc. can require the adoption of risky behaviour that may lead to falls, slips, unexpected shocks, loss of balance, etc. While almost always perceived as an "unforeseen event" that people attribute to their lack of skill or distraction, they can, in fact, be provoked by a wrong design of the opening and/or incorrect positioning of the handles, or by a failure to predict the actions and movements necessary for normal maintenance.

Similarly, the number of risk behaviours that can derive from a failure to correctly understand the function or functioning of an item are numerous and underestimated (see Sects. 4.5 and 5.4).

For example, the lack of understanding of the sense of opening a door (pushing instead of pulling, turning instead of pressing etc.),[14] the difficulty in comprehending the functions of a DVD player or the symbols and buttons placed on the dashboard of the car, which may be scarcely visible or too small, etc. generally lead to people seeking alternative methods, or proceeding by trial and error, like pressing the buttons randomly "to see what happens" (Fig. 5.2).

Finally, security in use is closely linked to the accessibility of products and environments (physical and virtual).

Accessibility is defined by Italian legislation as «the possibility, even for people with reduced or impeded motor or sensorial capacity, to reach the building and its individual real estate and environmental units, to enter it easily and to use it for spaces and equipment in conditions of adequate security and autonomy», and

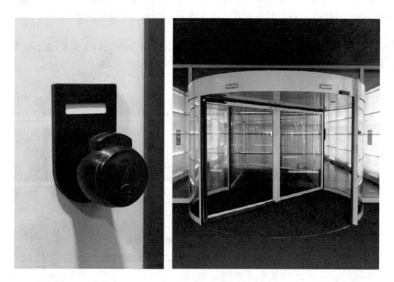

Fig. 5.2 Difficulties in use may derive from a lack of understanding of what to do and how to do it

---

[14] A detailed analysis of the errors that users can make due to the products and their operation being difficult to understand is reported by Norman (2013).

it means guaranteeing anyone the right to full and secure usability of the space in which he lives and in particular to the safety of use of the environments and products.

Incorrect or improper uses can derive from physical or perceptual limitations that make it difficult or impossible to use a product correctly. Many limitations on the ability to force and/or move fingers, visual perception and tactile perception,[15] the ability to pay attention, etc., can make it difficult to identify and/or grasp the opening handles correctly, insert and correctly rotate a key in the lock, locate and press the desired button, understand the function of the controls, etc.

For example, being forced to use a single hand to open a drawer or rotate a handle (because you are suffering from arthritis, using a cane, or simply have one hand full) can lead to loss of balance or poor control of the intensity and direction of the force being exercised. Handles, wardrobe doors and half-open drawers are often used as support points, such as shelves, tables and backs of chairs and armchairs. Similarly, the moderate limitations on visual capacity, such as a limited myopia or presbyopia, can limit the ability to read images and writing or make it difficult or slower to read road signs, even when wearing glasses.

For more information on accessibility issues, see Chap. 9 "Ergonomics and Design for All: Design for inclusion".

5.3 Usability

The term *usability* refers to the "adoptability and employability[16] of an object and is commonly understood as the "suitability of an object for the use for which it was designed and implemented" and the "ease with which people can use it".

Usability is therefore commonly recognised as the basic requirement of items used by people (which are made expressly to be used) and concerns—by definition—the quality of the interaction between people and the product. The requirement for usability is comprehensively covered in the technical legislation on ergonomics and, in particular, by ISO 9241/210: 2010, which deals with the issue of usability in the context of the Human-Centred Design approach and a global vision of users' experiences when using a product.

ISO standard[17] defines Usability as the extension with which a product or service can be used, within a system, by specific users to achieve specific objectives with effectiveness, efficiency and satisfaction in a given context of use. Where:

- **effectiveness** is the accuracy and completeness with which users achieve certain objectives;

---

[15] Both are characteristics of elderly people, but also of children, people suffering from myopia or other visual impairments and, as in terms of tactile sensitivity, anyone who is wearing gloves.

[16] See Devoto et al. (2016).

[17] ISO 9241-210:2010, Ergonomics of human-system interaction—Part 210: Human-Centred Design for interactive systems.

- **efficiency** represents the resources spent in relation to the accuracy and completeness with which users achieve their objectives;
- **satisfaction** is the freedom from discomfort and the adoption of positive attitudes towards the use of the product.

The definitions of usage context, user and task are particularly noteworthy.

The usage context is defined by the users, activities, equipment (hardware, software and materials) and the physical and social environments in which a product is used. The usage context is therefore defined as a complex system in which each element interacts with the others, and in which they play an integral part:

- **direct and indirect users**, that is, the people who interact—directly or otherwise—with the product;
- **tasks**, that is, the activities required to achieve a certain objective;
- **the product** in question;
- the physical, social and technological **environment**.

The framework provided by ISO 9241/210: 2010 proposes a highly pragmatic approach to the evaluation and design of usability. The latter, which is embedded in reality and in the specificity of the context, is variable, as are the conditions in which human-product interaction takes place and, finally, is closely linked to the complexity of external factors that can impact the specific usage context. In fact, the product cannot have intrinsic usability, as this depends on the context in which it is used and/or for which it is designed. To specify and measure the usability of the product, therefore, we must understand and describe the conditions in which the interaction takes place.

Every aspect of the ergonomic approach is summarised in the need to build on the knowledge and structured analysis of all the components of the usage context, that is, the characteristics of the users, the tasks required of them and, finally, the physical, organisational, technological and social environment.

On a conceptual level, the usability requirement can be defined in terms of *how* people carry out their activities to achieve personal goals (or required goals) and *what* the product must guarantee to allow these objectives to be achieved and ensure that the activities are safe, comfortable and satisfactory for the user.
Usability checks, therefore, represent an essential part of the iterative design process, as they allow the product to be evaluated in relation to the general usability objectives and the specific objectives required, at each step of the development process.

On an operational level, the definition of usability that includes one or more of the following factors is generally[18] accepted:

- *Utility*: pertains to the level with which a product allows the user to reach his goal. Utility is directly linked to the user's motivation to use a particular product. If a system is simple to use and its use is satisfactory and easy to learn, but it does not allow you to achieve a specific goal, or the user has no reason to

---

[18]See Rubin (1994), Stanton and Baber (1996 e 1998), Rubin and Chisnell (2011), Stanton et al. (2014).

use it, it will be deemed useless and rejected, even if offered at no cost. The usefulness of the product represents a necessary evaluation, since the first phase of product development, which must be established, even before its usability, safety and aesthetic pleasantness, is if the product will be desirable as it is useful and necessary for a purpose[19];

- *Ease of use*: generally, pertains to the possibility for a specific group of users to carry out and successfully complete a specific task. It is usually defined, in a quantitative sense, as "speed in the execution of a task" or "the absence of errors" and assessed on the basis of the percentage of people who can perform a task satisfactorily (e.g. 95% of people who regularly use a certain word processing program can use its new version after only 10 min of use).
- *Learnability*: the product or system should allow the user to reach an acceptable skill level within an acceptable, pre-defined time. On the other hand, learning may refer to the speed with which the user is able to use the product with an acceptable level of confidence after the planned training period.
- *Attitude*: pertains to how the user perceives and judges the product. Users generally tend to use a product that they find useful and satisfactory.
- *Flexibility*: the product should allow the user to carry out a wider range of tasks than just the primary use, and, moreover, should include usage methods beyond simply the "correct" one.

5.3.1 The Components of Usability

ISO 9241/210: 2010 provides the tools for measuring and evaluating the three components of usability: effectiveness, efficiency and satisfaction. The choice of the measures to be used, and their level of detail, depends on the objectives of the evaluation and on the importance that each aspect has in the usage context.

- **Effectiveness** is the accuracy and completeness with which users achieve specific objectives. In the case of products designed and used for simple activities, the effectiveness coincides with the accuracy and completeness with which the user can complete the requested task and obtain the expected result. In the case of environments or products designed for complex activities, the result may, on the

---

[19]It is clear that the need for, and desirability of, the product are not absolute values, but may derive from essential needs and induced needs. A chair is necessary to sit down because it responds to a primary need, but the last chair designed by a famous designer is not necessary, save for a need induced by taste or belonging to a certain social status.

The aspect that interests us here is not the moral value of the need for a given product, but how it can impact our choices, which are based on hierarchies of values that can be very different over time. The same person can privilege the aesthetic value in the choice of a dress with respect to its cost or its functionality (i.e. beautiful and desirable, though very expensive, or too tight, too short or too light) and its possible consequences (i.e. the chance of catching a cold or spending long hours in terribly uncomfortable positions) and, at the same time, buying an appliance with special functions, even if it is viewed as aesthetically unappealing.

contrary, be completely or only partially achieved, i.e. correspond to the correct execution of all or only part of the planned operations.

In a public office, for example, you can manage to achieve the desired design without finding the office we were looking for. Likewise, in the case of a window that can be opened many ways (vertical and pivoting windows) or a DVD player with various functions (copying a film, choice of scenes, restart from the last break), the user will be able to complete all the planned operations or only a part of them. In similar cases, the effectiveness can be judged in whole or in part, based on the number of operations that can be completed.

Similarly, the effectiveness with which individual objectives can be achieved will be more or less relevant based on the importance they have: not being able to reach the side entrance of a building quickly may be irrelevant if the goal is simply to satisfy your curiosity, but it can represent a dramatic problem if it is the emergency room entrance and we have just had an accident.

- **Efficiency** *refers to the resources spent in relation to the accuracy and completeness with which the users achieve the results (i.e. effectiveness).*

  Resources can include physical and mental effort, time, materials and the financial commitment necessary to achieve the required objectives: efficiency can therefore be measured in terms of effort required, in terms of time and in economic terms.

  You can have different efficiency evaluations based on the objectives or functions that are given more or less weight. If we are using a photocopier to make two hundred copies of a document, the efficiency can be evaluated based on the time required (the number of copies per minute), the material consumed (the number of copies per ink cartridge), the cost (the number of copies needed to pay off the purchase of the machine, or the cost of each copy in relation to the cost of the paper and ink needed, etc.).
- **Satisfaction** *refers to freedom from discomfort and positive attitudes towards the use of the product.*

  Satisfaction can be assessed based on the conditions of well-being or discomfort experienced by people using the product or an environment, the acceptability of the workload required by the use of the product or the extent to which effectiveness and efficiency is achieved in the performance of a given task. Further information can be obtained through long-term measures and assessments, for example, ones pertaining to the number of accidents caused by the product or similar products, to health problems expressed by users, to absenteeism detected in a given period of time, etc.

It is important to emphasise that the evaluation of usability (or one or more of its components: effectiveness, efficiency and satisfaction) can refer to the product as a whole (for example, a photocopier is generally usable, i.e. for all its functions and for all his parts), or only partially (a photocopier is completely usable for making single copies or on the front, but it cannot reduce, enlarge or print in colour; or, it is perfectly cost-effective for ink consumption, but it is not for the number of copies per minute).

The definition of usability can be further investigated by considering other aspects related to *learnability* and *memorability*.

As Polillo writes (2010, p. 70), "*it is necessary to consider the evolution that the user may experience over time in his relationship with the product or system. At the beginning, he does not know everything (beginner), then he starts using it (novice user), eventually becoming competent and, in some cases, an expert on the product. In this learning process, the user can encounter various difficulties, depending on the characteristics of the system. Even products with very similar functions can have very different learning profiles, in fact*".

The frequency with which people come into contact with, or use, a product is important here. As stated in Sect. 4.5 regarding the frequency of use, an occasional user will have a radically different relationship with the product than a continuous user, i.e. a person who uses the product on a daily basis, establishing habits based around it.

It is important that the methods of use are easily memorised, particularly for occasional users. If this is not the case, the user will have to re-learn how to use it with each new use.

Nielsen (1993, p. 26) defines usability on the basis of five attributes[20]:

- Learnability: The system should be easy to learn so that the user can rapidly start getting some work done with the system.
- Efficiency: The system should be efficient to use, so that once the user has learned the system, a high level of productivity is possible.
- Memorability: The system should be easy to remember, so that the casual user is able to return to the system after some period of not having used it, without having to learn everything all over again.
- Errors: The system should have a low error rate, so that users make few errors during the use of the system, and so that if they do make errors they can easily recover from them. Further, catastrophic errors must not occur.
- Satisfaction: The system should be pleasant to use, so that users are subjectively satisfied when using it; they like it.

Finally, the definition of **User Experience (UX)**, "includes all the users' emotions, beliefs, preferences, perceptions, physical and psychological responses, behaviours and accomplishments that occur before, during and after use" is essential (ISO 9241/210:2010). It is important to note that, in defining the standard, the User Experience includes "all emotions, beliefs, preferences, user perceptions, physical and psychological responses, behaviours and achievements that occur before, during and after use", i.e. it considers the interaction with the system as a whole. This definition appears to be of particular interest for design, in which the reference to the "use of the product" is necessarily limiting. The definition of UX as a "consequence of brand image, presentation, *functionality, system performance, interactive behaviour and assistive capabilities of the interactive system, the user's internal and physical state*

---

[20]See Nielsen (1993, p. 26), quoted by Polillo as it pertains to usability evaluation criteria.

resulting from prior experiences, attitudes, skills and personality, and the context of use" is also of specific interest for the field of design.

Usability, in fact, *"when interpreted from the perspective of the users' personal goals, can include the kind of perceptual and emotional aspects typically associated with user experience. Usability criteria can be used to assess aspects of user experience".*

The definition of usage experience thus refers to the **overall quality of the interaction between people and the products/systems** with which they interact and a design vision that is based on the understanding and interpretation of all its aspects—physical, sensorial, emotional, cultural, etc.

5.4   Making the Relationship with Products and Systems Understandable

In this paragraph, a "translation in terms of design" of what was extensively treated in the texts that analysed "human error"[21] and the cognitive processes that govern the actions of individuals is explored, in spite of the case limitations.

The cognitive aspects of the interaction between people and systems are dealt with in Chaps. 13 and 14; the necessary bibliographical references are also indicated there.

The concepts of *visibility, mapping, feedback, affordance, meanings and usage constraints* are summarised and reworked in terms of design, starting from Norman's text, *The Design of Everyday Things*[22] which, as mentioned in the introduction, was a pivotla step in the formation of the area of research and experimentation around "Ergonomics and Design".

Making the relationship with products and systems understandable, along with the physical or virtual environments, means designing them so that the people they target, and who can then use or relate to them, can easily and quickly understand how they work, the function of their parts or components, the sequence of actions to be carried out, etc.

It means, in other words, *ensuring ease (and safety of use)* for people who may enter into a relationship with the product.

In the case of consumer products, as well as equipment and building components, this means focusing attention on the ease with which the object of interest will be perceived and recognised for the function it performs, and ensuring it is possible to identify the parts that make it up, the usage methods, the outcome of our actions, etc.

In the case of buildings, ease and safety of use depend on the clarity and logic of the planimetric layout, on the availability and readability of the internal signs, and translate into the ease with which people can find their bearings and become familiar

---

[21] In terms of "human error", we limit ourselves to some bibliographical references that are essential for design. These include: Norman (1993), Reason (1990), Mantovani (2000).

[22] See note 11 in Chap. 2.

with the place, along with the safety of the interaction. In both cases, formal solutions that denounce the function of a specific element, the use of colour and contrast to favour the visibility of the parts and the command and handling mechanisms, different surface treatments to differentiate between vertical and horizontal planes, fixed parts instead of mobile ones and, in general, highlighting of the elements of interest[23] are essential.

Firstly, consider the *visibility*[24] of the parts or components of the object of interest, which allow us to recognise the function. Designing visibility means making information relating to the functioning of an object, its fixed moving parts, the command and handling mechanisms and, in the case of an environment, the articulation of paths, directions, entry and exit doors etc. clear, through the use of signals with clear and unambiguous information (Figs. 5.3 and 5.4).

> Visibility, that is the easy identification of the entrance of a building and then of the paths and internal entrances, allows us to quickly orient ourselves. The visibility of the commands and/or parts of an object allows us to understand how it is used, what parts to act on, what can be moved, opened, rotated. It can be said that visibility is the foundation of usability.

Norman refers to *natural signals* as sets of information that can be interpreted immediately, without the need for instructions or references. This information uses consolidated codes within a given group of people and a logical-spatial correlation between the form or operation of the command or adjustment tool and the action that must be performed.

The former exploit codes of behaviour or language that commonly used by a specific segment of the population. For example, the use of English on-off terms that are now universally recognised in all Western countries as indicating on-off, or the arrangement of car driving pedals (clutch, brake, accelerator) from left to right, which is recognised by anyone with a driver's licence, etc.

The latter are based on immediately associating the commands to the actions that can be carried out and to the results that can be obtained through their use.

---

[23]For example, a moving door can be made easy to identify through the use of colour. Similarly, the handles and the opening systems must stand out because of their shape and size and must be familiar and recognisable by sight.

A dark-coloured handle on a dark door is obviously hard to make out, as is a lock or a button that does not contrast from the surface on which it is placed.

The use of colour or different surface treatments can be used to differentiate between vertical and horizontal surfaces, to distinguish doors and drawers and to make handles and opening systems visible.

[24]See Chaps. 13 and 14, which are dedicated to cognitive ergonomics, in particular, 13.4 "Designing mental models".

Fig. 5.3 Visibility

When we use a new object our actions are guided by a series of questions to which we must respond in order to use it:

Which parts are mobile and which parts are fixed?
Where do you grab the object? Which part should be handled? What should you grip? If it is voice activated, where do you talk?
What kind of movement is possible: pushing, pulling, turning, rotating, touching, feeling?
What are the relevant physical characteristics of the movments I have to perform? How forcefully should the object be handled? How far should we expect it to move? How do you judge the success of the handling?
Which parts of the object are supporting surfaces? How much weight and size can the object support?

For, or within, an unknown or changed environment, the questions will be:

Which is the entrance?
Which are the most important parts (for example, the reception in a hotel, the ticket office and the waiting room in a station, the reception in a hospital, the information desk in all public buildings)? Where are the stairs, elevators, etc.?
Which room am I looking for and where is it?
What are the paths I have to follow? Where are the directions and what exactly do they mean?
If I found the directions: is the path I am following right? Will it take me to the place I'm looking for? And in what time?
Where is the exit? How do I get there?
In addition:
Who can I ask for information?
If I get lost: where is there a map or an indication that will lead me back to the entrance?
If the situation is dangerous or urgent: where are the emergency exits, stairs, elevators, etc.?

Fig. 5.4 Visibility

The interpretation of the correlation between command and action, or the mapping[25] process, allows us to establish an immediate relationship between the parts of an object and what they allow you to do.

The information can exploit the position of the controls in relation to the position of the objects which we will act upon: an example is the position of the switches with respect to the arrangement of the light points, or the arrangement of the ignition and adjustment knobs with respect to the position of the stoves (see Fig. 5.5).

---

[25]*Mapping* literally means "cartographic projection" and indicates the graphic projection of spatial relationships and, by extension, the correspondence between commands and functions, see. Norman (2013). See Chaps. 13 and 14, which are dedicated to cognitive ergonomics, in particular, 13.3.1 on the concept of mapping.

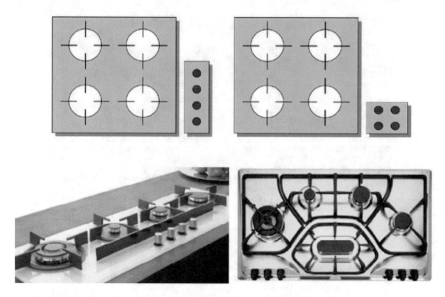

Fig. 5.5 Mapping. The two diagrams in the figure above are reworked from Norman (2013)

Object mapping can also take advantage of physical links: for example, push the control lever up to open the boot of a car, turn the steering wheel to the left to rotate the tires and the car in the same direction.

The ease with which environments and products can be used, and the ability to perform actions without making mistakes, is based on people's ability to make generalisations, stemming from skills previously developed with other products.[26]

The classic arrangement of the ignition and adjustment knobs (top left) does not provide any indication on the correspondence between knob and stove. An arrangement based on spatial correlation (top right) allows instead the immediate association between command and commanded object.

The information that the environment or product can provide also relates to what actions we can reasonably perform with that product or within that environment, and the effect that our actions have produced or are producing. In the first case, the information derives from the affordance of the product, that is, what the product allows us to do; in the second case, from the feedback that follows our actions.

---

[26]Previously acquired skill can obviously be an advantage or, in some cases, a disadvantage: it is possible to drive any manual car with relative ease if we have owned or used a car of this type for some time, but it will be much more difficult to have to relearn to drive an automatic.

Fig. 5.6 Affordance

The **affordance**[27], or literally invitation or authorisation, indicates the set of real and perceived properties of material things that provide us with information on how they can be appropriately used.

A chair authorises and invites you to sit or support something; *it is made* for sitting or supporting weight. A glass *is made for* looking through but also breaking.

Finally, Norman defines the concept of a *signifier*, defined as the part of the product that communicates where the action should be performed. The handle that indicates where to place the hand to open the door is significant, but also the indication "push" that indicates what we must do (Figs. 5.6 and 5.7).

A chair or an armchair, formed by a flat element about 50 cm from the ground sup- ported by three or four sufficiently stable legs, backrest and eventual arms, tells us the authorization—the invitation—to sit down, or to support a bag or a parcel. A chair or armchair from the 1700, exhibited in a museum, will tend to communicate the same invitation to us; however, we will have to resist by virtue of the constraint—in this case physical and cultural—which prevents us from using it as a "chair" and instead requires us to observe it as museum piece.

---

[27]See Chaps. 13 and 14, which are dedicated to cognitive ergonomics, in particular, 13.3.2 on the concept of affordance.

Examples of physical feedback

- The minute hands moving on a watch dial
- The key that turns without getting stuck inside the lock
- The silent click that can be felt by pressing a button
- The soft click up to îend of strokeî and the soft return of the keys on the computer keyboard that indicates the insertion of the letter
- The handle that turns all the way down (and the door that unlocks)
- And, of course, all the actions that have obviously come to fruition (which make us say or think: this works): the water that comes out of the tap, the fire lit on the stove, the car that starts, the hinge that closes, the door that opens, and so on.

Examples of light feedback

- The red alarm light
- The red hold light
- The flashing light of a printer when no colour ink remains
- The flashing light of the indicator (on the car panel)
- The light of any color that indicates on or off in any electrical appliance (television, iron, radio, etc.).

Examples of found feedback

- The clicking when a door lock closes
- The sound of a zipper moving
- The hollow sound of a door closing badly
- The rumble of a crushed muffler
- The rattle of pieces that are incorrectly placed
- The whistle of a kettle when water boils
- The click of the toaster when the slice pops out
- The sharp sound of a fan when it jams
- The indescribable altered noise of a complex machine when it begins to fail
- The ticking of the indicator after it has been activated.

Fig. 5.7 Examples of physical, light and sound feedback

Information concerning the effect produced by our actions constitutes the **feed-back**,[28] or literally "return information", that is, the signal that tells us if the action ended, if it had a follow-up and, above all, if it was "the right one". The information on the outcome of an operation obviously allows us to understand whether or not the action carried out is the correct one, and also allows us to make the subsequent choices (for example, continue the planned sequence of actions, or go back and start again). The clarity and unambiguity of the information depends, on the ease with

---

[28] See Chaps. 13 and 14, which are dedicated to cognitive ergonomics, in particular, 13.4 "Designing mental models".

which it will be possible to understand the operation of a given product and the possibility of using it correctly and without making mistakes.

Without going into the merits of the analysis of human error, which we refer to in Chaps. 13 and 14 and the previously cited texts, it is useful to recall the difference between lapses and the errors that may occur through human actions.

"Human error" is defined as any deviation from appropriate behaviour.

- *Lapses* arise from a behaviour that leads us to do something when in reality we wanted to do something else. We can take one action instead of another (action-based lapses), or forget to take action (memory-lapses). Lapses occur more frequently in the actions we normally carry out and with greater competence. Lapses are easily identifiable, even by the user himself, but only in the presence of adequate forms of feedback that allow us to verify the sequence of the action during its execution.
- Mistakes arise from conscious decisions, when the objective is
  wrong, or we do not understand the actions we have to perform or when we understand them incorrectly. *"Mistakes have three major classes: rule-based, knowledge-based, and memory-lapse. In a rule-based mistake, the person has appropriately diagnosed the situation, but then decided upon an erroneous course of action: the wrong rule is being followed. In a knowledge-based mistake, the problem is misdiagnosed because of erroneous or incomplete knowledge. Memory-lapse mistakes take place when there is forgetting at the stages of goals, plans, or evaluation"* (Norman 2013, pp. 171–172).

Many errors originate from a wrong interpretation of the necessary action sequences or from a wrong choice among the available alternatives (see Sect. 13.5 "From perception to action" and "The Norman model of action" shown in Fig. 13.6). Errors also occur when habitual actions take place. The same person, who is perfectly capable of performing a certain action in normal circumstances, can instead make mistakes or forget basic steps when he is in an abnormal situation, or in situations with personal or environmental stress. We all routinely and safely drive our car, carrying out the actions we need to move through traffic, or on poor roads or the motorway, but we can perform wrong actions or skip essential steps (for example, changing gears, choosing between the brake pedal and the accelerator, etc.) in conditions of fatigue or distraction, of environmental stress due to a traffic jam, in panic conditions and so on. Likewise, in situations that differ from the norm, we can go the wrong way, not see a red light or a stop sign, press a wrong switch, forget to complete a procedure (for example, do not save the job we just finished on our computer, leave the house keys in the lock), etc. (Figs. 5.8, 5.9, 5.10 and 5.11).

Examples of physical constraints include the size and position of the holes provided for mounting shelves or other furnishing elements or the common keys used for doors and windows, but also for turning on the car, for opening covers, etc. The keys generally have a direction (that is, they can be inserted

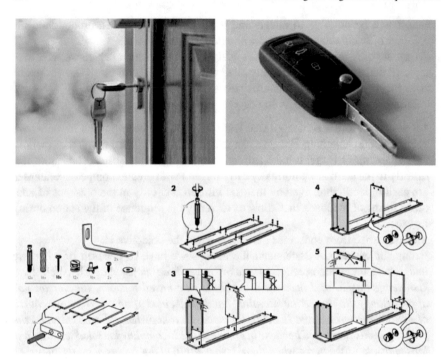

Fig. 5.8 Examples of physical constraints

Fig. 5.9 Examples of invitations and cultural constraints

Fig. 5.10 Examples of cultural invitations

Fig. 5.11 Examples of logical invitations

in a vertical slot or in a horizontal slot) made evident by the shape of the lock, but they do not have a signal indicating the orientation with which they must be inserted (upwards or low, right or left). A well-de- signed key should work in both directions (many ignition keys for cars of more recent production are designed so that the orientation is indifferent).

The top of a table covered with paper or cardboard, or a top in plywood, invites any person accustomed to drawing to use it as a huge sheet of paper. In this case, the obligation not to do so is strictly cultural (at least in the absence of controls) and obviously not always effective.

Design makes it possible to signal and invite users to carry out the appropriate actions, and to prevent inappropriate or risky actions, through the provision of invitations to use and constraints of use.[29]

Invitations to use refer to the previously described concepts of affordance, signifiers and mapping.

Affordances can signal how an object can be moved, what its function is, which parts to fit together during assembly and so on. Signifiers indicate where we must perform our actions. Mapping suggests what the relationship between the commands, the actions to be carried out and their possible effects are.

Constraints limit the possibilities of action. The knob can only be turned to the left, a button can only be pressed downwards, the steering wheel of a car can be rotated in both directions, but not tilted forward or pulled towards the driver, etc.

In other words, invitations to use suggest what actions I can perform with a given object, while constraints limit the number of alternatives and prevent erroneous actions.

Constraints can be divided into:

- *Physical constraints*: these are the physical limitations that circumscribe the number of possible operations so that, to make an object work, it is not necessary to have instructions or training because the object can only be used in that particular way. For example, a large pin cannot enter a small hole, an electric plug can only fit into a socket suitable for its size, etc. Physical constraints are useful and effective if they are easy to see and interpret because in this way the range of possible actions narrows before they are even executed; otherwise the physical constraint prevents the wrong action only after it has been tried.
- *Semantic constraints:* these are the constraints that rely on the meaning of the situation to circumscribe the set of possible actions. To drive a car, you need to sit looking forward, to write you must point the tip of the pen or pencil down and so on. Semantic constraints rely on the user's knowledge of the situation and sharing the same knowledge with the majority of potential users. In different cultural environments, the same codes of interpretation may not apply. Some operations, which are obvious for those who are used to driving for a long time and have experience with cars and different controls, may not be so for those who are learning to drive or for those who are using a new car for the first time. For example, the actions required to operate the reverse gear provided for in many car models (raise the gear locking ring or, instead, press the knob downwards) or, similarly, use the mouse with a double click or drag on the icon in PC and Apple systems respectively.
- *Cultural constraints:* these are the constraints that refer to accepted cultural conventions. A cultural convention would be that writings are made on purpose to be read and that the letters must be arranged in a straight and vertical position.
  Each culture also provides for a series of permitted or excluded actions in different social situations. We know how to behave differently at a party or during

---

[29]The definitions of physical constraints, semantic constraints, cultural constraints and logical constraints are taken from Norman (1988 and 2013), and by Hall (1996).

a university lecture; we know how to behave at a restaurant or in the post office queue and how to move around a subway station or shopping centre, etc. because we have acquired the basic rules of behaviour through family and social education, or because we have already experienced similar situations and we expect the signals coming from the outside world. An unknown environment, a brand new experience, or personal situations (anxiety, haste, tiredness, distraction, or other impediments of any kind) diminish our ability to control the environment, and may prevent us from acting in the right way.

- *Logical constraints*: these are the constraints related to "the only logical option" (the only piece left or the last free accommodation).

5.5 The Perceptual and Emotional Quality of the Product

The perceptive and emotional aspects of the interaction between people and products refer to the collection of sensations and emotions—both positive and negative—that the aesthetic and sensorial qualities of the products evoke.

Talking about the overall quality of the interaction obviously cannot limit us only to the relationship dynamic with the product, but presupposes evaluating and designing the set of factors that determine the interaction experience, which includes the sensory and emotional aspects, the aesthetic appeal and, finally, the visceral reaction that allows the product to "hook" the user.

As Norman writes, in addition to the components of Design that are strictly related to use, i.e. usability (or its absence), aesthetics and practicality, there is (naturally) "*also a strong emotional component to how products are designed and put to use. (…) the emotional side of design may be more critical to a product's success than its practical elements*". We can describe "*three different aspects of design: visceral, behavioural, and reflective*" (Norman 2004, p. 5).

As we will see in Chap. 14, visceral design involves people through the shape and symbolic aspects of the product.

Behavioural Design concerns the use of the product, its effectiveness and the pleasure of using it.

Reflective design concerns the overall judgment on the product and the value that each of us rationally attributes to it. As Caratozzolo and Parlangeli write in Chap. 14, "*an object that offers us an emotionally positive experience positively impacts our future intentions of use*".

The definitions of "pleasure in use" and "appeal", introduced by Jordan (1998 and 2002) and by Buti (2008), are also of considerable interest in this regard.

"Pleasure in use", defined as the dimension that "*goes beyond usability*" to respond to the more strictly subjective needs of the user, approaches the needs that do not derive only from the use of the product, but from the meaning that each individual attributes to it when they come into contact.[30]

---

[30]See Green and Jordan (2002), Jordan (2001).

Buti defines appeal as the set of sensations (tactile, prehensile, functional, thermal, chromatic, acoustic) that derive from the relationship with the product, its shape, its surface and its consistency (Buti 2008, pp. 158–161).

The evaluation of sensations can be partly conducted in an objective manner, but it mainly requires a subjective evaluation that can collect people's impressions and assessments with respect to the texture of a surface, the combination of colours, the shape, weight, etc. of the product.[31]

The emotional dimension of the interaction between people and products, the dimension linked to use and, finally, the reflective dimension, if considered as a whole, unified part of the *experience* of people who interact with the product, define the approach that most closely resembles design and is more able to provide answers to the problems posed by design.

The definition of the User Experience (UX) as a "person's perceptions and responses resulting from the use and/or anticipated use of a product, system or service" which "includes all the users' emotions, beliefs, preferences, perceptions, physical and psychological responses, behaviours and accomplishments that occur before, during and after use", includes the sensory and emotional sphere as an integral part of the interaction between people and products. It is important to note that UX is also a "consequence of brand image, presentation, functionality, system performance, interactive behaviour and assistive capabilities of the interactive system", and that "the user's internal and physical state resulting from prior experiences, attitudes, skills and personality, and the context of use".[32]

The study of appeal, in fact, involves every element of an aesthetic evaluation—that is, subjective taste and preferences, and social, cultural and individual factors that can influence them—and the sensory and perceptive elements that can also shape our appreciate, annoyance or repulsion when we come into contact with a product or an environment.

Appeal and, in general, the subjective aspects of the individual environment-product interaction are covered by the area of ergonomic research that we now know as New Human Factors, NHF, or new human factors. The NHF are aimed at investigating the motivations that push the user to appreciate or prefer a given product, the trends and preferences that can be recognised in the universe of potential users/buyers, the relationships between the formal and sensory properties of the product and the sensations and emotions that they can arouse.

It is important to note that the sensations and emotions aroused by the product, as well as its aesthetic evaluation, refer to a relationship with the product, that is still linked to its use. Works of art, in fact, are not of interest for New Human Factors, as aesthetic evaluation obviously takes on another, more complex dimension there,

---

[31]For this subject, see the SEQUAM method, Sensory Quality Assessment Method, described by Buti (2008, pp. 167–177).

[32]See the definitions of ISO 9241-210: 2010 set out in par. 5.3. In terms of the relationship between User experience and Usability, note 3 of the User experience definition also indicates that: "Usability, when interpreted from the point of view of users' personal goals, may include aspects of the User Experience".

but rather consumer products, i.e. the objects with which we establish a relationship stemming from their function.

The research and experiments carried out in this sector are myriad and the interpretations that can be given to the definition of "pleasure in use" are just as numerous. In this field, we can spot some main components:

- *the emotional and perceptive component*, i.e. the ability to recognise, use and appreciate the sensory properties of a product or an environment, the information (but also the emotions) provided by the surfaces, colours, smells, sounds, the feeling of discomfort or of well-being that the isolated or combined actions of those stimuli produce on the individual;
- *the model and social behaviours component*, i.e. the set of socio-cultural, generational and individual character influences—that is, characteristics related to one's own personal experience—from which we can derive models that the behaviours of the subjects and their abilities and relationship methods refer to.

In both cases, the subjective aspects of the interaction between people and products represent the shift in ergonomics towards themes that are less and less linked to the functionality and use of the product, and increasingly tied to its effect on the user in psychological and emotional terms. The goal of ergonomics—and, of course, design—is to evaluate the user's judgement, that is, to evaluate the value that the user attributes to their relationship with the product and translate this value into design tools. The quality of the product thus becomes the quality that the individual experiences, what positive (or negative) feeling the product arouses with its presence, the value that can be attributed to it in terms of sensations and emotions and, finally, the judgment with which we value or reject that product. Norman himself considers the concepts of usability and experience inseparable. In fact, the user experience concerns every aspect that is related to the interaction of the user with the product, i.e. "*how it is perceived, learned, and used. It includes ease of use and, most important of all, the needs that the product fulfills (…) both in terms of function and aesthetics, and to ensure that the products are easy to understand and to use*" (Norman 1998, pp. 47–48). "To evaluate the value", or evaluate, writes Jordan (1999, p. 209), "*the emotional, hedonic and practical benefits associated with products*" means paying attention to all the factors—cultural, social, individual—that can influence the psychological and emotional reactions of the subject.

This requires us to:

- understand and interpret users' needs;
- link the properties of products to the user's emotional responses;
- develop methods for investigating and quantifying enjoyment. The aspects at stake are obviously numerous and pertain, in addition
  to the strictly functional sphere, to every level—psychological, emotional and sensorial—that we can break our relationship with the product down into.

The contribution of Ergonomics remains essentially methodological in nature, aimed at analysing the user's needs in a structured manner and translating the information relating to the user-product interaction into design tools and data, where the user is not only the one who "uses the product to carry out certain activities" but is the person who comes into contact with the product and expresses needs both in terms of "use" and expectations and preferences.

References

Buti LB (2008) Ergonomia olistica, il progetto per la variabilità umana. FrancoAngeli, Milano

Murphy's law and other reasons why things go wrong. Price Stern Sloan, Los Angeles, 1977

Devoto G et al (2016) Il nuovo Devotino. Vocabolario della lingua italiana, Le Monnier-Mondadori Education, Milano

Green SG, Jordan PW (2002) Pleasure with products, beyond the usability. CRC Press, Londra e New York

Hall ET (1996) La dimensione nascosta. Bompiani, Milano (1st ed.: Hall ET (1966) The hidden dimension. Doubleday, Garden City)

Jordan PW (2001) New century supertrends: designing plesurable future. In: Helander GM et al (eds) Affective human factors design, proceedings of the international conference 27–29 June 2001 in Singapore, Asean Academic Press, Londra

Mantovani G (2000) Ergonomia, lavoro, sicurezza e nuove tecnologie. Il Mulino, Bologna

Nielsen J (1993) Usability engineering. Academic Press, S. Francisco

Norman DA (1988) The design of everyday things. Doubleday, New York. (1st ed: Norman DA (1988) The psychology of everyday things. Basic Books, New York)

Norman DA (1993) Things that make us smart. Addison-Wesley, New York

Norman DA (2004) Emotional design. Basic Books, Cambridge

Norman DA (2013) The design of everyday things. Basic Books, New York

Polillo R (2010) Facile da usare, una moderna introduzione all'ingegneria dell'usabilità. Apogeo, Milano

Reason J (1990) Human error. Cambridge University Press, Cambridge

Rubin J (1994) Handbook of usability testing: how to plan, design and conduct effective tests. Wiley, New York

Rubin J, Chisnell D (2011) Handbook of usability testing: how to plan, design, and conduct effective tests. Wiley, Indianapolis (1st ed.: Rubin J (1994) Handbook of usability testing: how to plan, design, and conduct effective tests. Wiley, New York)

Stanton NA, Baber C (1996) Factors affecting the selection of methods and techniques prior to conducting a usability evaluation. In: Jordan P et al (ed) Usability evaluation in industry. Taylor & Francis, Londra

Stanton NA, Baber C (1998) A system analysis of consumer products. In: Stanton NA (ed) Human factors in consumer products. Taylor & Francis, Londra

Stanton NA et al (2014) Guide to methodology in ergonomics: designing for human use. CRC Press, Boca Raton (1st ed.: Stanton NA, Young MS (1999) A guide to methodology in ergonomics: designing for human use. Taylor & Francis, Londra e New York)

UNI 8290 (1983) Edilizia residenziale, sistema tecnologico

UNI 8289 (1981) Edilizia. Esigenze dell'utenza finale. Classificazione residenziale,sistema tecnologico

UNI EN ISO 12100 (2010) Sicurezza del macchinario. Principi generali di proget-tazione. Valutazione del rischio e riduzione del rischio

Chapter 6
Human-Centred Design—User Experience: Tools and Intervention Methods

6.1 Introduction

The specificity and innovative value of the ergonomic approach to design lies in its ability to evaluate the range of variables that define the interaction between people and what they come into contact with (that is, the characteristics and abilities of users; the characteristics of products and the activities for which they are, or can be, used; the characteristics of the physical, social and organisational context), their reciprocal relationship and what could change with time, by periodically identifying and interpreting the needs and expectations that people express, or can express, regarding this interaction.

The design intervention, then, is based on the ability to understand, interpret—and imagine—the different situations and scenarios in which people can, or could, come into contact with the system, by identifying the range of variables involved and the complexity that defines their reciprocal relationship. This is achieved by defining the system of requirements that are needed for the product and the necessary parameters and criteria for evaluating and designing the quality of the product.

Ergonomics' intervention methods allow us to define and manage these relationships, organise the data collected during the design process and define the integrated intervention requirements.

In particular, the methods for evaluating usability and safety of use, as well as the methods for evaluating the User Experience, are based on the collection of information about the interaction between people and products and systems they come into contact with within a given usage context. They also allow us to identify and analyse the behaviour of people (or users), their needs (needs, expectations and desires) and, finally, the type and frequency of the errors that they can make while performing the required tasks.

This chapter was co-authored by Francesca Tosi (Sects. 6.1; 6.4.1; 6.4.2; 6.4.10), Alessia Brischetto (Sects. 6.2; 6.3) and Mattia Pistolesi (Sects. 6.4.3–6.4.9).

© Springer Nature Switzerland AG 2020   111
F. Tosi, *Design for Ergonomics*, Springer Series in Design and Innovation 2,
https://doi.org/10.1007/978-3-030-33562-5_6

As we saw in Chap. 5, the conception, design and development process of the product/system or service starts with knowledge and interpretation of the specificity of the usage context—or contexts—for which said product/system or service is intended. The usage context is described as the set of variables that define the interaction between people and the elements of the system they come into contact with, of which said people are an integral part, that is: the people, the predicted tasks, the physical, technological, organisational and social environment.

In the case of the industrial design and production process, the variables to be considered also include client requests and the limitations imposed by corporate organisation and the production process (costs, time, materials, available machines and processes, etc.).

As we will see in the following section and the Appendix of this book, the process of conceiving and developing the design, and the subsequent manufacturing of the product, typically stem from the requests of a client and an exploration phase in which we must define "what" the new product must be, "why" it must be produced and what sector of the market it will target, at least on an indicative basis.

This exploration phase is follow by the definition of the design brief and—within the framework of the HCD/UX approach—phases for figuring out, specifying and interpreting the usage context(s) and, subsequently, identifying (and interpreting) the expressed and unexpressed needs of potential users and defining the project requirements.

The design phase, and the subsequent verification of the design proposals, starts from the previous investigate phases, then, and will develop based on the design requirements.

Naturally, the evaluation process for existing products/systems or services has a similar approach.

6.2 Human-Centred Design Methods

"Anthropocentric" design, or Human-Centred Design (HCD), is an intervention philosophy that aims to develop products/systems or services that can satisfy people's needs, so that interacting with them is characterised by a high level of usability and ease of understanding and can offer a user experience that is positive and satisfying.

As part of this vision, the products must meet a series of requirements that are based on careful evaluation of the users' needs and, naturally, respect for the limitations imposed by production and marketing needs.

As noted in the ISO 9241-210: 2010 standards, once the need to develop a system, product or service has been identified and, as a result, the reference problem, there are four essential steps that must be followed to integrate the requirements for usability into the product/system-development process (Fig. 6.1):

(a) understand and specify the context of use;
(b) specify the user requirements;

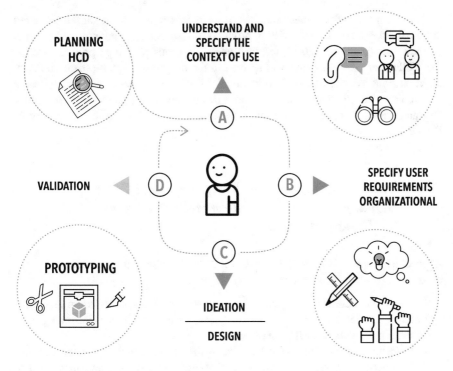

Fig. 6.1 The human-centred design (HCD) process

(c) produce design solutions to meet these requirements;
(d) evaluate the designs against requirements.

The first two phases are part of so-called "User research", which is dedicated to "understand and specify the context of use"—i.e., defining the profile of the users (reference target), the real and potential context of their interaction with the new product and explicit and implicit user needs—and to "defining the requirements" of the design.

The second phase for defining the requirements is a result (or, rather, the output) of "User research".

In the third phase of producing "design solutions", we instead move to the actual design of solutions based on the requirements (phase a–b). In the fourth and final phased, which aims to "evaluate the designs against requirements", the technical-functional aspects and User Experience aspects will be examined using an iterative evaluation process for the design solutions that have been produced and any modifications to them.

Within the HCD process, the involvement of the users plays a fundamental role in phases a and b ("understand and specify the context of use" e "specify the user requirements") and phase d ("evaluate the designs against requirements").

The users (or, rather, select samples of users that represent the target audience) are involved via trial uses of existing products, questionnaires and interviews, direct observations (phases a and b) and trial uses of various prototype products (phase d). While users were not always involved in the design-only phase (c) in the past, today, various methods are used to involved users in the phase for producing design solutions: these include the methods that developed in the fields of Participatory Design (originally called Co-operative Design) and Co-design (see Chap. 7 by A. Rinaldi).

6.2.1 Understanding and Specifying the Usage Context; Defining the Design Requirements; Producing the Design Solutions

According to the ISO 9241-210: 2010 standard, understanding and specifying the usage context must include a description of:

- the users' characteristics: physical characteristics, abilities, skills, training, habits, preferences, etc. In some cases it will be necessary to classify users into categories, for example, by their level of experience.
- the tasks they will have to perform: analysing the tasks that may impact the usability of the product or system, for example, indicating their frequency and duration. The physical stress and safety levels should also be checked.
- the physical, technological and social environment in which users will use the product/system: this involves defining the characteristics of the physical environment, the available technology (e.g. the availability of electricity, water, IT services, tools and equipment, etc.), the social organisation, including the tools (e.g. technology, organisational structure, work procedures, standards and habits and behaviour, etc.).

The description and understanding of the usage context and its components allows us to a collect a considerable amount of information that is necessary for identifying the needs of users and subsequently defining requirements.

Typically, the collection phase for the needs of users (both expressed and unexpressed) employs a series of investigative methods, the most notable of which include: individual questionnaires and interviews, focus groups, field observations and spontaneous suggestions from users.

As has already been noted, many methods have already been investigated regarding the HCD approach. The following table lists some of the main evaluation methods and their usability in the different phases of the HCD process.

These must be accompanied by the evaluation of competitive products that are already on the market (competition analysis) and the best practices.

The methods noted in Fig. 6.2 (and the variety of others that are available) allows us to collect the information we need to identify the types of activities that users

Planning HCD	Context of use	Requirements	Design	Evaluation
• Usability planning and scoping • Usability cost benefit analysis	• Identify stakeholders • Context of use analysis • Survey of existing users • Field study/user observation • Task analysis	• Stakeholder analysis • User cost-benefit analysis • User requirements interview • Focus groups • Scenarios of use • Personas • Existing system/competitor analysis • Task/functio n mapping • Allocation of function • User, usability and organizational requirements	• Brainstorming • Parallel design • Design guidelines and standards • Storyboarding • Affinity diagram • Paper prototyping • Prototyping	• Participatory evaluation • Assisted evaluation • Heuristic or expert evaluation • Controlled user testing • Satisfaction • questionnaires • Assessing cognitive workload • Critical incidents

Fig. 6.2 Human-centred design support methods. Reworked from Maguire (2001)

perform regularly, the objectives associated with these, the context they are performed in and the motivation that leads people to act in a certain manner. The methods can be combined and used in a "creative way", especially when dealing with highly complex systems.

Combining them allows us to identify various types of requirements in parallel. Furthermore, after the phase for identifying the needs and related requirements, it will be possible for us to draft a document of the results that have emerged and define the initial design ideas or identify new activities to validate. Once the usage context has been defined (either through prediction or hypothesis), the team of designers and/or experts must draft a document of the requirements during the first phase of the process.

Said ISO 9241-210 standard suggests that the following aspects can be identified (this is an indicative list; some points will be considered rather than others as it relates to the product or system investigated):

- the services required by the new system (or product) in relation to operational and economic objectives;
- the relevant legislative and regulatory requirements, including those for health and safety;
- the communication and co-operation between users and other relevant figures.
- the activity of users, including the division of tasks, their well-being and their motivation;
- the services required by the various tasks;
- the design of workflows and organisation;
- the management of changes that occur due to the new system, including training activities and the staff involved;
- the feasibility of the various operations, including maintenance works;
- the design of work stations and the human-computer interface.

The conclusion of the first two phases occurs with the definition of the requirements and, on an operational level, the creation of a document that specifies the requirements for the future product. This will be shared with the client (where one exists) and the work team and, once it has been discussed and revised, will allow us to plan subsequent phases, particularly the design phase.

This document will define what the new product must be and why and how it must be produced. The requirements document will never be definitive within the iterative process, but is instead subject to amendments and additions throughout every phase of the design process.

The next phase is that of conception (c), which aims to "produce design solutions" on the basis of the requirements identified in the previous phases (a–b). Next, we will move to the phase for verifying the design solutions that have been produced (d), in relation to the technical and functional aspects and the aspects that relate to usability and the user experience.

6.2.2 Verification and Validation Phases

The evaluation phase is an essential part of the HCD approach. Thanks to the use of prototypes (physical and/or virtual), we can quantify the safety, usability and user experience standards of the solutions that have been developed, with reference to the initial design requirements. The term "evaluation" can refer to two very different activities (Polillo 2010, p. 352):

- checking that the product is consistent with what is expected in the requirements documentation. This type of evaluation is referred to as verification;
- checking that the product effectively satisfies the needs for which it has been conceived. This type of evaluation is referred to as validation.

The difference between verification and validation is that the requirements documentation is not always exhaustive as it pertains to the expectations and needs of the users.

The validation phase is very complex, as it focuses on the actual quality of the product: it involves checking that the prototype is able to effectively satisfy the needs (both expressed and unexpressed) of users and the client. For this reason, the involvement of users is crucial.

There are many tests for this phase, and these vary based on the type of product/system or service being evaluated. Literature typically speaks of two branches of methods:

- expert evaluations (which are typically performed by usability experts) without any user involvement, which are known as inspections. These correspond to heuristic evaluations;
- evaluations carried out with user involvement. Evaluations of usability and the User Experience belong to this category.

6.3 Usability Tests

Usability tests are based on collecting information about the way in which user-product interactions take place within a given usage context and allow us to identify and analyse people's behaviour, needs (both expressed and unexpressed) and, finally, the frequency of errors that people can make while performing their required tasks.

According to Wilson (1995), usability tests can be divided into direct evaluation methods and indirect evaluation methods.

Direct evaluation methods are based on techniques for observing and evaluating the behaviour of users when interacting with a product (environment, system or service). They are defined—as objective methods are—based on their ability to supply objective information and consist of collecting information about the performance of users.

On the other hand, indirect observation methods allow us to collect information about users' interpretation of what they are doing. Indirect observation methods are also defined as subjective methods and refer to the creation of reports about users' attitudes and behaviour. The subjectivity of these refers to the fact that the information produced is filtered through the judgement of the observing party.

Generally speaking, usability tests aim to extract concrete indications on how to improve the product or system. The person who conducts the test observes and analyses the users' behaviour and tries to identify the critical factors, "what" caused the user to make the error and "when", and the nature of the difficulties they encountered.

In addition to the involvement of the user, the test should be conducted by at least two people: a facilitator, who manages and guides the test, and one (or more) observer(s) who attends the testing sessions and takes notes of the relevant information. The evaluators' roles are critical and require in-depth knowledge of the activities needed to perform the tasks; a good practice that can be adopted by the observers is to verify first-hand the tasks that will be assigned to the users, at some time prior to the test itself.

6.4 Summary of the Main Methods of Investigation for Human-Centred Design

Multiple investigative methods that can be used in accordance with the Human-Centred Design (HCD) approach are found in literature; these can be used to evaluate usability and the User Experience. Each method has its own characteristics, advantages and disadvantages that make it appropriate (or not) for the type of product that must be evaluated and/or designed and the phase of the evaluation and/or design process in which they can be used.

The UNI 11377-2: 2010, "Usability of industrial products" standard, clarifies the objectives of the main evaluation methods and, in particular, their applicability to the industrial products sector.

The selection of the evaluation methods depends on five factors:

- accuracy of the methods;
- aspects to evaluate;
- acceptability and suitability of the methods;
- skills of the designers involved in the process;
- cost-benefit analysis of each method.
- As highlighted in the previous section, the HCD methods can be divided into methods that require the direct involvement of users and methods that do not require this (expert evaluations).

As previously emphasised, a wide variety of methods are available to-day and we can say that each discipline and each research group has developed its own methods and/or combination of existing methods.

In the conception field, particularly the field of Design, investigative methods aimed at the specificity of the industrial design and production process are used and allow us to examine the complexity of the interaction between the user and the product, the system or the service (which can often be physical and virtual) and the variety of limitations that are found in the production processes.

What follows is a summary of some of the most commonly used methods in the field of design for evaluating the safety and usability of a product/system or service.

6.4.1 Task Analysis—TA

Overview

The Task Analysis (TA) is based on the breakdown of the activities that a user must perform to achieve their stated goal (or goals), such as the performance of a task. The tasks to be completed are indicated by the experts who design the test in relation to their initial evaluation, for example: high-risk tasks for the safety of the users, or that highlight more critical use or a lower level of understanding, etc.

The TA is typically structured into five fundamental steps:

1. collection of data/information;
2. description of activities;
3. selection of the most important activities;
4. breakdown of the activities to identify and describe tasks, objectives and end goals;
5. organisation of the activities and sub-activities in order of their importance or suitability.

Procedure

The test is carried out with certain actual or potential users who are in line with the target audience. These are invited to perform some tasks that are typical of using the product; these have been previously codified by the experiment team into high risk, high priority, frequent and/or critical. Team members assist the users, interpreting what they say (direct questions and inviting them to "think aloud") and their non-verbal cues, thus identifying the criticisms and strengths of the product.

When the various phases of analysis come to an end, the debriefing is carried out to share the results. This allows us to identify solutions for the criticisms that arose, in terms of (Fig. 6.3):

- new features to implement, because they are missing;
- existing features to remove, because they are useless or redundant (duplicates found);
- features to modify, because they are critical (high risk of causing errors) or complex (long and/or repetitive). The set of conclusions provides the guidelines for subsequent phases of product design.

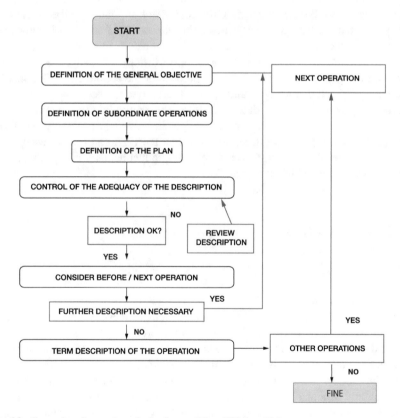

Fig. 6.3 General outline-task analysis. *Source* Oliva (2005, p. 114)

6.4.1.1   Hierarchical Task Analysis—HTA

Overview
The Hierarchical Task Analysis (HTA) is based on the breakdown of activities, objectives and plans into a sequence (or hierarchy) of required tasks and sub-tasks. The HTA allows us to structure a hierarchical sequence of tasks that must be carried out to complete the required action and to examine the functional and cognitive logic that determines whether or not the objectives are achieved.

Procedure
We start by defining the main objective of a task for a specific activity. Said objective is the highest level of the hierarchy.

As you go into the description of the activity, you gradually add sub-activities until you have a set of tasks that are organised in a sequence, which includes all the steps necessary for the achievement of the goal. The last point described will be the final phase, that is, the actual performance of the activity. For each objective, and for each corresponding task sequence, the different results that can occur are taken into consideration (not achieving the goal, only partially achieving the goal, fully achieving the goal, misunderstanding the purpose of the action, etc.). This method requires a very detailed analysis and does not require long execution times, unless a particularly complex task is described. Its use is required during the preliminary design phase, in the form of an a priori study that clarifies all the necessary operations to be performed. It is equally true, however, that the HTA can also be used in the evaluation phase of a project, to verify that the product actually allows the user to complete the intended objective (Fig. 6.4).

6.4.2   Link Analysis

Overview
The Link Analysis represents the sequence in which the components of the product are used in a given scenario and according to the indicated operations. The sequence indicates the links between the elements and the product interface. This could be used to determine if the relationships between the elements are optimal in terms of the sequence of activities they perform. The recorded-time data, which is related to the duration of attention, could also be recorded to determine if the elements are efficiently arranged. The data link could also be used to evaluate a range of alternatives before the project that is considered to be best can be accepted.

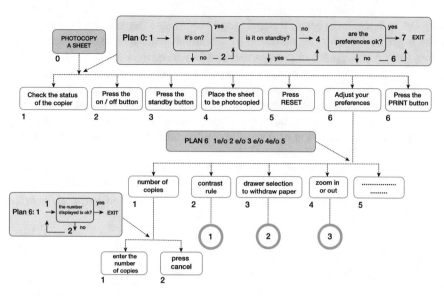

Fig. 6.4 HTA reworking of a professional photcopier. *Source* Oliva (2005, p. 118)

Procedure

For example, this methodology can be used to study the usability of a photocopier. In this case, the fundamental action consists of identifying the physical relationships that exist between the machine's various buttons and their spatial setting.

The objective of this analysis method is to verify whether or not there is a logical consequence to the layout of the keys and their links with respect to the functions that can be carried out. The links indicate the path of the user's eyes and hands when carrying out the various operations that are required to operate the product correctly.

6.4.3 Layout Analysis

Overview

The Layout Analysis works by analysing the graphic and display interfaces, using functional criteria on the importance of the contents, the sequences and the frequency of use.

This method analyses an existing system and suggests improvements (Stanton et al. 2014, p. 39).

Procedure

The Layout Analysis can be applied, for example, to the study of a product's interface (whether it is physical or virtual) by firstly analysing the spatial layout of the buttons, in order to identify potential improvements that could be implemented.

On an operational level, the first step is to arrange the keys that are found on the interface that is being examined by functional groups, taking into account the frequency and sequence of use by the user when doing the task.

In the same way, when evaluating the physical space of a room, it is best to start by defining the activities that take place therein and grouping them according to the type and number of people who perform them, the interactivity with which they are repeated. This is done by evaluating that there are no overlapping spaces or people that could cause confusion or unease or, on the other hand, that there are no free spaces that could be used for other activities.

This method requires a relatively detailed analysis of the activity and the amount of time required depends on the number of people involved (if it refers to a physical space) and the complexity of the actions that must be carried out to achieve the objective.

6.4.4 Heuristics

Overview

The Heuristic method is probably the simplest option available, as it requires the researcher (evaluator) to sit down and evaluate, based on their own experience and skill, whether a product is usable, if it causes errors and if it is safe and well designed. This method can be used at every step of the design cycle of a product (from concept to prototype, etc.). To perform the evaluation, the usability expert should consider one Heuristic rule at a time, and examine the functions of the system in detail, in order to ensure its compliance. Furthermore, Heuristics can be used to evaluate other aspects of a product's usability, such as predicting errors and the time needed to perform a task.

Procedure

A thorough analysisi will result from having spent some time familiarising oneself with the device and any accompanying documentation (e.g., manual) (Stanton et al. 2014, p. 22).

Then the researcher (or analyst) should try to perform an exhaustive set of tasks in order to explorer every aspect of device functioning.

During the analyses, it is important that the researcher remains as impartial as possible to ensure the objectivity of the results.

6.4.5 Checklists

Overview

Checklists are a series of pre-defined points that the researcher (evaluator) can use to verify a product or a design.

This method is used to identify the actual usability and related problems, as well as checking what errors the user has made or could make. There are many models for checklists in the relevant literature and it is wise to choose the most appropriate model based on the case being evaluated.

Procedure

The application of this method consists of a preliminary inspection of the product or the design being evaluated, with reference to each point of the checklist. If this refers to an existing product or a prototype, it is important for the researcher to familiarise himself with this first-hand before starting his evaluation. The points (checklists) can be adapted on the basis of the needs for the analyses, with this variability in the analyses required a certain skill from the designer.

6.4.6 Questionnaires

Overview

Questionnaires represent an effective data-collection tool that can be used to get users' opinions when using a given product. Naturally, a large amount of questionnaire is needed to obtain a statistically relevant quantity of data. The so-called Likert scale is a very commonly used technique. The questionnaire is composed of a series of questions and/or affirmations, with five possible answers for each. A scale from 1 to 5 is associated with these, where 1 corresponds to "strongly disagree" and 5 corresponds to "strongly agree".

Procedure

A questionnaire is a method that can be used in:

- the design phase. The questionnaire can be administered as a preventative measure to explore the hypothesised criticisms of the user and obtain useful feedback from their opinions on the product (can also be submitted for evaluation in the form of a prototype) for future optimisation;
- in the development phase of the product. The questionnaire can provide data regarding on-going changes, based on the skills and experience that the surveyed users have already acquired, in order to determine the sequence for a global improvement plan for issues that are believed to be heterogeneous or contradictory.

6.4.7 Interviews

Overview

The most commonly used technique is the individual interview, which allows the researcher to interview a user by analysing each aspect of their criticism. This test's positive outcome depends heavily on the experience and empathy of the evaluator.

The interview may be carried out in three contexts: in a lab, in a pre-established scenario or in a real context. The test may be rendered ineffective in the first two cases, due to the embarrassment experienced by the user in an artificial situation. The third case offers easier conditions for the user and allows us to collect data about their actual experience, as well as their "spontaneous" observations.

Procedure

To ensure the reliability of the interview, it is important to choose the sample of users to be analysed carefully and calmly. The application phases call for more or less structured interviews depending on the case. A brief description is found below:

- unstructured interview (exploratory survey): open and free questions between the researcher (interviewer) and the interviewee;
- structured interview: similar to questionnaires with pre-set questions;
- semi-structured interviews: these are used when the researcher has a reasonably clear idea of the components to be evaluated and tested.
- Interviews, whether structured or unstructured, are probably the most widely used tool for collecting information in different disciplines. The term "structured" implies that the content of the interview is based on pre-determined questions, while "unstructured" refers to interviews with open questions that are typically used in the early investigative stages.
- Interviews are frequently used to support Task Analysis and can also be applied in the early phases of problem analyses or in verifying the accuracy of information that has been collected via other methods.

6.4.8 Observation

Overview

This technique is useful for evaluating physical tasks and usability, as users are not always able to explain how to use the product in detail (whether they normally use it or it is the very first time) and they may have a distorted view of how they behave in various real usage situations.

There are currently various types of observation techniques, including direct observation, indirect observation and participant observation.

Generally, two people are required for any type of observation: one or more researchers and the participant.

Procedure

The observation phase begins with a scenario. The researcher provides the user a product and established the tasks he must perform. The researcher (observer or evaluator) sits and observes the user as he interacts with the product. Some parameters for observation may include usage errors and the time taken to complete the pre-established activities.

The information that emerges from the observation phase can be useful during the design phase of a new product.

It is wise to use a camera and a computer during the observation.

One of the major concerns with this method is the its intrusiveness, as the observed party may change their behaviour when under observation.

6.4.9 Thinking Aloud

Overview

While using the Thinking Aloud method, the researcher asks the user to express what he is doing and what he thinks out loud while performing set tasks. The researcher will have the task of noting the important statements made while the task is being performed, the difficulties that are experienced and the uncertainties and mistakes that are encountered.

This technique is very useful, since it allows the researcher to gather and obtain information that would be otherwise unobtainable, relating to the strategies employed by the user when performing the tasks, the difficulties encountered during the test, his thoughts and his expectations. Recording audio and video is advised, so everything that happened during the test session can be reviewed at a later date.

Procedure

The researcher who accompanies the user during the activity is tasked with asking some questions to understand the choices they make and actions they perform. The questions asked must be simple, clear and open; this is because their objective is to explore the actions of the user and understand their cognitive process.

To correctly employ the Thinking Aloud method, the researcher is advised to film the observed party's face with a camera as they perform the test, recording what he says and, at the same time, what he is doing with another camera. This is very useful for a successful evaluation, as facial expressions can be as revealing as words and actions.

Reviewing the two video recordings simultaneously is recommended when collecting and analysing data (Fig. 6.5).

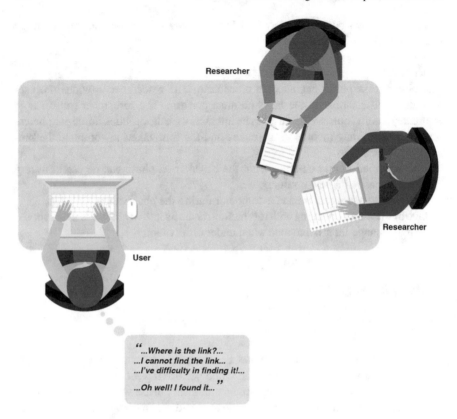

Fig. 6.5 Thinking aloud technique

6.4.10   Scenario

Overview

The Scenario method realistically describes the sequence of actions that a person carries out when using a product/system or service in one or more specific usage contexts, typically via images.

Scenarios are a very helpful tool, because they allow the designer or the design team to explore how a product must work in detail (it allows us to consider the desired characteristics of potential users, their tasks and their environment and allows us to simulate possibility usability issues), in order to ensure a good User Experience and satisfy the pre-established objectives.

Procedure

The representation of a usage Scenario is at the discretion of the designer and the design team, who can choose to describe the development of a scenario via images (see Fig. 6.6), stories, videos, diagrams and more.

Fig. 6.6 Example of a usage scenario for sports activities performed on a piece of gym equipment

Once the means of representation has been chosen, it is wise to outline the target user or category of user (in this phase, it is crucial that the players described are people with a specific identity, as the use of simplified or fictitious characters increases the risk of lacking specificity and losing sight of the needs of real users).

After identifying the characteristics of the users, we proceed as follows:

1. identify the objectives;
2. list the actions required to achieve the objective;
3. list the questions that the user could ask while performing the action;
4. describe the usage context's characteristics (physical-temporal characteristics).

References

ISO 9241-210 (2010) Ergonomics of human-system interaction—Part 210: hu-man-centred design for interactive systems

Maguire M (2001) Methods to support human-centred design. Int J Hum Comput Stud 55(4):587–634

Oliva S (2005) Valutare le attività: "Task Analysis" – Analisi dei compiti. In: Tosi F(ed) Ergonomia progetto e prodotto. FrancoAngeli, Milano

Polillo R (2010) Facile da usare: una moderna introduzione all'ingegneria dell'usabilità. Apogeo, Milano

Wilson JR (1995) A framework and a contest for ergonomics methodology. In: Wilson JR, Corlett EN (eds) Evaluation of human work. Taylor & Francis, London

Chapter 7
Co-design and Innovation: Tools, Methods and Opportunities for the Generation of Innovation Through User Involvement

7.1 Introduction

Design is a strategic innovation factor that can be used by people and society; it allows us to identify possible scenarios and innovative solutions in every field of application.

Design walks hand in hand with the concept of change; without Design, there is no progress or innovation (Antonelli 2014).

The Commission Staff Working Document on "Design as a driver of user-centred innovation" analyses the contribution of Design to innovation and competitiveness.

> Though Design is often associated entirely with the aesthetic qualities of products, its application is actually much broader. The needs, aspirations and abilities of the user are jumping-off points and the focus of Design activities, with a potential to integrate environmental, safety and accessibility considerations into the products, services and systems. (Commission of the European Communities 2009)

As Norman writes, Design is the deliberate molding of the environment to comply with the needs of the individual and society at large, across all disciplines, from the arts and sciences to the humanities, engineering, law and business management (Norman 2008).

Today, the role of the designer is expanding and moving from that of a problem solver—called upon to optimise production processes and improve products from a functional and aesthetic point of view—to that of a problem finder, that is, a person who identifies and interprets the needs of people and society in a critical and innovative way, offering visions of possible and plausible futures. It is necessary, therefore, to develop a greater awareness of the complexity of the role that Design plays and to expand our view of the discipline.

Many people still believe that Design is only about products and that it focuses on the aesthetic value of these, but Antonelli suggests that this approach is, in fact, limited. In reality, Design is very speculative: it involves imagining possible and

---

This chapter was authored by Alessandra Rinaldi.

© Springer Nature Switzerland AG 2020
F. Tosi, *Design for Ergonomics*, Springer Series in Design and Innovation 2,
https://doi.org/10.1007/978-3-030-33562-5_7

plausible futures and understanding how today's choices will affect tomorrow; it is about translating the major technological innovations, which may happen in the field of science or history, into objects that we can use daily (Antonelli 2011).

The designers of today are not solely focused on shapes, materials and finishes. They "conceive", playing a creative and strategic role that involves them understanding what the reactions to changes could be through the use of almost artistic tools.

Once upon a time, design was affirmative; today, it is critical and also broaches human issues; it must not only solve problems, but define them as well. One of the fundamental tasks of Design is to help people to understand change. Designers stand between revolution and everyday life. In the wake of this change, Design has embraced the concepts of "generosity" and "sharing", which were not so clearly delineated in the past. The designer must create a space for the person who will use their design.

From this point of view, open source also becomes a form of Design, because, as an expression of collective effort, it gives everyone the chance to change something, to create new spaces that are otherwise unavailable and to create changes. Dunne and Raby propose a type of Design that is used as a tool to create not only things, but also ideas. For them, Design is a means to ruminate on how things could be, to imagine alter-native visions of the world and possible futures. This is not a typical type of prediction or forecasting, which identifies and extrapolates trends; Critical Design poses "what if" questions, which aim to stoke debate and discussion on what kind of future people want. In a period in which existing systems are reaching their limits, Critical Design suggests an expanded and renewed role for Design.

The result is a series of scenarios that help to illuminate moral, ethical and aesthetic issues, which go beyond corporate and social limitations and existing technological approaches (Dunne and Raby 2013).

The evolutionary paradigms of Design have been summarised in the table drafted by London designers Dunne and Raby, the originators of Critical Design.[1]

Their provocative manifesto is a very significant contribution to the search for the new direction of Design.

It is thus clear that Design, and the role of the designer, will have to expand as the world expands. They cannot be limited to designing objects "to be owned".

In this scenario, what are the skills that the designer needs to function in a world that is increasingly characterised by the complexity of interactions? What tools and strategies does Design have to handle innovation? (Fig. 7.1).

---

[1] Anthony Dunne is a Professor at the Royal College of Art, London and Head of Interaction Design. He is also a partner in the Dunne and Raby studio. Fiona Raby is Professor of Industrial Design at the University of Applied Arts, Vienna, and Reader in Design Interactions at the Royal College of Art. Their work focuses on Critical Design.

DESIGN TODAY - DESIGN TOMORROW

affirmative - critical

problem solving - problem finding

provides answer - asks questions

design for production - design for debate

design as solution - design as medium

in the service of industry - in the service of society

functional functions - functional fictions

for how the world is - for how the world could be

change the world to suit us - change us to suit the world

science fiction - social fiction

futures - parallel worlds

the "real" real - the "unreal" real

narratives of production - narratives of consumption

applications - implications

fun - humour

innovation - provocation

concept design - conceptual design

consumer - citizen

makes us buy - makes us think

ergonomics - rhetoric

user-friendly - ethics

process - authorship

Fig. 7.1   Design today and tomorrow. *Source* Dunne and Raby (2013)

7.2   Design for Innovation: From Participatory Design to Co-design

In the previous chapters, much was said about Human-Centred Design, User Experience and Driven Innovation as approaches to Design and innovation that focus on the individual, his needs and his experience when using a system, product or service.

The HCD approach, which developed in America in the 1970s, allows companies and designers to define the characteristics of the services and products to be designed, based on the needs and expectations of future users.

Designers, in their role as design experts, consider users to be passive objects ("user as a subject to be observed"), to be observed and inter-viewed to collect their options on design concepts that others have gene-rated, particularly as they relate to the usability of products and services.

In parallel in Europe, particularly the northern countries, a collective approach to design creativity has been adopted, in what is known as "Participatory design". Users are considered to be a partner in this; they are seen as experts due to their experience and are involved from the conceptualisation phase for design ideas. Both of these approaches are starting to influence each other (Fig. 7.2).

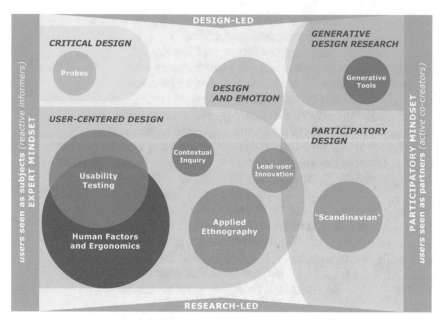

Fig. 7.2   Emerging trends in design research. *Source*   Sanders and Stappers (2012)

Manzini explains that, while designers initially felt the need to focus more closely on people, their behaviours and contexts, through a one-way activity in which the designer was the observer and the user was merely the subject to be observed, this asymmetrical relationship between active observer and passive observe has subsequently been questioned. It was realized that relations with context, given their complexity, require active involvement on the user's part. This reflection led to the definition of the user as an "expert actor", the holder of knowledge that only he, thanks to his direct experience, could possess. The user's experience as a new area for analysis has led to the creation of numerous tools that aim to collect motivations and meanings behind people's actions, such as participatory observation and storytelling. Numerous experiments have shown the effectiveness of these methods, but also some limitations to them. The observation of what is there, that is, of what exists, cannot tell us much about what could be, that is, about what human creativity and technological innovation could create. This brings us to a phase in which user-based research meets the need to propose something that has not yet been thought of, and certainly not experienced or tried (Manzini 2009).

This is the source of the attempt to transform the end user into a de- signer, or the concept of Co-design. This is a new approach that sees the user becoming a co-designer and the designer becoming a facilitator and a mediator in the co-designing process. The term Co-design is used to indicate a creative endeavour that is shared between the designer and the end user, who work together as equals during the process of generating and conceiving ideas and the subsequent design phases (Rizzo 2009).

According to Sanders and Stappers (2008), designers have becoming increasingly closer to the future users that they design for over the last six decades.

The evolution of user-centred design and innovation methods is leading to an increasingly active engagement of users in the design processes. The co-design phenomenon has become an essential component of design, for products, services and product/system services, and favours the generation of design ideas and identifying innovative solutions based around the needs of users.

In the field of Participatory design, the concepts of co-creation and co-design have been widely disseminated. Today, they are often confused and/or seen as synonyms. Opinions about who should be involved in these collective creativity endeavours, as well as when and what their role should be, vary.

Sanders (2006) uses co-creation to refer to any act of collective creativity, which is, in a very broad sense, shared by two or more people. The same author refers to Co-design as collective creativity applied to the entire span of a design process. Therefore, Co-design is a specific example of co-creation. While for some, Co-design refers to the collective creativity of designers who collaborate with each other, it is now commonly agreed that the term Co-design implies a creative process between designers and those who are not experts in design, who work together to generate ideas and concepts for a new product and/or service.

Sanders also notes that, as designers becoming increasingly closer to the users they design for, the emphasis that is placed on the initial phase of the design process (known as the front end), that is, the collection of activities that are performed to inform and inspire the exploration of open questions, is growing.

The front-end is often defined as "fuzzy", due to its characteristic ambiguity and chaotic nature. *"In the fuzzy front end (FFE), it is often not known whether the deliverable of the design process will be a product, a service, an interface, a building, etc."* (Sanders and Stappers 2008). *"Considerations of many natures come together in this increasingly critical phase, e.g. understanding of users and contexts of use, exploration and selection of technological opportunities such as new materials and information technologies, etc."* (Sanders 2006). *"The goal of the explorations in the front end is to deter-mine what is to be designed and sometimes what should not be designed and manufactured. The fuzzy front end is followed by the traditional design process where the resulting ideas for product, service, interface, etc., are developed first into concepts, and then into prototypes that are refined on the basis of the feedback of future users."* (Sanders and Stappers 2008).

The Human-Centred Design approach, which has proved very useful in the design and development of consumer products, becomes less effective in dealing with the scale or complexity of contemporary challenges, in which products are no longer simply designed for users, but for the future experiences of people, communities and cultures that are now connected and informed in ways that could not have been imagined even 10 years ago (Sanders and Stappers 2008).

As Rizzo (2009) writes, the broadening of the subject matter of design towards a series of areas removed from the traditional ones, from services to product/service

The traditional design disciplines focus on the designing of products	The emerging design disciplines focus on designing for a purpose
Visual communication design	Design for experiencing
Interior space design	Design for emotion
Product design	Design for interacting
Information design	Design for sustainability
Architecture	Design for serving
Planning	Design for transforming

Fig. 7.3 Traditional and emerging design practices. Reworked from Sanders and Stappers (2008)

systems, health, well-being etc. involves an increased interest in including the user in the design process, encouraging their participation in the inspiration, creativity and conception phases.

The Table in Fig. 7.3, developed by Sanders, highlights how we are moving from designing categories of "products" to designing for people's purposes. The traditional Design disciplines on the left are focused on products or technology. Here, the designer acquires the skills necessary to design and produce products, such as brand identities, interior spaces, buildings, consumer products, etc.

Emerging design practices (on the right) focus on people's needs and the needs of society and require a different approach, as they need to consider more perspectives and address broader fields of research.

The emerging design practices will change what we design, how we design, and who designs. (Sanders and Stappers 2008)

The transition from Human-Centred Design to Co-design is having an impact on the roles played by individuals in the design process. In the classic user-centered design process, the user is a passive object of study, while the designer acquires knowledge from theories and develops more knowledge through observation and interviews. The designer then adds an understanding of technology and the creative thinking needed to generate ideas, concepts and techniques to develop the design. In Co-design, the roles are mixed: the person who will eventually be affected by the design process is given the position of 'expert of his own experience', and plays an important role in the development of knowledge, in the generation of ideas and in the development of the concept. The designer, meanwhile, acts as a facilitator, providing tools for generating ideas, expressing them and giving shape to ideas.

In this way, users, if their creativity is stimulated, can provide important ideas and suggestions. Co-design, therefore, is an emerging approach, one that is more interested in research and experimentation with users for the generation of ideas that can inspire Design, rather than validating results.

Co-design is an approach to Design that is characterised by the involvement of non-designers in the generation of ideas and in the actual design process, which involves the use of so-called user-driven innovation research. This approach to design is aimed more at "discovering" than at "answering", at imagining new opportunities that become visible over the course of the process. In this context, designers/researchers will play the role of facilitators and mediators in the design and will use their skills as a tool to stimulate and direct creativity and focus on the needs of the participants, as end users to whom the design is targeted.

What is the co-designer/user's role?

As has been mentioned, Co-design implies the inclusion of non-designers in the process of generating ideas and developing the project. In this sense, all users are given the opportunity to generate and conceive design ideas, but their degree of involvement depends on their level of skill, passion and creativity. Sometimes, "users" can play co-creation roles throughout the design process, that is, become real co-designers, but this is not always the case. All people are creative, but not all people become designers. Sanders and Stappers (2008) argues that *"four levels of creativity can be seen in people's lives: doing, adapting, making and creating"*.

"These four levels vary in terms of the amount of expertise and interest needed", depending on the level of experience and interest they may have in a given usage context, for a certain type of product or service. The Table shown in Fig. 7.4 shows how *"expertise, interest/passion, effort, and returns grow with each level. People live simultaneously at all levels of creativity in different parts of their daily lives. For example, they may be at the creating level when it comes to cooking but at the adapting level when it comes to the use of technology products. People with a high level of passion and knowledge in a certain domain who are invited to participate directly in the design process can certainly become co-designers"* (Sanders and Stappers 2008).

"Users can become part of the design team as 'expert of their experiences'" (Sleeswijk Visser et al. 2005), but in order for them to take on this role, they must be given appropriate tools for expressing themselves.

Rizzo (2009) poses two questions: how to select the people who be- come part of a design team, so as to obtain the greatest possible degree of creativity, and whether or not expressive skills and visualisation skills are an element to be considered in this choice. Fischer (2002) proposes considering the users and the role that they can

Level	Type	Motivated by	Purpose	Example
4	Creating	Inspiration	Express my creativity	Dreaming up a new dish
3	Making	Asserting my ability or skill	Make with my own hands	Cooking with a recipe
2	Adapting	Appropriation	Make things my own	Embellishing a ready-made meal
1	Doing	Productivity	Getting Something done	Organising my herbs and spices

Fig. 7.4   Four levels of creativity. *Source* Sanders and Stappers (2008)

play on a range from active to passive, scaling their involvement along a scale of possibilities in terms of the aims of Design and its associated skills.

In the last ten years, research groups at a number of academic institutions, Design consultants and Design research groups in companies have explored all of the Co-design tools, techniques and processes that can be used to apply them. Interest in co-designing tools and techniques has also been growing rapidly.

To date, as Rizzo writes (2009), a systematic elaboration of the Co-design methods has not yet been developed, but other co-design experiments can be studied instead, that is, design experiments aimed mainly at understanding useful tools that are necessary for co-design and for supporting creative collaborations and generating design and innovation processes.

And what role does the professional designer play when future users are co-creating tangible visions for new products and/or services?

According to Sanders and Manzini, new design scenarios reveal that design skills will become even more important in the future. Designers will always be in great demand since the utility of Design thinking is recognised as a driver of innovation in the face of the challenges of global and systemic problems.

In fact, the Design Thinking model is identified as a new paradigm for tackling problems in various professions, particularly in the business and management sectors, and addressing and managing complex challenges. Designers, who deal with open and complex problems, create frameworks and develop ad hoc methods and practices; this is why certain organisations are interested in studying how designers work and adopting some of their practices. Having to deal with open and complex problems leads to a particular interest in the ability of designers to create frameworks, and towards the ability of companies to use frameworks in the fields in which they practice (Dorst 2011).

Designers will be needed, because they possess highly developed skills that are relevant at the higher complexity levels of the design process. Most of the designers acquire skills in visual thinking, in guiding creative processes, in searching for missing information and in being able to make the necessary decisions in the absence of complete information through their training. In the near future, designers will find themselves involved not only in product design, but also in the design of complex environments and systems.

As the scale and complexity of design problems increases, we will need the specific skills and expertise of designers to find the way forward. We will use the idea-generating aspect of Design to confront change. The use of generating design tools allows us to look ahead towards possible futures in which people will live, work and play. Designers will have the task of exploring the potential of generating tools and bringing Co-design codes into their work. They will be an integral part of the creation and exploration of new tools and methods for the thought-generating aspect of Design and will have to create tools that can be used by 'non-desi-gners' to express themselves creatively.

Finally, designers must maintain a role in the Co-design teams, because they provide the specialist knowledge that other stakeholders do not have. Designers will identify, in a professional way, existing, new and emerging technologies and will have an overall view of production processes and company contexts. This knowledge will always be relevant throughout the Design development process (Sanders and Stappers 2008).

7.3 Co-design Tools for the Conception and Development of a Design

Co-design requires designers to develop and be aware of specific tools and methods that involve users in the design process, from the conception phase to the development phase, and for efficient communication with all stakeholders with an interest in the project. The role of the designer in this complex and articulated process is to mediate between different interests, extend individual interests to common interests, and facilitate the generation of ideas by the participants. Designers must create frameworks and tools that support the participation of non-designers in the design process. The *Design game* for example, is seen as an effective method to support the participation of every member of a multidisciplinary team, as it facilitates communication and the generation of ideas. Visual representations and mock-ups can very helpful to the generation of ideas, stimulating the creativity of the participants through tactile and tangible experiences Ehn and Kyng (1992). Argue that the strength of the mock-up is in its abstract "draft" nature, which is far from the finished object, as opposed to the prototype which has a much higher level of detail. In this sense, mock-ups help people to develop creative thinking and to use this as a design tool.

Debate still rages regarding what tools can favour the engagement of the user as a source of information, as a generator of idea and as a designer within the design process.

A systematic elaboration of Co-design methods has not yet been developed, but other Co-design experiments can be studied instead, namely design experiments aimed mainly at understanding useful tools that are necessary for Co-design and for supporting the creative collaboration and generating of Design and innovation processes. However, each co-design experience involves experimenting with custom co-design methods based on the attitudes, interests and skills of the participants, who are used to encourage the generation of ideas that can inspire the Design.

Experienced designers, starting from the methods and tools borrowed from Human-Centred Design, or Design focused on the person and the context in which the interaction occurs, and from User-Driven Design, that is, Design based on the experience of people, are experimenting with new avenues. What is clear is the shift in perspective that this methodological approach entails: the user/person is no longer considered as a stereotype that represents a class of physical, cognitive or other characteristics, but as an individual with thoughts, attitudes, values, a culture, and, most importantly, unparalleled expertise regarding their own experiences.

Based on the experiments conducted so far, including those within companies, a series of original instruments have emerged, such as Design probes and Co-design workshops, which primarily support Design for users at present.

7.3.1 Cultural Probes

The cultural probes (Design probes) are one of the most innovative methods developed to date to generate ideas in the design process, through the active involvement of users. The probes were developed by Design for experience, which, starting from the development of User-Centred Design methods, such as participant observation and story collection, generated a tool to collect and record inspiring data from people's lives, values and thoughts in real time. As Rizzo writes, there is no noun that allows the meaning of the English word probe to be expressed in Italian. Its etymology derives from the Latin probare, translated into Italian with test, examine, try, force and trial (Rizzo 2009).

Probes are small kits that can include any type of artefact (such as a postcard, a camera, post-it notes, pencils, a diary, open questions), which are designed by the Design team and given to the participants in the Design process to allow them to record specific events, feelings or interactions. The goal is to collect data and information from people, in order to better understand their culture, their thoughts and their values, and thus stimulate the designer's imagination (Gaver et al. 1999).

Probes are a type of exercise that the design team assigns to the participants and they can have at least three different functions: guiding the user to record information about his experiences; stimulating the user to visualise their thoughts, concepts and ideas; highlighting possibilities that do not exist at present, but are possible in the future.

The recording in a diary, by the user, of all the things that occur in daily life or limited to a daily activity, with respect to participant observation or ethnographic investigation, is varied. According to Carter and Mancoff (2005), the observer in the field should not influence the events and cause the observed party to change their behaviour due to the presence of an observer. Furthermore, the data collected by an observer who does not belong to the observed context may not be completely reliable, as they are filtered. Finally, tracing experiences in the context in which they occur solves the problem of the reliability of memories, which instead emerges in storytelling.

Cultural probes, which were first developed by a research group at the Royal College of Art for a design research project of the 1990s, funded by the European Union, are now considered to be an empirical research method for analytical experimentation and are used with the aim of gathering information to inspire designs.

Mattelmaki (2005) divides the possibilities of applying probes, based on their purpose.

- For inspirational purposes: probes are a tool to inspire and guide the design towards new experimental situations and to imagine new hypotheses and new future users. Interesting topics and discussion points can emerge from the involvement of the user. The probes are a valid support for collecting information about the user experience, but the challenge is represented by the designer's ability to structure them so as to allow the user to identify the critical issues and indicate possible solutions.
- For information purposes: probes can contain the description of a current situation in people's lives and stimulate the user to analyse and interpret his own experience. In this case, designers use the probes to focus on the problems of a context in which they are operating.
- For participatory purposes: probes can be used to stimulate the user's active participation in trialling new tools and equipment.
- For dialogue purposes: in this case, probes become a tool for users to communicate their emotions and experiences and for de-signers to engage a process of empathy, putting themselves in the user's shoes.

The application of the probes in the Co-design process implies the definition of the design kit to be implemented, based on the objectives to be achieved and the users that it intends to engage. Design kits are not available but must be designed in an adaptive manner that depends on the operating context. Sanders and Stappers (2008) has attempted to design libraries of probes and to define cues for their construction; however, these indications can only constitute a trace that can be used to facilitate their definition.

Rizzo (2009) writes that the probing process can be guided as a sup-port tool to connect Design thinking with the world of user experience. This process facilitates the collaboration of users, designers and researchers in a multidisciplinary manner. In addition to dialogue, the probes inspire designers, can help to collect information about the user and the subject of the design: they help designers to have access to visions of future experiences. The probes contain elements of uncertainty and have an experimental nature, which makes them difficult to organise in libraries of pre-determined instruments. They are part of the repertoire of tests that must be designed each time, in the same manner as an experiment in any of the disciplines that investigate social phenomena.

7.3.2   The Co-design Workshop: From Conception to Prototyping

The Co-design workshop consists of activities for creation, visualisation, experimental prototyping and the conception of innovative, interesting and viable solutions, which engage end users as active actors in the design process.

These activities can be mainly divided into three parts (Sanders 2012):

- Cognitive and exploratory activities, which can refer to the "Say" phase, during which people recount their experiences, document them through annotations, images, photographs, using different tools and methods made available by designers.
- Co-conception activities, which can refer to the "Do" phase; once the previous phase has been analysed and a co-design brief has been identified with the participants, we move to the discussion and collaboration phase, which involves empathy with the co-designers and generating ideas, concepts and potential creative solutions.
- Development and prototyping activities, which can refer to the "Make" phase. In this phase, we start the creation, visualisation, experimental prototyping and conception of the solutions that have been identified as the most interesting and viable. In this phase, based on the explanation of how people would do things, designers/design researchers will provide the most suitable tools to facilitate communication and the formalisation of ideas.

The Co-design workshops that have been tested to date can be traced back to the task of drafting future uses for products and services. The idea behind this design methodology is that users, when allowed to express themselves freely and work with prototypes or construct new ones, can help designers to not only understand their explicit needs and expectations, but also their latent and future ones.

The method of the Co-design workshop is participatory; the designers who use it believe that working with users in an active manner allows them to design products and services that are more useful and meaningful to people. The limits linked to this methodology are, on one hand, the possibility of inhibiting the participants with the presence of the stakeholders, and, on the other hand, the difficulty of supporting creativity. In the Co-design workshops, envisioning prototypes are used to indicate simulations designed by designers to be manipulated by the participants, to express their ideas about future situations, or created during the workshop by users/participants who become the creators of the products they use, the places they live, so as to generate ideas and inspiration for the Design.

As Rizzo (2009) writes, the Co-design workshop is currently becoming a new user-centred design tool that focuses on how people build artefacts based on the functions they are familiar with, on the activities they would like to do with them and on their symbolic meanings. This tool allows you to access the contents exchanged during the co-design activity and to use this information as additional knowledge that can be applied to the project or used to inspire it.

References

Antonelli P (2011) Talk to me: design and the communication between people and objects. MOMA, New York

Antonelli P (2014) Design e Musei del Futuro. Lecture a Meet the Media Guru. www.meetthemediaguru.org/lecture/paola-antonelli/

Carter S, Mankoff J (2005) When participants do the capturing: The role of media in diary studies. Proceedings of the SIGCHI conference on Human factors in computing systems, Portland, pp 899–908

Commission of the European Communities (2009) Design as a driver of user-centred innovation, Brussels

Dorst K (2011) The core of the 'design thinking' and its application. Des Stud 32(6):521–532

Dunne A, Raby F (2013) Speculative everything: design, fiction, and social dreaming. MIT Press, Boston

Ehn P, Kyng M (1992) Cardboard computers: mocking-it-up or hands-on the future. Design at work. ACM, New York, pp 169–196

Fischer G (2002) Beyond 'couch potatoes': from consumers to designers and active. FirstMonday 7(12). http://firstmonday.org/issues/issue7_12/fischer/index.html

Gaver W et al (1999) Design: cultural probes. In: Interaction, vol 6, no 1. ACM, New York, pp 21–29

Manzini E (2009) Preface. In: Rizzo F. Strategie di co-Design. Teorie, metodi e strumenti per progettare con gli utenti. FrancoAngeli, Milano, pp 7–10

Mattelmaki T (2005) Applying probes, from inspirational notes to collaborative insights. In: Co-Design, vol 1, no 2. Taylor & Francis, pp 83–102

Norman DA (2008) Il Design del Futuro. Apogeo, Milano

Rizzo F (2009) Strategie di co-design. Teorie, metodi e strumenti per progettare con gli utenti. FrancoAngeli, Milano

Sanders EBN (2006) Design serving people. In: Salmi E, Anusionwu L (eds) Cumulus working papers. University of Art and Design Helsinki, Copenhagen, pp 28–33

Sanders EBN, Stappers PJ (2008) Co-creation and the new landscapes of Design In: CoDesign, vol 4, no 1. Taylor & Francis, pp 5–18

Sanders EBN, Stappers PJ (2012) Convivial toolbox. Generative research for the front end of design. BIS Publisher, Amsterdam

Sleeswijk Visser F et al (2005) Contextmapping: experiences from practice. In: CoDesign, vol 1, no 2. Taylor & Francis, pp 119–149

Chapter 8
Design Thinking and Creativity: Processes and Tools for New Opportunities in People-Centred Innovation

8.1 Introduction

This volume reviews how project-oriented Ergonomics, and in particular Design-oriented, in its evolution from User-Centred Design to Human-Centred Design to User Experience, represents a key strategic factor for product/system innovation.

This HCD approach is continuously evolving, regularly taking on new forms in relation to the socio/cultural, technological context and the increasingly specific demands aimed at achieving radical innovation. This leads to a discussion about Design Thinking and a *People-Centred* approach, given that Design needs to explore increasingly detailed and ambitious application fields and scales of intervention, passing from product design to interactive systems to services and social business strategies, adapting methods and tools in order to accommodate the contributions of other disciplines and other actors.

It is in this context that the Politecnico di Milano IDEActivity Center research group was born, active in innovation studies driven by creativity through design, and aimed at focusing on the changes taking place. The IDEActivity method is founded on some models of Design Thinking, creativity as a "lateral thinking" ability, a *People-Centred* approach which puts people at the centre of the design process, and a method that entails co-participation.

The aim of this method is to be a flexible tool able to meet specific needs of small or large companies, or public bodies with very different objectives and configurations.

This chapter was co-authored by Marita Canina, Laura Anselmi and Carmen Bruno.

© Springer Nature Switzerland AG 2020
F. Tosi, *Design for Ergonomics*, Springer Series in Design and Innovation 2,
https://doi.org/10.1007/978-3-030-33562-5_8

8.2   From Human-Centred to People-Centred: The Role of Design in the Evolution of Design Processes

As previously mentioned, Design has been exploring progressively complex scenarios, leading to a transformation in design culture and an evolution in the approach taken to projects, substantially changing their methods.

A fluid approach to design is emerging in response to economic, cultural and social changes, and reflection is necessary to understand the changes that occur in design under the current trends and new boundaries.

As reported in *Red paper 02: Transformation design* (Burns et al. 2006): in the first decade of the twenty-first century, design underwent two major changes—changes *where* engineering skills are applied, and changes in *who* practices design.

In recent years, new spaces for creativity and innovation have emerged, such as FabLabs and Creative Labs, as well as new figures, the makers, who in a short time have created a movement that celebrates the Do-It-Yourself approach, democratising digital technologies, making them accessible to a wider pool of people. The people involved in this movement use all types of available technology that enable them to build whatever they want to create. The concepts of community, sharing, creativity, problem-solving and social change form the manifesto of a new society profoundly shaped by the exponential development of digital technology over the last decade. This has given rise to the new attitude of people to "do" and "create", based on collaboration and sharing of knowledge, which can create opportunities for social and technological innovation with a bottom up approach. It has been estimated that 80% of scientific instrument innovations arise from "amateurs" (Von Hippel 2005, p. 175). In this scenario of diffused creativity, there is a tangible change in the approach to the project and in the democratisation of the designer's role. Design must then analyse current transformations, and identify a new role and new areas of intervention, which move in the direction of enabling the creativity of people, educating them to implement a collaborative design process in order to generate innovation in and for the community.

When these emerging trends are taken into consideration, design can, and must adopt a *People-Centred* approach in which the individual as a society and a community is put at the centre of the project in a holistic manner. Therefore, there is a clear need to identify different methods of analysis and new forms of expression of Design and creativity that can accommodate the social changes that are emerging in connection with technological development.

8.3   The Evolution of Creativity and Design as Key Elements of Innovation

The market in which today's companies compete has a high level of complexity due to the dynamics created by the globalisation process, in which new competitors constantly come into play, and new needs and markets emerge. Moreover, the exponential development and diffusion of digital technologies offers the opportunity to experiment with new business models that often compete with more traditional ones. Consequently, innovation has become a fundamental requirement in the current socio-economic context for all companies that want to be competitive in the market.

> Therefore, it has become strategic for companies to adopt a new approach, that includes a new mental approach, with which to face challenges in innovative and ongoing ways, which enable the pursuit of a model geared towards radical innovation, based on the *Disruptive Innovation* identified by Christensen et al. (2015), professor at Harvard Business School.
>
> According to his research, *Sustaining Innovation* "takes steps" based on progressive improvement measures of a product, service or process, while *Disruptive Innovation* "takes leaps", constantly generating innovative solutions and creating new markets. With the Disruptive process, companies are guaranteed the ability to differentiate themselves in the market by adopting a modus operandi (mindset) driven by strong creativity and out-of-the-box, unconventional thinking applied to a deep understanding of the needs (manifest and latent) of potential users and end-users. It allows the pursuit of "new paradigms," "new meanings" of product/service/system that generate value for the market and that, consequently, enable the company to differentiate itself or to assume a leadership role among competitors.
>
> Creativity precedes innovation. Following the definition provided by Amabile (1988), Creativity is in fact the production of new ideas (unique compared to others available in the company) that are useful (if they have the potential to create value in the company). These ideas, in terms of products, services, and processes, become innovation once they are adopted, integrated and shared by the company and by the community.

Creativity is thus a primary source of innovation, and companies that want to increase their competitiveness on the market must exploit the creative potential of their resources. Innovation is driven by human creativity i.e. spontaneous act that is often strongly propelled by intrinsic motivations, through which the individual improves themself and their world. It has become necessary to place the human creative impulse at the centre of change, as the sole agent able to effectively and efficiently intervene in problem-solution, capable of providing a company with the flexibility and openness necessary to achieve a competitive advantage. Generating a creative climate within a company thus allows it to effectively utilise a single "human resource" by putting it in a position to contribute towards the growth of its working environment.

Creativity and creative thought, together with Design, become an important driving force for innovation, insofar as they are in close correlation (Von Stamm 2008, p. 165). From this emerges the need for a new culture that aims to stimulate, encourage and welcome the opportunity for the development of personal and group creative skills as a source of innovation strategies.

Creativity is widely regarded as a "spark of genius" and as the privilege of a few talented people. However, some research shows that the majority of people of average intelligence possess creative thinking skills, i.e. dealing with problems from solid knowledge, but also adopting new perspectives, producing ideas that break away from "traditional patterns", with the aim of finding innovative and effective solutions whatever the scope.

This style of thinking is expressed in a process that does not always possess linear trends, and consists of collecting, selecting and reconfiguring information from all available options, and identifying *useful connections to generate new conclusions*. Poincaré (1924) speaks of Creativity as the ability to combine existing elements into new, useful combinations. He considers Creativity to be a skill that is based on the ability to identify knowledge bases to "disconnect" and reconnect by association, in never-used-before patterns. With Poincare we also find the first references to a multi-phase structure of the creative process. In fact, it was his reflection on using the creative thought process to solve mathematical problems that inspired Wallas (1926) to divide the creative process into four phases: preparation, incubation, illumination and verification.

This classification served as a springboard for research on movements in the field of Creativity in Design that was studying new models to better describe the different phases of the process.

The psychologists J. P. Guilford and E. De Bono, who conducted studies on the human mind and its mechanisms, subdivide the thought into convergent and divergent (Guilford) or vertical and lateral (de Bono 1970), identified by the way in which you deal with a problem. The first type of thinking—convergent or vertical—is carried out in linear sequences: cause and effect, premise and consequence. It essentially engages the left hemisphere of the brain, where the language centres are situated. The second type of thinking—the divergent or lateral—occurs in a non-sequential manner, but according to similarities, analogies and differences, symmetries and asymmetries. It mainly involves the right cerebral hemisphere, where the vision centres are located.

If a problem is confronted with the first type—rational method of thought—correct results are achieved, but these results are limited, due to the rigidity of traditional logic models. Instead, when a truly different and innovative solution is required, the pattern of reasoning must be changed, and one must see things in a different way. Therefore, convergent or vertical thinking, i.e. based on logical deductions, must be abandoned in favour of the laterality of creative thinking, which stimulates a greater divergence of views.

However, it can be said that both styles of thinking, vertical and lateral or convergent and divergent, are essential to the production of ideas, and are complementary: vertical thinking exponentially increases the effectiveness of lateral thinking by making proper use of generated ideas. Everyone can then be creative. Knowledge and information are the basis of creativity.

Unfortunately, average creativity skills steadily decrease from childhood onwards, due to social pressures and a lack of training of creative faculties in everyday life. It is possible to stimulate innovation and creative thinking on both individual level and in groups, using tested and proven techniques and methodologies which aim to promote and generate creativity, break established schemes, stimulate the imagination and, regardless of the method, improve the conditions in which creative ideas are produced. Tools and methods have also been developed to overcome the conceptual blocks, i.e. "mental walls" that block a person from correctly understanding a problem and devising an appropriate solution. Although the primary cause of conceptual blocks is the lack of mental flexibility, many other types of blocks have been identified, among which some are emotional, environmental, cultural, and intellectual (Roth 1973).

These elements are key to building new strategic scenarios of project and training, fit for the specific needs of companies, institutions and organisations. When one starts from the idea that creativity is what underpins innovation, not as a single event, but as a systematic approach, knowing and stimulating creative propensities become strategic in the project planning framework.

When defining Creativity and innovation, it is essential to take the cultural and socio-economic context in which we live into account. In fact, with human evolution, the manifestation of creative actions affects more and more new fields, always fluctuating, adapting and re-calibrating its definition. The interactive map "25 years of Creativity Research"[1] (Williams et al. 2016) shows how the research focus has been put on Creativity during the last 25 years.

In recent decades, particularly with the emergence of Information Technology, and more recently with the spread and democratisation of digital production and sharing technologies, society has begun a transformation process, becoming increasingly interconnected. Online presence is associated with very high levels of connectivity and unprecedented potential for sharing information, connecting people and ideas that used to have the tendency to be kept separate, and facilitating collaboration within the digital world and beyond (Literat and Glăveanu 2016). These features not only have an impact on Creativity as a phenomenon but redefine it as a creative collaboration process with others, mediated by digital technology. Some examples of digital environments that support mass collaboration and social production are wikis (Wikipedia), virtual worlds (Second Life), and media sharing (Flickr, YouTube). Although it has often been thought throughout history that creative individuals were working in isolation, research in fact shows that intelligence and creativity stem in large part from interaction and collaboration with other individuals and artefacts (Fischer et al. 2005; Csikszentmihalyi 1996). The skills and experiences of an individual can significantly influence the success of a collaborative project. Individual and social expression of creativity must balance and complement each other. Monitoring these changes is critical to Design in order to identify new spaces and project and action scenarios, guided by human evolution.

8.4   Design Thinking: Process or Mindset?

The word Design is not exclusively used to indicate a product, but also to describe a process. In the broadest sense of the term, it is to be understood as a lever which

---

[1] For inspection at the following link http://app.mappr.io/play/CreativityResearch1990-2015.

can push change, starting from listening to people. When implemented and applied in a synergic manner to organisation (Design Thinking), it takes the form of an innovative process, which is a methodology that connects the typical creative approach to traditional thinking which focuses on a rational and logical ways of approaching problems.

Design Thinking—which cannot be regarded as a linear analytical method, but a new way of thinking, a new design culture characterised by a creative-logic approach to problem solving, and Human-centred—aims to generate a process of innovation. It entails experience, real needs and materialised innovation process models run by multidisciplinary teams for the realisation of new products, services and profoundly innovative business models.

This method is not only a driving force for innovation, but also offers new process models and tools that help to improve, accelerate and visualise any creative process, making it accessible not only to designers but also to multi-disciplinary teams and every type of organisation.

Born in the 2000's in Stanford, California, Design Thinking is a process of articulated thought fit for conceiving new realities and managing complex problems that can be analysed and solved thanks to the creative vision and outside-of-the-box thinking that typify the Design culture and its methods. The term was coined in the 80's by Professor Rolf Faste from Stanford University, a pioneer in the Human-centred approach, known for his contributions to the creative practice of Design, or "Design Thinking" and his experimentation with a personal approach to problem-resolution based on perceived needs.

However, his colleague David Kelley and associate Tim Brown—founder of IDEO's design studio (ranked thirty-fifth in the *Forbes'* most innovative companies at global level in 2010)—have implemented Design Thinking in fields such as business innovation, kick-starting its dissemination. From the United States, Design Thinking spread rapidly to Canada, Asia and Australia. It has also been present in Europe for several years, starting from Germany, where, thanks to the HPI-Stanford Design Thinking Research Program, set up by Stanford University and by Hasso-Plattner-Institut (Potsdam, Germany) was formalised and has been studied scientifically since 2008.

Today, Design Thinking is not just a cognitive process or mindset, but a tool that fosters greater "democratisation" of innovation within corporate organisational charts, connecting typical creative Design approach to traditional business thinking. "A powerful methodology for innovation" has emerged from the research results of the HPI-Stanford Design Thinking Research Program, collected in the Springer series "Design Thinking Research". "This methodology integrates human, business and technological factors in *problem-forming, problem-solving, and -design*: Design Thinking (Fig. 8.1). It is a Human-centred methodology that integrates Design, social sciences, engineering and business skills. It connects the focus on the end user and their needs with multidisciplinary collaboration and iterative improvement to produce product innovation, service and systems" Meinel and Leifer (2010).

Meinel and Leifer (2010, pp. XIV–XV) include the following four "rules of Design Thinking" in the results of the studies conducted by the research program:

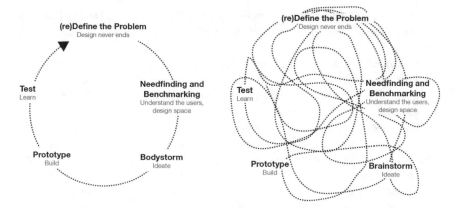

Fig. 8.1 Design Thinking is commonly viewed as an iterative series of five main stages. On the left you see the standard form. To the right you see something closer to reality. While stages suffice, the appropriate expertise required to choose the right inflection points is an activity of great intellectual effort that requires practice and is learnable. Readapted from Meinel and Leifer (2010)

- **The Human Rule**: All Design activities are essentially social in nature. According to some studies, successful innovation through Design Thinking activities always emerges from a "Human-centred point of view". This is imperative to solve technical problems that aim to satisfy human needs and to make the human element known to all technologists and managers.
- **The Ambiguity Rule**: Design Thinkers must maintain ambiguity. There is no possibility of "discovery" if a box is hermetically sealed, restrictions are enumerated excessively, and the fear of failure always kept on hand. Innovation requires testing the limits of our knowledge, the limits of our ability to control events and the freedom to see things differently.
- **The Re-Design Rule**: All design is re-design. The human needs we seek to satisfy have been with us for millennia. Through time and evolution there have been many successful solutions to these problems. As technology and social circumstances are constantly changing, it is imperative to understand how these needs have been addressed in the past so we can apply "forward-thinking tools and methods" to best estimate the social and technical conditions we will encounter 5, 10 or 20 years from now.
- **The Tangibility Rule**: Making ideas tangible facilitates communication. Interestingly, this is one of our latest discoveries. While conceptual prototyping was a key activity during the whole period of research, it is only in recent years that "prototypes are a means of communication" has been realised. The rule of "making it tangible" is one of the most important findings of the research program on Design Thinking.

Therefore, Design Thinking is a process and a mindset. But it is much more: applied to daily work, the potential of Design Thinking problem-solving finds its expression in a profound form of innovation culture.

After identifying the principles, rules and distinctive features of Design Thinking, it is now time to go into more detail and try to understand how it works and is implemented in practice.

One can summarise Design Thinking by highlighting some of the main distinctive features of the methodology[2]: the *Human-Centred approach* (A deep understanding of human behaviour), *multidisciplinary* and *collaboration*, *Creativity* and the penchant for *wild ideas* that allow you to go beyond the limits of knowledge.

Taking these characteristics into account, and the definition given by Brown (2008): *"Design thinking is an approach that uses the **designer's sensibility and methods** for problem solving to meet people's needs in a technologically feasible and commercially viable way. In other words, design thinking is **human-centred innovation**"*, you can understand why research has an interest in identifying the mental strategies of designers during project development. This combination of *design* and *thinking* offers the opportunity to apply Design tools to other contexts such as business forms, services and processes (Tschimmel 2012).

The starting point is *"Be empathetic. Try to figure out what people really value"*, this basically describes the most profound and valuable part of this methodology.

In the traditional method of troubleshooting, you often find a problem, define the steps and tools to use to reach a solution, then adhere to the plan and hope for the desired result. Instead, Design Thinking begins with observation and an understanding of culture and context (what people need). This approach explores the strategic dimensions in which the product ranks and uses tools and techniques that integrate imagination, creativity and intuition with logic, analysis and planning.

Despite Creativity being the main vehicle of Design Thinking, it is necessary to provide a precise scheme within which this creativity is stimulated and decoded. Starting from observing people, we face a process divided into different phases. The subdivision and the respective visualisation of the different phases of the process depend above all on the methodological paradigm with which the creative process is analysed and described (Dorst and Dijkhuis 1995; Tschimmel 2012).

Among the many process models that have been defined, two of the best known were chosen for this text: "3 I" model (Brown and Wyatt 2010) developed by IDEO, and the Design Thinking model of Stanford d.school. They are presented and discussed below.

8.4.1   The "3 I" Model

In April 2015, IDEO launched a new evolution of the HCD toolkit, the Field Guide to Human-Centred Design.[3] The model included in this design kit is named after the process steps: *Inspiration, Ideation, Implementation*. It was developed in the context of social innovation, with the idea that adopting a Human-Centred approach means

---

[2]Citation: Definition by David Kelly.

[3]http://www.designkit.org/resources/1.

believing that all problems, even those seemingly unsolvable such as poverty, gender equality and clean water, are solvable and that people who face these problems on a daily basis are the ones who hold the key to their solution. Not only does it focus on the creation of products and services that are centred on the individual, but the process itself is profoundly human. *The Field Guide to Human-Centred Design* defines the process, and the mindset that design should address the social sector. The kit includes 57 design methods to understand the people for whom you are planning, to have more effective brainstorming, to prototype ideas and finally come up with more creative solutions, and a complete list of worksheets and project cases showing the Human-Centred Design in action.

Obviously, the model is based on the experience of IDEO in the field of business innovation, therefore it has been applied to case studies in the social sector, but it is usable in every area.

To summarise, in *inspiration* you learn how to better understand people by observing their lives, listening to their hopes and desires, and focus on the challenge by identifying the opportunity and potential of the project; *ideation* is to make sense of everything you've heard, generate lots of ideas, identify the most promising project opportunities, test and refine your solutions; *implementation* is the design phase which brings the solution to life, to consider the idea in relation to the market, and to build financial resources and models that ensure the solution is implemented in the best way possible, and can be sustained long-term. Implementation is an iterative process that requires many prototypes and pilot tests to refine the solution and support the system. These phases are not always carried out in a sequential manner. In fact, the projects may go back through *inspiration*, *ideation* and *implementation* more than once to allow the team to refine their ideas and explore new directions (Fig. 8.2).

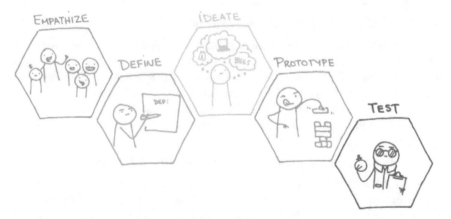

Fig. 8.2   The five steps of Design Thinking defined by Stanford d.School

8.4.2 The d.School Model

Another Design Thinking model, developed in a university context, is that of d.school at Stanford University and IDEO. It is a research program that aims to understand the approach of designers, turning it into a scientific method. In this model, also based on the experience of IDEO, the Design Thinking process is visualised in five steps (Empathise, Define, Ideate, Prototype, Test), which are all interconnected.

EMPHATHISE: *"To create meaningful innovations, you need to know your users and care about their lives"*

Empathy is at the core of a Human-Centred design process. It is the stage in which to understand the physical and emotional needs of people in the context of the design challenge. To empathise it is necessary to: observe users and their behaviour, involve them by active interaction, and full immersion in their frame of reference.

DEFINE: *"Framing the right problem is the only way to create the right solution"*

The definition is based on clarity and attention to the area of design with the development of a point of view built on the results of the empathic phase. We must therefore define the challenge based on what we have learned about the user and the context, giving sense to the information gathered. The goal is to define the problems they want to solve, and the opportunities to be grasped, in a meaningful and feasible way.

IDEATE: *"It's not about coming up with the 'right' idea, it's about generating the broadest range of possibilities"*

Ideate is the method of the design process that focuses on idea generation. Mentally, it is a process that could open new perspectives and require a new iteration of the exploratory and defining phases. Ideation provides the starting material to produce prototypes and obtain innovative solutions.

PROTOTYPE: *"Build to think and test to learn"*

The prototyping phase is the iterative generation of artefacts intended to answer questions that bring us closer to the final solution. In the early stages, a crude prototype offers the possibility to try out various alternatives before achieving the optimum result, and to collect useful feedback from colleagues and other people involved. In the later stages, both the prototype and the demand could become more refined. A prototype can be anything that a user can interact with: a post-it wall, a role-playing game, an object, or even a storyboard.

TEST: *"Testing is an opportunity to learn about your solution and your user"*

The Test phase is aimed at obtaining feedback for perfecting prototypes and solutions, and to learn more about the user. It tests the effectiveness of ideas through the feedback of the participants. As for the first phase, in this case it is necessary to observe and listen to people, to allow them to experience and experiment with the prototype by themselves. By observing this interaction, you can gather important information, review prototypes by fixing some usability problems, or generate additional insight that necessitate the repetition of some previous stages.

8.5   The IDEActivity Method: Design Thinking in the Italian Context (or European Context)

Starting from a knowledge base of Design Thinking patterns, the IDEActivity Center research team of Politecnico di Milano—active in innovation studies driven by creativity through design—has developed its own methodology that is based on exigencies in the national context.

Using the study of various most significant existing models as a springboard—the 3 I model (Brown and Wyatt 2010), the Double diamond model by the British Design Council, the Service Design Thinking proposed by Stickdorn and Schneider (2010) and a key reading in Human Centred Design (HCD, IDEO 2011)—the IDEActivity method was developed, in which the user is seen as a partner throughout the entire creation process (co-design) and is recognised as having significant creative potential.

The method is aimed at promoting the factors that allow for the pursuit of radical innovation, by ensuring that this approach is transformed into a permanent attitude within a company, even one of small/medium size, of an institution or of an organisation.

IDEActivity uses the sensitivity and methods of Design to find solutions that meet the needs of people in a technologically feasible and commercially viable manner.

Creativity as a "lateral thinking" capability and its application through a Human-Centred Design and co-participation approach, allows companies to be competitive. Therefore, the knowledge of some creative techniques plays a strategic role within the organisation. The goal is to be able to develop creative potential by activating the lateral thinking of each individual. Creativity, in fact, is not an innate ability and, like all other faculties, can be stimulated, amplified and channelled in specific areas (Fig. 8.3).

Fig. 8.3 Creative approach that facilitates the processes of innovation through design. Edited from T. Brown

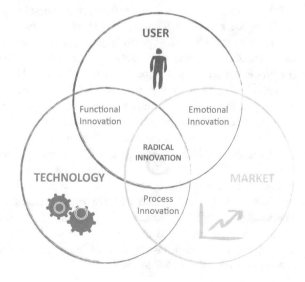

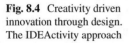

Fig. 8.4 Creativity driven innovation through design. The IDEActivity approach

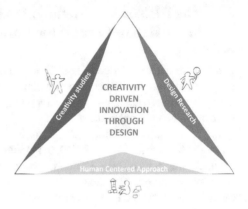

IDEActivity has an innovative methodological approach that, by combining methods and *tools* of different disciplines (Creativity and Design), allows design with the user's proactive participation (Canina et al. 2015).

It becomes a strategy to make the process that leads to the formulation of new ideas more efficient, stimulating creativity and generating radical innovation.

The method is designed to be a fluid and flexible tool that adapts to meet the needs of companies with different objectives and configurations, all in order to create a fertile climate in which Creativity, Design and innovation are interdependent.

Combining research in Design, studies on Creativity and a people-centred approach makes it possible to achieve innovation driven by Creativity through Design (Fig. 8.4).

The structure of the IDEActivity method integrates and amalgamates several known techniques and others which were developed ad hoc.

It is characterised by fundamental "play", meaning to get involved, to collaborate, to form a team and to look at things from another point of view with the help of others.

The process on which it is based requires a participatory co-design approach, based on the active involvement of potential users, and is structured on the development of a dedicated *Toolkit* which offers techniques and methods to guide participants through the process leading to creation and new solutions. The *Toolkit* has been designed to make tools and guidelines readily available, in order to stimulate creativity and enable you to identify both the design challenge and the final objectives.

A key feature borrowed from Design Thinking models is communication through visual, perceptive and communicative dimensions, typical of the language of Design.

The tools pertaining to Design and Creativity are divided into the three phases that make up the IDEActivity method: *Explore, Generate* and, transversally, *Set-Up* (Canina et al. 2013) (Fig. 8.5).

The creative process is divided into two major phases: *Explore* and *Generate*. Prior to conducting a creative session, it is necessary to consider the goal to be achieved, the available group, the location where the session will take place, and choosing techniques to use while taking all these elements into consideration. This is why *Set-Up* is at the heart of the process.

Fig. 8.5 The process that leads to creation and the achievement of new design solutions

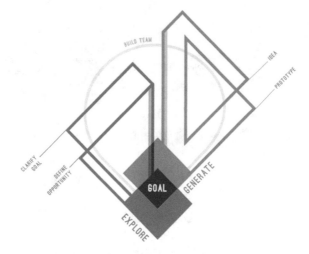

8.5.1 Explore

Through specific instruments, the *Explore* phase promotes and facilitates the analysis of the context of the market and the people, (re) defining a clear objective and building design scenarios that provide a glimpse of new opportunities. It is extremely important to define your goal, and to define it in a precise manner by devoting yourself to the *Clarify Goals* phase. In the *Define Opportunity* phase only the aspects of the objective that seem promising are highlighted as possible project opportunities.

8.5.2 Generate

The *Generate* phase is designed to make ideas tangible by generating appropriate solutions in line with the context and objectives of the project. Once you have discussed the basis for good design you move on to the *Idea* phase, the moment of generation and choice of ideas.

Lastly, the state of abstraction during the ideation phase passes to the physical with the *Prototype* phase, reducing the uncertainties of a project and abandoning other alternatives.

Each phase of the method is always constituted by a first phase of divergence, which is followed by a classification, and finally convergence to arrive at the definition of the problem or a solution.

The *Creative Diamond* (Tassoul and Buijs 2011) is characterised by a diamond shape and its phases have specific rules (Fig. 8.6).

The idea is to transfer, through *learning by doing*, those skills of approach to the project that can break old patterns and undertake new strategic roads in terms of product, service and/or system, which are replicable in different needs of innovation.

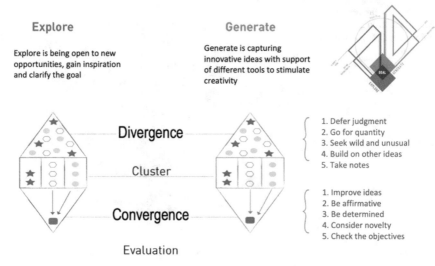

Fig. 8.6 The Creative Diamond representing the divergent and convergent phases (Tassoul and Buijs 2005)

In fact, the IDEActivity method is set up in such a way that project activities are always flanked by a flexible training structure, customisable and adaptable to the specific needs of the contexts to which it is applied on a case-by-case basis.

This activity, which sees the integration of *Design Knowledge* and *Creative Thinking*, seeks to consolidate creative behaviour and lateral thinking: that is to fix a creative "mindset" in the individual and facilitate the creative behaviour of the group.

Micro-objectives also remain key features of this method, such as: elimination of factors inhibiting creative thinking; recognition, management and overcoming of creativity blocks; identification of potential areas of development of creativity through Design in the organisation (fluidity, flexibility, originality, elaboration); techniques for the development of tactical creativity and strategic creativity that foster thinking and therefore the creative attitude.

8.5.3 How Design Thinking Affects Real Projects

The basic research on Creativity, carried out with the aim of contributing to structure the IDEActivity methodology, and the study of the scientific production on related topics (such as Creativity, Design Thinking, Service Design, Co-design, Creative Leadership, etc.) has allowed the mapping of the fundamental stages of the creative process, by identifying specific techniques and creating ad hoc tools that can be used to stimulate visualisation and retention areas in the Co-design project. The realisation of the first tools was the necessary step to create "proprietary products" such as *IDEActivity Toolkit*, the *Service Design Thinking for Educators* and *Co-design in the*

DiDIY scenario. Tools that organise and integrate various known techniques with the techniques and tools created specifically to facilitate Co-design activities, throughout the entire process, targeting innovation in areas such as digital technology, education and production.

The different aspects of the toolkit, with relevant guidelines, are the result of adapting the process and the techniques to the technical scope in which you are working. The *IDEActivity Toolkit*[4] is a customisable tool, configurable to the specific needs of a project, which allows the practical application of the method. It was brought to life during a project with the Enterprise Network "Rold Research",[5] in which the goal was to facilitate a group of companies to identify a common field of design application in which the specific competencies of each company bring strong added value in relation to that of the others. It contains a set of documents with the relevant tools and guidelines, an integration of well-known techniques and ad hoc designed tools, created specifically to facilitate co-design activities throughout the entire process (Canina et al. 2013).

From the toolkit, the *Service Design Thinking Guide for Teachers*[6] was developed as part of the "Innovative Design of the educational process in schools", created to meet current training needs of teachers of all levels and order.

This Guide, written collaboratively with teachers,[7] contains process and design methods specifically adapted to the formation context. It offers new ways of being collaborative in the design stages, and it allows teachers to create high-impact solutions. The guide offers a variety of teaching methods to choose from, with concise explanations, useful tips and suggestions. It is an ongoing project, which has been enriched by pilot experiences conducted in the classes of innovative teachers, and later also by other teachers who have practiced the method after dissemination activities through the creation of an online platform (Canina et al. 2016).

From the start, the IDEActivity toolkit was specifically created as part of the European project *Digital Do It Yourself*,[8] *Co-design in the DiDIY scenario*. The *Toolkit and guideline* provide support to people who generate innovative solutions in their professional field by applying the basic features of Digital DIY. The toolkit provides guidance on the application of a strategic planning approach to the use of digital production and sharing technologies, which then becomes a means to enable new opportunities and innovative ideas (Canina and Bruno 2018).

---

[4]Download from http://www.ideactivity.polimi.it/toolkits/ideactivitytoolkit/.

[5]Site of the Network of Companies http://www.roldresearch.org/.

[6]The project "Innovative Design of the educational process in schools" funded by Telecom Foundation was developed in collaboration with ANP (National Association of managers and the high professionalism of the school) and MIP Politecnico di Milano. The guide can be downloaded from http://www.ideactivity.polimi.it/toolkits/servicedesignthinking/.

[7]Available at www.innovazioneinclasse.it.

[8]Official site of the project DiDIY http://www.didiy.eu/ where you can download the toolkit.

8.6   Conclusion

In recent years, Design Thinking has been gaining momentum among Italian companies, with more and more managers and entrepreneurs welcoming this alternative approach to innovation. In fact, as previously said, this method generates a change of perspective that can support innovative business models. In this sense, and in order to cope with a socio-economic environment such as the current one, several small and medium Italian enterprises have started working to renew themselves internally regarding their approach to and management of projects, relying on consultancy and training groups created to transfer their know-how of the design process in areas of application such as business reality. The aim is to provide a method that allows an SME to differentiate itself in the market by leveraging on the creation of a working environment based on a strong team play, the sharing of ideas and enhancing the creative potential of each internal resource through a broader perspective of the Human-Centred approach. In this sense, research shows that Design Thinking can support not only creative processes, but also the execution phase and acceleration of innovation processes.

This method is not only a driving force for innovation but offers new process models and tools that help improve, accelerate and visualise any creative process, making it accessible not only to designers but also to multi-disciplinary teams and to all types of organisation, even in the Italian context.

References

Amabile TM (1988) A model of creativity and innovation in organizations. Res Organ Behav 10(1):123–167

Brown T (2008). Design thinking. Harvard Business Review. June 2008

Brown T, Wyatt J (2010) Design thinking for social innovation, stanford social innovation review. Leland Stanford Jr. University

Burns C, Cottam H, Vanstone C, Winhall J (2006) Transformation design, Red paper 02 Report, Design Council

Canina M, Anselmi L, Coccioni E (2013) Design training plans in creativity techniques for companies. E & PDE 2013—The 15th international conference on engineering and product design education conference, Dublin, 5-6 September.

Canina, M., Bruno, C. (2018) Discovery DiDIY. An immersive gamified activity to explore the potentialities of digital technology, Design Research Society 2018 (DRS2018), Limerick, 25-27th June 2018

Canina M, Coccioni E, Anselmi L (2015) Creativity and design tools as an emotional approach to learning, Cumulus Milan 2015 Conference. June 2015

Canina M, Salvia G, Bruno C (2016) Design and creativity enhancing innovative educational processes. The 18th international conference on engineering and product design education (E&PDE 2016). Design Education: Collaboration and Cross-Disciplinarity. September 2016

Christensen CM, Raynor ME, McDonald R (2015) What is disruptive innovation? Harvard Bus Rev 93(12):44–53

Csikszentmihalyi M (1996) The creative personality. Psychology Today 29(4):36–40

De Bono E (1970) Lateral thinking: creativity step by step. Harper & Row, New York

Dorst K, Dijkhus J (1995) Comparing paradigms for describing design activity. Design Stud 16:261–274. Elsevier

Fischer G et al (2005) Beyond binary choices: integrating individual and social creativity. Int J Hum Comput Stud 63(4–5):482–512

Literat I, Glăveanu V (2016) Same but different? Distributed creativity in the internet age. In: Creativity, theories–research-applications, vol 3, no 2, pp 330–342

Meinel C, Leifer L (2010) Design thinking. understand, improve, apply. Springer, Heidelberg

Poincaré H (1924) The foundation of science. Science Press, New York

Roth B (1973) Design process and creativity. In: Beer-Sheva, Israel: Mechanical Engineering Department, University of the Negev. https://dschool.stanford.edu/resources/bernie-roth-treatise-on-design-thinking

Stickdorn M, Schneider J (2010) This is service design thinking. Basic—tools—cases. BIS Publisher, Amsterdam

Tassoul M, Buijs J (2011) Clustering, from diverging to converging in CPS process. pp 1–16

Tschimmel K (2012) Design thinking as an effective toolkit for innovation. In: ISPIM conference proceedings, the international society for professional innovation management (ISPIM), Barcellona

Von Hippel E (2005) Democratizing innovation. MIT Press, Cambridge. https://web.mit.edu/evhippel/www/books/DI/DemocInn.pdf

Von Stamm B (2008) Managing innovation, design and creativity. Wiley, West Sussex

Wallas G (2014) The art of thought. Solis Press, Londra (ed. originale: Wallas G (1926) The art of thought. Harcourt, New York)

Williams R, Runco MA, Berlow E (2016) Mapping the themes, impact, and cohesion of creativity research over the last 25 years. Creat Res J 28(4):385–394

Chapter 9
Ergonomics and Design for All: Design for Inclusion

9.1 Introduction

The results of design research and experimentation in the field that is now known as *Design for inclusion* offers a largely new framework than that of times past, one that exists in the context of the profound social changes that have occurred in industrialised countries and which involves a progressive advancement of the contents and objectives of Ergonomics/Human-Centred Design and Design for All.[1]

In fact, the social changes of recent years offer a radically altered outlook, both in terms of the demographic shifts in industrialised countries and the shifts in lifestyles and expectations of their populations.

First consider the demographic shifts, which are, on one hand, characterised by ageing populations and, on the other, the growing presence of immigrant populations from southern countries in industrialised countries, particularly Europe.

Demographic shifts stemming from the growth of elderly populations are bolstered by the immigration-based shifts on one hand and the decline in birth rates on the other. Together, these are leading to a profound change in family structures and the dissemination of new ways of cohabiting and living and, naturally, new expectations and needs in terms of the level of independence offered by living environments and the ability to participate socially.

In recent years, we have seen social policies promoted by the Europe Union, and European funding for research, in the field of inclusion and social innovation.

The "Europe 2020 leading strategy"[2] sets the goal for a smart, sustainable and inclusive economy, based on developing social inclusion and innovation, defining the latter in this way: *"Social innovation can be defined as the development and implementation of new ideas (products, services and models) to meet social needs and create new social relationships or collaborations. It represents new responses to*

---

[1] For more information on the topics covered in this chapter, see: Buti (2008), Steffan (2014), Tosi (2014), Spadolini (2013), Di Bucchianico and Kercher (2016).

[2] European Commission, Guide to Social Innovation, European Union (2013).

© Springer Nature Switzerland AG 2020

F. Tosi, *Design for Ergonomics*, Springer Series in Design and Innovation 2,

https://doi.org/10.1007/978-3-030-33562-5_9

pressing social demands, which affect the process of social interactions. It is aimed at improving human well-being. Social innovations are innovations that are social in both their ends and their means. They are innovations that are not only good for society but also enhance individuals' capacity to act".

The chief elements of social innovation are:

- identification of new/unmet/inadequately met social needs;
- development of new solutions in response to these social needs;
- evaluation of the effectiveness of new solutions in meeting social needs;
- scaling up of effective social innovations.

The new approach to the subject of *inclusion* of the more neglected segments of the population *and social inclusion* highlights the diversity of the target population and brings attention to all of the conditions, however obvious they may be, that lead to exclusion from the ability to carry out daily activities and actively engage with society.

As Elton and Nicolle write (2015, pp. 300–301), *"There are many different approaches to design. The approach selected is often dependent on the type of value the product and/or service intends to deliver to the end-users. (...) the inclusive design approach, which aims to deliver 'mainstream products and/or services that are accessible to, and usable by, people with the widest range of abilities within the widest range of situations without the need for special adaptation or design' (BS 7000-6 2005). Accessibility and usability are the key criteria of this approach. Accessibility refers to allowing users access to the features of products and/or services through their sensory, physical and cognitive capabilities.*

In essence, inclusive design can be categorised as a specific type of human-centred approach to design. The inclusive design approach specifically focuses on understanding the needs, capabilities and attitudes of people who have some form of impairment and then applying this knowledge to main-stream design. Thus, ergonomics/human factors play a significant role in the inclusive design approach".

The themes of inclusion and social innovation involve very broad user profiles, which are brought together by the need to safeguard and improve their levels of personal independence in their living environment (domestic, urban, communal) and participate actively in life and society.

In this matter, where the focus is on the needs (needs, expectations, desires)[3] of people within their living context/environment, the contents and objectives of Ergonomics/Human-Centred Design and Design for All converge, along with, more generally, Design for inclusion.

In fact, certain innovative elements characterise both the objectives and the set-up of the research areas and the design experiences that have developed in recent years, in both the fields of HCD and DfA and Design for social inclusion.

Innovative objectives, aimed not only at guaranteeing adequate levels of care and assistance to the most vulnerable segments of the population, but at creating living environments, products, services, and, in general, social and environmental

---

[3]For the meaning of the term "needs" used in this book, see Sect. 1.1.

conditions aimed at the well-being and autonomy of the person, to facilitate and promote real participation in life and social integration.

Innovative approach to the analysis of the demographic and social framework that characterises all European countries. This approach is based on an organic assessment of the complex framework of social changes, linked not only to the conditions of disability and reduced autonomy, but also to the complex phenomena of ageing populations, immigration and, finally, the shift in family structures.

These changes result in the creation of large segments of the population that are characterised by specific needs for the safety and usability of living environments and active participation in life, and require the definition of solutions that can favour and maintain independence and participation in people's social life over time, as an essential condition not only for individual well-being, but also for the sustainability of the health and social system of European countries.

9.2 Ergonomics, Design for All, Design for Social Inclusion: The Birth of a Common Language

The relationship and points of convergence between the areas of research and intervention in Ergonomics/HCD and Design for All start with their common focus on the specificity and complexity of each intervention case—despite their different perspectives—starting with the identification of the specific needs that people express, or can express, in terms of their relationship with a product[4] or a system.[5]

Today, this is the basis for discussing *Design for inclusion and social sustainability*, subjects that are recognised as a field of research and experimentation common to Ergonomics/HCD and Design for All and a central theme in many international conventions,[6] as well as a specialty field for researchers and professionals in these two research areas.

However, let's start from the evolution of the two approaches to what is now defined as Design for inclusion.[7]

Attention to the needs of the most vulnerable segments of the population, and the definition of design criteria that can guarantee adequate security and accessibility to people with disabilities in the use of products, environments and services, is the aim of many design research areas that have developed since the Second World War, in both the United States and Europe.

Barrier Free Design started in the United States in the 1950s, with the aim of developing and spreading design principles that were based on accessibility. This allowed the reintegration of wounded soldiers following the Second World War and, later, the Korean War and Vietnam.

---

[4]For the meaning of the term "product" used in this book, see Sect. 1.1.

[5]See Chap. 1, notes 2, 3 and 4.

[6]See Chap. 1, note 9.

[7]On this subject, see: Steffan (2014).

Barrier Free Design, Inclusive Design,[8] and, in general, numerous research areas aimed at "accessible design", have, with time, provided a vast heritage of design principles and intervention criteria, which, while representing fundamental contributions to culture and design practice, remain focused on a marked specialisation on designing "for the disabled" and, in particular, the needs of those who are physically disabled.

It is through the Design For All approach, and, in part, the Universal Design[9] approach, that we overcome the traditional "disabled designs" specialisation.

The Stockholm Declaration of 2004, in defining Design For All, says "*is design for human diversity, social inclusion and equality. This holistic and innovative approach constitutes a creative and ethical challenge for all planners, designers, entrepreneurs, administrators and political leaders. Design for All aims to enable all people to have equal opportunities to participate in every aspect of society. To achieve this, the built environment, everyday objects, services, culture and information—in short, everything that is designed and made by people to be used by people—must be accessible, convenient for everyone in society to use and responsive to evolving human diversity. The practice of Design for All makes conscious use of the analysis of human needs and aspirations and requires the involvement of end users at every stage in the design process*".[10]

With Design For All, the focus shifts *from a marked specialisation approach*, aimed at responding to the needs and expectations of people with disabilities or physical, sensory or cognitive restrictions, *to an entirely inclusive approach*, which

---

[8]The British Standards Institute (2005) defines Inclusive design as *"The design of mainstream products and/or services that are accessible to, and usable by, as many people as reasonably possible (...) Without the need for special adaptation or specialised design"*.

[9]Universal Design is defined as the *"design of products and environments that can be used by everyone, with the maximum possible coverage, without the need for adaptations or specialised design"*.

The objective is to create products that are can be fully used by people with reduced autonomy and that are aimed at the widest possible range of users in which users with disabilities are not seen as a separate segment but an integral component. In 1997 the Center for Universal Design developed the 7 principles of Universal Design, which are aimed at addressing the design of environments, products and communication systems.

1. Fair use: Design is useful and saleable to people with differing abilities. 2. Flexibility in use: the design appeals to a wide range of individual preferences and abilities. 3. Simple and intuitive use. The use of the design is easy to understand, regardless of the user's experience, knowledge, language skills or current level of concentration. 4. Perceivable information. The design effectively communicates the information needed by the user, regardless of the environmental conditions or sensory capabilities of the user. 5. Fault tolerance. The design minimises the risks and negative consequences of accidental or unintentional actions. 6. Low physical effort. The design can be used efficiently, comfortably and with a minimum of effort. 7. Size and space for approach and use. Appropriate sizes and space are provided to approach, reach, manipulate and use them (objects and environments), regardless of the size, posture or mobility of the user. See: Center for Universal Design, College of Design at North Carolina State University (NCSU)—http://www.ncsu.edu/ncsu/design/cud/index.htm.

[10]EIDD, European Institute for Design and Disability, Stockholm Declaration 2004.

starts from the needs of specific user groups and aims to create products whose image, functions and methods of use can be aimed at the entire population.

It is in this area that the relationship between Design For All and Ergonomics has developed and consolidated in these years, which identifies the person and their specific needs as the jumping-off point and the central objective of any evaluation or design intervention.

As Steffan[11] writes, it is, in fact, the concept of "usability", understood as the possibility to fully enjoy the good (product, environment, service), that denotes the shift from a design aimed solely at ensuring accessibility (physical or perceptive) of places and objects to a fully inclusive conception of design intervention, one that is aimed at ensuring the effective well-being of people, regardless of their skill level.

Thus, the Design For All approach opens up a field of research and intervention in Ergonomics that allows the needs and rights of the most vulnerable segments of the population to be studied and interpreted, as a jumping-off point for a kind of design that is more conscious and attentive to the various needs of all possible users. The principles, theoretical processes and experiments that have developed in the field of DFA supply a wealth of very valuable experiences that can be applied to every area of design.

Ergonomics, and in particular, Ergonomics for design, deals with the specificity and complexity of each intervention case—be it evaluation or design of a product, an environment or a service—starting from the identification of the *specific needs* that the user expresses, or can express, with respect to the relationship with that product, according to the same *specific context conditions* in which this relationship takes place, of their reciprocal conditioning and their variability over time.

On the cognitive level, Ergonomics supplies a wealth of multidisciplinary knowledge that allows us to evaluate and design the compatibility of the product with respect to the characteristics and capabilities of the users for whom the product is intended.

As a methodological approach, Ergonomics offers a structured set of evaluation methods aimed at identifying, interpreting (and imagining) the real needs of users in their relationship with the product, the possible sources of discomfort and/or frustration, expectations and possible wishes.

Ergonomics, therefore, may represent the necessary pragmatic contribution to Design For All, as well as to many other research sectors that have focused on accessible and inclusive design for years, whose contents have developed on a chiefly theoretical level to this point, through the development of an exhaustive and structured framework of definitions, objectives and intervention criteria.

---

[11] See Steffan (2014).

9.3 The Reference Framework

To identify and understand the needs[12] of the people targeted by the design for inclusion, we must firstly focus on what is meant by the definitions of *population ageing* and *disabilities,* and, more generally, the difficulty and fragility surrounding the physical and social environment.

Population ageing refers to the phenomenon of the average age rising, caused by the growth in life expectancy and the parallel reduction in birth rates, which is leading to the gradual consolidation of population groups, (and as we shall see, the market) characterised by a greater need for security and usability when using products.

In quantitative terms, the increase in the elderly population has been a growing figure in all industrialised countries for several decades. The trend, already evident since the 1980s, has become the most relevant and analysed demographic phenomenon of the last few years, the impact of which, as is well known, affects the structure—and stability—of pension and health systems in all Western countries; it is also the source of the need to move care and assistance activities towards home care.

Life expectancy in European countries, which was 45 at the beginning of the 1900s, has progressively grown throughout the 20th century, reaching the current figure of 84 years for women and 80 years for men.

The most notable forecasts regarding ageing population indicate, at a global level, an increase from 11 to 22% in the percentage of people over 65 between 2000 and 2050, and from 900 million in 2015 to 1400 million in 2030 and 2100 million in 2050. As for Europe, the percentage of people over the age of 60 is expected to be 34% by 2050.[13]

The Istat 2017 data[14] on national demographic forecasts between 2016 and 2065 shows a steady growth of the elderly population emerging against a parallel decrease in under-14s. The age groups over 65 rise from about 22% in 2016 to over 33% in 2045, maintaining a percentage between 32 and 35% in the years that follow. At the same time, forecasts indicate a progressive decrease in the population under 14 from the current 14–12% in 2030 and mostly stable in the following years. The period around 2045 is indicated as the most critical moment, when the active population would fall to 53% of the total and the average age of the population would rise from the current 45 years to 49.7.[15]

---

[12]For the meaning of the term "needs" used in this book, see Sect. 1.1.

[13]World Health Organisation Executive Board, Multisectoral Action for a Life Course Approach to Healthy Ageing: Draft Global Strategy and Plan of Action on Ageing and Health, World Health Organisation (2015).

[14]ISTAT, Report 2017: The demo- graphic future of the country. Forecasts of the resident population as of 2065, ISTAT 2017. See also: ISTAT, Annual Report 2017, the situation in the country, ISTAT 2017.

[15]Part of the ageing process that is occurring can be explained by the shift of the baby boom generation (1961–75) between the late active age (40–64) and the senile age (65 and over). The peak of this ageing will hit Italy in 2045–50, when there will be a quota of over 65s close to 34% (See ISTAT Annual Report 2017).

The percentage increase of the very old (i.e. of people over the age of 85), passed from at 2% of the population in 2005 to 7.8 in 2050.

The forecasts on immigration, which see the absolute weight of migrations on the variation of the population in the period 2016–2065 as being around 10.2 million individuals is also significant.[16]

It should be emphasised that the phenomenon of immigration is obviously concentrated in the youth and adult ages, with an equally evident effect on the predictions of births, which will tend to partially offset the reduction in the new births of the current Italian population.[17]

The growth of the elderly population and the reduction of births have brought about profound changes in the family structure, with important implications linked to the need for care and assistance for elderly family members and, in parallel, the presence of an increasingly small number of kids and infants (Fig. 9.1).

The last few years have seen a significant change in the traditional family, with an increase in single-person families and so-called "extended families", i.e. cohabitation models that are partly independent from age, and motivated by mainly economic reasons and the need for work. They are also partly linked to the ageing of the population and to the growing number of elderly people living alone.

The typical family is today composed of parents and one or two children, and there are often elderly relatives who are assisted by the family and by so-called "informal caregivers", even if they do not share living quarters.

This has repercussions on organisation that impact not only elderly people but all age groups and everyone who is directly or indirectly involved in care.[18]

The term disability defines "any limitation of the ability to act caused by a state of disability/impairment", followed by a disadvantageous condition or *handicap* experienced by the person in terms of the actions, activities, behaviours or attention-response abilities that required by the physical and social environment in which they live.

A disability, therefore, is a limiting condition that causes personal difficulty, which translates into a disadvantage, or handicap, that manifests itself in their interaction with the physical and social environment when these prevent or limit that person's autonomy.[19]

---

[16]7.7 million directly and 2.5 indirectly as an additional effect on the dynamics of births and deaths (See ISTAT report 2017).

[17]In the predicted population estimate of residents in Italy, a significant contribution comes from the forecast of migrations abroad. The migratory balance with foreign countries is expected to be positive, on average, more than 150,000 units per year (133,000 as of the last recording in 2015), though this is marked by strong uncertainty. We do not exclude the possibility, though the probability is low, that it may become negative in the long run (Cf. ISTAT, Annual Report 2017).

[18]On these subjects, see the book: Tosi and Rinaldi (2015b).

[19]In Italy, the principles of "accessible design", which have already been present in the legislation regarding the elimination of architectural barriers since the end of the 1960s, developed in the two decades that followed, culminating in the enactment of law 13 of 1989 "Provisions to favor the overcoming and elimination of architectural barriers in private buildings" and the subsequent

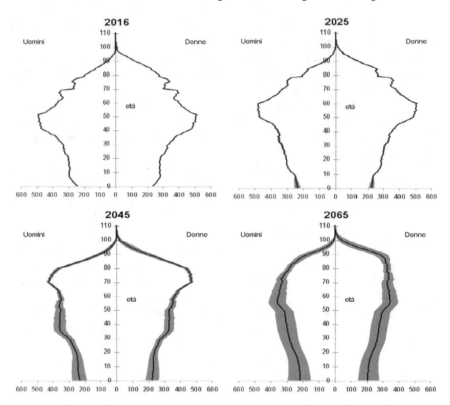

Fig. 9.1 Resident population pyramid—median scenario and 90% confidence interval. Italy, 2016–2065, 1 January, data in thousands. *Source* ISTAT report 2017: the demographic future of the country. Forecasts of the resident population in 2065

This reduced capacity not only concerns the conditions of a comprehensive dis-ability, but the whole range of more or conditions that are defined by a more or less marked or evident *distance* from what is commonly referred to as normal physical, perceptive and cognitive capacity.

This distance can relate to the different aspects of movement skills, visual or auditory capacity, but also learning skills, the ability to control yourself in daily activities and in relationships with others and the ability to decode information from the physical and social environment social.

---

implementation of decrees 236/1989 and 503/1996, which provide detailed design information for the construction of the works in residential buildings and public buildings respectively.

The legislation for the elimination of architectural barriers, which is universally considered to be one of the best European regulations, even if it is still poorly applied today, has been flanked by numerous national and regional legislative instruments since the end of the 1980s, which concern the construction of residential facilities for the elderly and, subsequently, programs for the adaptation of private housing and the safe-guarding of the autonomy of housing for the elderly.

Among the latter are the difficulties of placing events and objects in space and time, attention difficulties, difficulties in relating to people (for example in verbal or behavioural expression, in understanding language, etc.), reduction in risk perception, i.e. the difficulty of predicting the possible risk factors related to the use of objects and equipment or situations that require attention, such as travel, transportation, etc., and, in general, problems with orientation.

The conditions of the individual's inability to deal with the physical and social environment may appear unique, or be the consequence of the other. They can concern not only situations of complete disability, but also a much wider range of people who find themselves in a precarious condition when compared to what is required by the physical and social environment in which they live, leading to a disadvantage in the performance of common life activities daily.

Even if we limit ourselves to physical and sensory disabilities, weak conditions with respect to the environment concern children, pregnant women, most people over 65, people with vision and hearing impairments, etc. Even temporary conditions of weakness can derive from contingent situations: carrying a heavy suitcase or a child in your arms, having to cross rough terrain without the proper equipment, etc. can transform everyday activities into obstacles that are grueling to overcome.

The need for greater legibility of road signs, as well as written information (on the information leaflets for medicines, on the labels of food products, instruction booklets, etc.) is, of course, an issue for people with significant visual ailments, but also for most short-sighted or presbyopic people, or those suffering from other visual issues that affect approximately half the population.

Designing safe and easy-to-use products, therefore, means responding to the essential needs of the most disadvantaged categories of users, but also to needs that touch very large segments of the population, meaning the greater usability of products, environments and services will result in conditions of greater well-being, savings in time and energy and a general improvement in quality of life.

Difficulties in integration and participation in social life also concern all those who do not possess, or do not yet possess, codes of conduct, language, and, in general, the ability to address and respond in an appropriate manner in a given social environment. People who have immigrated from countries other than the one they live in, people who are not part of the social group with whom they have to, or want to, enter into a relationship (by age, educational level, geographical or cultural affiliation, or simply by social background) and, of course, people with disabilities, are in fact excluded or limited in terms of integrating and participating in social activities.

An example of particular relevance to the subject of this chapter is represented by problems in relating to the physical environment, services and products of use, and the consequent social exclusion, caused by the dissemination of communication and interface systems based on digital language.

The widespread use of electronics and information technology has led, as is well known, to a relatively rapid disappearance of mechanical controls and their progressive replacement with a computer-based interface with increasingly complex and extensive performances, the use of which requires the use of digital language and logic and a method of learning new functions and procedures, based on the ability to establish immediate analogies with similar dialogue systems (i.e. on the habit of using mobile phones, vending machines, computer programs, etc.)

The dissemination of communication and interface systems based on digital language is now so widespread that it touches almost all of our daily activities.

In the space of just over 20 years, we have shifted from a world of predominantly mechanical and manual objects to a world in which most of our daily activities require interaction methods that presuppose knowledge of digital language and interaction methods. This forces us to continuously learn new dialogue logics, procedures of use and, more recently, new modes of communication, social interaction and new ways to physically interact, thanks to the advent of touch screen.[20]

This phenomenon has also occurred an unprecedented speed, imposing not only the need to learn new languages, but also to keep abreast of continuous evolutions and to do it with a speed that can only be fully achieved by younger generations and, to some degree, somewhat older people who can (or must) use prior knowledge of IT systems; for example, many years of work, the use of next-gen mobile phones and then smartphones and digital devices for listening to music, going online, communicating, and learning how to use and interact with them over time.

Ironically, a population that is increasingly made up of elderly people is addressed with languages and ways of interaction designed for the generation of digital natives, proposing systems of communication and participation, but also ways of using essential daily use products, that pre-suppose radically new skills with respect to the past and the acquisition of skills typical of younger generations.

ICT technologies also offer great opportunities.[21] The use of environmental control and regulation systems based on digitally controlled technologies, the use of voice-activated communication and information systems and assistive technology for environmental control and personal safety are examples of how computer technology can ensure the safety and protection of people and allow access to information, even in situations of reduced mobility or perceptive capacity.

---

[20]It is important to highlight, especially for those who are now starting out as a design—or as a design student—that the difficulty of learning new tasks and new languages is not only linked to the gradual decrease in the ability to learn and memories new tasks and procedures that characterises the third age, but also to the ability, or lack thereof, to associate new requirements and procedures with prior knowledge.

The inability to exploit similarities with previously assimilated usage procedures, and previously used interfaces, can make learning extremely difficult, long before we reach old age, with the consequence that the design of digital interfaces, and of their usage methods, cannot be aimed at generic "age classes", but rather at varying skill levels and customs regarding the use of languages and dialogue procedures that are specific to digital interfaces.

[21]On these subjects, see Chaps. 3 and 16 by Brischetto and see: Spadolini (2013), Tosi and Brischetto (2016), Brischetto (2017).

9.4   The Knowledge of the *Context of Use* as the Basis of the Design

The goal of a fully inclusive design is to assume the needs, expectations and desires of people with a reduced capacity (motor, perceptive, cognitive) as a regular component of the collection of needs to which the design must respond. In these needs, we can recognise—permanently or temporarily—people of different ages, health conditions, economic possibilities and cultural level.

Reduced capacity or ability are related to the conditions of disability and, as we have seen in the previous sections, the entire range of conditions of *distance* from what is commonly viewed as a normal skill.

The concrete effects of this distance—that is, the consequences on daily life and relationships—depend largely, as we have seen, on the contextual conditions in which the person lives.

The ergonomic approach to assessing the needs of people presupposes, as we have seen, the evaluation and interpretation of the complexity of variables that define the *interaction context* between the people and the systems in which, and with which, they come into contact. Among these are the characteristics and capabilities of the person, the activities they must and/or want to carry out, their objectives and their expectations and, more importantly, the characteristics of the physical, social and technological environment.

The ICF "International Classification of Functioning, Disability and Health" (international classification of functionality, disability and health) has a similar outlook. ICF describes the limitations or disabilities of the individual according to the pathologies and/or impairments from which they can originate and from the limitations of the individual and social activities that these can provoke.[22]

The ICF classification, published by the World Health Organization in 2001, constitutes, together with the ICD 10 (International Statistical Classification of Diseases and Related Health Problems) the evolution of the International Classification of Impairments, Disabilities, and Handicap (ICIDH), which was published for the first time in 1980 and contains the definitions of:

- impairment: any loss or abnormality of a psychological, physiological or anatomical structure or function;
- disability: the limitation or loss, due to impairment, of the ability to perform an activity in the manner and to the extent considered normal for a human being;
- handicap: the discrepancy between the efficiency and the state of the subject and the expectations of efficiency and status coming from the environment (social and physical) and from the subject himself. The handicap, therefore, is a limitation on the ability manifested in the relationship with the environment (physical and social) in which the person lives and is the consequence of disability and not the disability itself.

---

[22]See WHO, ICF International Classification of Functioning, Disability and Health, www.who.int/classifications/icf/en/.

The ICF, therefore, provides "a frame of reference that allows the measurement of the role of the environment and the health outcomes of people and, at the same time, more effective monitoring of the socio-health interventions, transversally and in an interdisciplinary manner" (Leonardi 2003).[23]

The ICF classification integrates and extends the definitions of impairment, disability, handicap, relating them with functional states associated with the state of health, and with the different dimensions in which they can be evaluated, namely the *body dimensions* (bodily functions, body structures, impairments), the *dimensions of individual activities*, i.e. the range of activities that the individual can perform (defined as the performance level with which the individual performs a task or action) and the *dimension of participation and social activity*, i.e. the areas of communal life in which the individual may be involved (i.e. the level of participation and restriction to daily life).[24]

The three dimensions allow us to define the individual's levels of functionality or disability and are studied starting from the contextual factors—divided into environmental factors and individual factors—based on the relationship model shown in Fig. 9.2.

In the last decade, ergonomic they have been developed some standards specifically aimed at design for the global user base. In particular, ISO 20282: 2006, Ease of operation of everyday products—Part 1: Context of use and user characteristics; part 2: Test method.

ISO 20282:2006 defines *ease off use* based on its effectiveness, or the "*percentage of users who reach the main objectives of use of the product interface with completeness and accuracy*", focusing on daily use products that are defined as "*those products that a person must use in his daily life, that a person can buy, rent, or use, regardless of whether they are owned by public organisations or individual property*".

Daily use, mechanical and electrical products with mechanical or digital interfaces are the products necessary for carrying out normal daily life activities, whose simplicity of use represents the discriminating factor between the possibility or the impossibility of using them independently.

The ISO/IEC Guide 71: 2014, Guide for addressing accessibility in standards, is similar and is aimed at providing basic information and knowledge on the needs of

---

[23]See Leonardi, ICF. The classification of the functioning of disability and health by the World Health Organization. Work and discussion proposals for Italy, www.silsismi.unimi.it/SILSISMI/Indirizzi/Indirizzi_doc/Sostegno/Leonardi271103.pdf.

[24]The contents of the UN Convention on the Rights of Persons with Disabilities, published in 2007, which states "respect for the difference and acceptance of people with disabilities as part of human diversity" is of particular interest.

In particular, the UN Convention introduces "a framework based on human rights, absent in the ICF model. (…) Whatever the cause of the functional limitation and whatever the nature, this is attributable to human diversity. In reality, a person with a spinal cord injury that has caused paraplegia cannot be described only on the basis of their functional limitations, and the latter, while producing conditions of dependence on third parties (e.g. for dressing, moving, washing, etc.) do not automatically produce a dependency. (…) It would be more correct to define functional impairment as one of the characteristics of the person and not "the characteristic from which to start, otherwise we risk reducing that same person to that single characteristic". See Griffo (2009).

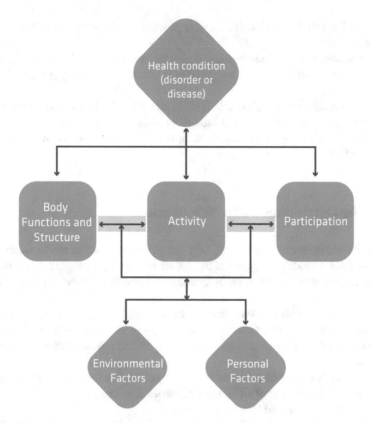

Fig. 9.2 Reworked from ICF "International Classification of Functioning, Disability and Health", World Health Organisation

people with different skill levels. The contents of the Guidelines are addressed, in particular, to the drafters of regulatory standards and, therefore, aims to disseminate the tools necessary to "include" the needs of the elderly and, more generally, of people with "a reduced level of ability" in the criteria and objectives of general rules.

In both cases, the aim of the regulatory tools is to extend the meanings and fields of application of usability, applying and reinterpreting the framework of definitions from Human-Centred Design to the sector of daily use products and design "for the maximum number of users".

Both regulatory instruments have been an essential reference in the drafting of UNI 11377, Usability of industrial products—part 1: general principles, terms and definitions; part 2: methods and intervention tools,[25] published in September 2010 and currently in the process of being updated, which defines the general principles

---

[25]The UNI 11377: 2010 standard is the first standard in the field of ergonomics that is specifically aimed at the evaluation and usability of design in the Product Design sector. The standard was prepared by the "Usability of industrial products" working group established within the "Ergonomics" Commission of UNI, GL "General Principles", and composed of: Francesca Tosi (coordinator),

for the design and production of everyday products, and the terms to be used in the field of Ergonomics for design.

The standard also provides the basic definitions and theoretical contents for usability and its components, which are applied to the design sector and, in particular, that of product design.

The standard provides the basic definitions and theoretical contents of for usability, which are applied to the industrial products sector, and presents some of the main methods for evaluating the ergonomic quality of products, which can be applied in the field of Ergonomics and Design.

9.5   The Development of a New Market

The social changes of recent years, and the growing attention to the needs—and rights—of the population segments at risk of exclusion, have gone hand in hand with an equally growing attention by companies towards the "new" third generation market.

This aspect is obviously very relevant for Design for inclusion, which, like any other research and design area, needs availability and interlocutors both within companies and public administrations, so that projects and research breakthroughs can be made.

The ageing of the population, and the very significant consequences for the demographic and social policies of all Western countries, are accompanied by two radically new phenomena compared to the recent past.

Thus, we see some aspects that characterize and also give rise to this new market sector.

The first is the shift in time of entry into the real age of old age and the consolidation of a segment of the population that carries specific security and usability needs, with respect to the use of environments, products and services, and are increasingly attentive and aware of their needs and rights.

So-called "old age" today corresponds to ages that greatly exceed those of the recent past. If the demographic data set the date of entry into the third generation at 60 or 65, the sixty-year-olds are defined, even in journalistic and commercial terms, as *young elders*, while the immediately higher age groups are labelled as *active unemployed, elderly in good health,* etc.

Today's seventy-year-olds (but also those in their eighties and nineties) are often active people, capable of managing and planning their existence and, moreover, boast a lot of free time. In fact, these people are the target for travel offers and wellness packages offered by hotels, spas and gyms, daily use products advertised for their simplicity and safety of use, products for the care of body and agents who propose

---

Lucio Armagni, Paola Cenni, Luigi Bandini Buti and Laura Anselmi. In addition, Barbara Simionato, Fiammetta Costa and Sabrina Muschiato collaborated in the endeavour. The standard is currently being updated.

(and promise) a physical image that is not "young" but "active and efficient", referring to an image of old age that is certainly not a "phase of decline" but rather one of freedom from the burden of work and family, with time and financial resources to dedicate to one's own well-being.

A second phenomenon, which derives in part from the first is the spread of greater awareness of people's own needs and rights, including safeguarding their personal autonomy. Compared to the younger population, the elderly person also has a more reflective attitude towards the choice of new products, and a greater attention to quality, reliability and safety, as well as ease of use, familiarity of the product, aesthetic appeal and so on.

Within the same social group, the elderly buyer generally has a greater ceiling for expenditure and, above all, economic stability compared to younger generations. Retirement income, which represents the economic security of the majority of people over the age of 65, is also, in many cases, the main source of steady income for the family group. Instead, younger people can rely on forms of temporary work that often do not allow them to engage in purchases—and life projects—of any real value.

Companies are now turning to the new market for safe and easy-to-use products (so-called "user friendly" products), proposing solutions potentially aimed at very large groups of possible buyers, that is, to all those who are attentive to safety, well-being and ease of use. It can be said that the exponential growth of the new market represented by the elderly has overcome the traditional subdivision between "normal" products and "disabled" products, which, though there are many exceptions, has sadly characterised production in this sector until relatively recently.[26]

---

[26]The legislation for the elimination of architectural barriers and the financing of protected structures for the third generation (see note 19 of this chapter) gives rise to a new sector of interest in the late 1980s and early 1990s, for designers, construction companies, furniture manufacturers, aids, assistance equipment and, even if only marginally at the offset, for household products companies (kitchens and appliances, furniture, bathroom furnishings, etc.). Since the 1990s, numerous studies have been carried out in the field of architectural design, along with the construction of buildings and the creation of exceptional design solutions. In the field of Design, research fields and professional specialisations are developing in the fields of furnishings, healthcare products and assistance, and fully accessible household and daily use products. The companies that are beginning to produce furniture and equipment for residential facilities for the elderly are equally numerous, and also specialise in "home furnishings for the elderly".

In many cases, these are specialised products (furnishings, aids and equipment for assistance), that are directed—unequivocally—to the defined market (with a very unfortunate association of terms) "for the elderly and disabled".

Accessible design is aimed at the sector of disabled people with accessible kitchens, equipped with aids and facilities for the accessibility of worktops and containers, equipped bathrooms that comply with the provisions of law 13, are equipped with support handles, maneuvering spaces, and sanitary facilities that are suitable (in terms of position, height, available space) for the movement of the wheelchair. Similarly, design studies and experiments are being developed to meet the needs of the blind and visually impaired. Without going into the merits of the "history of accessibility", we can say that numerous, and, in many cases, considerable design solutions, technologies and experiments have been developed and trialed in this sector. The objectives, however, remain directed towards the "specialisation of the design" and towards a substantial separation between the universe of "normal" design (for people without difficulties with normal needs and normal expectations) and that of

Furthermore, it is not only the elderly who have caught the interest of companies. Products for early infancy, furniture for children and clothes to "play and move safely" are aimed at a market that is willing to spend more on this than in other sectors, and one that is acutely aware of what safety and reliability requirements they expect for their children, who are often the "only children" of families of three and often four generations.

Moreover, attention to the safety and usability of products is increasingly widespread in the market, where buyers are attentive to their purchases and increasingly aware of their needs. The economic crisis of recent years has made this trend even more evident: household appliances and home furnishings, but also office equipment, TVs, music players, landline and mobile phones, must ensure security and ease of use for themselves and their families, reduction of energy consumption, reliability over time.

9.6 Design for Inclusion

As we have seen, the jumping-off point for any evaluation and/or planning intervention for Ergonomics is identifying and interpreting the needs of the people who actually enter—or can enter—into a relationship with the product or system, within a specific usage context, and the translation of these needs into the design requirements.

Both the definitions and the operating principles of Design For All, which indicate "Along the same lines there are both the definitions and the operating principles of the Design For All which indicate as the objective of the project"[27] as the objectives for design, are similar.

This approach allows—and requires—us to shift attention *from the specific* (traditionally defined on the basis of age, specific physical or cognitive characteristics or specific limitations) to a design approach based on the identification of *needs profiles*, that is, a necessity but also a collection of expectations, attitudes and desires related to the relationship with a specific product or system.

In other words, the traditional evaluation of the "skill levels" or "inability" of users (which were rigid by necessity and often insufficient to define the different and multiple realities of the human condition) is replaced by a design approach that is

---

planning "for the elderly and disabled", who are characterised by specific needs for accessibility and security.

[27]Definition taken from the website of the IIDD, Italian Institute for Design and Disability. The same website states that the objectives of Design for All are achieved through:

- the design of products, services and applications that can be readily used by most users, without having to make any changes;
- the design of products that are easily adaptable to different users (for example, by changing the user interface);
- the use of standardised interfaces that are compatible with specialised equipment (e.g. assistant technologies).

centered on the reality of needs and expectations shared by different people (in terms of age, level of autonomy and health conditions), and can relate to each individual at different stages of his life.

In operational terms, it is a matter of passing from a design "for disabled people", "for elderly", "for the blind" etc., to a design aimed at ensuring and/or enhancing the usability and handling of products, the simplicity and comprehensibility of their usage methods, and the visibility and legibility of the components, written indications, symbols and icons used to understand them, etc.

This conceptual and methodological shift from design "for user profiles" and "for levels of skill or disability" to design aimed at responding to the needs and expectations of users highlights the contribution that Ergonomics for design can offer to the Design For All approach and Design for inclusion.

The central aspect of this step is to assume the needs, expectations and desires of people with reduced capacity (motor, perceptive, cognitive) are a normal component of the collection of needs that the design must satisfy; we can recognise, in these needs—permanently or temporary—people who are diverse in terms of age, health conditions, economic possibilities, cultural level.

To make daily use products that are completely usable, safe and pleasant to use, therefore, means not adapting products that were originally designed for optimal capacity and "average" needs and making them accessible to people with reduced capacity (of movement, perception, etc.) but—conversely—restructuring the design process, starting from the highest level of needs and expectations, and from the lower level of ability, to create products that are pleasant and easy to use for everyone.

A classic example, which is usually posed to students in anthropometry classes, applied to the design of furniture and everyday products is the criterion of "maximum accessibility", which can be used to define the height of shelves or kitchen wall units: in this case, the hand grip height of the 5th percentile of the female population (i.e. the grip height of the smallest person of height, which also corresponds to the grip height from the sitting position) is used as a reference, in order to guarantee accessibility in height to this user profile and to anyone taller. Conversely, to make children and dangerous objects inaccessible to objects, for example, in an elementary school or at home, reference will be made to the 95° hand grip height of children in this age group, storing all dangerous items above this height (see Chap. 10, "Elements of anthropometry").

Concrete examples of a fully inclusive approach to design are any solutions aimed at crafting products based on the specific needs of people with reduced capabilities, which can be easily used by any user: designing traditional telephones with keys that are sized to be easily identified by sight and touch and that can be easily used without pressing two or more at the same time; the use of writing that can be read by shortsighted or presbyopic people; attention to the correct ratio of colour and contrast between the figure (for example, symbols and warnings written on the outside of household appliances, televisions, telephones, etc.) and the background they are placed on; the use of shapes that are easy to handle and manipulate for handles and systems used for opening and the provision of doors and walkways that are accessible for people with wheelchairs or walking aids too, etc. These solutions represent design focuses that can respond not only to the needs of specific categories of users, but also simply make it easier, safer and more enjoyable for anyone to use everyday products.

The objective of "maximum inclusion" has also become, as we have seen, a commercial objective and, as such, must actually be achievable. If it is, in fact, clear that the objective of aiming the design at "all users" clashes with production needs, which must be based on containing production costs and on the exact identification of the target audience, companies that turn to the market for the most vulnerable segments of the population will propose safe and easy-to-use products that can meet the needs of clued-in buyers who are attentive to their needs, and, which are, in many cases, adaptable, without difficulties or additional costs, to the needs of people with reduced abilities (motor, visual, hearing, etc.).

The adaptability of products and systems represents a further, central aspect of inclusive design.

If one of the basic principles of Design for All is the design of products that can be used by the maximum possible number of people without the need for complex and/or expensive aids or adaptations, it is clear that we cannot satisfy all the different and specific needs that can derive from equally specific and different individual situations with a single product.

In fact, designing "everything for everyone" means confronting objective technical difficulties or unsustainable costs and/or production times for companies.

In 1998, the results of the TIDE research, published in *Improving the quality of life for european citizen,*[28] was already addressing the feasibility of design solutions oriented towards maximum inclusion. The problem was posed with great lucidity, identifying the feasible production process for products that can be effectively used by very wide groups of users.

Considering the wide variety of people within each generic user group (e.g. elderly, disabled, children), it becomes difficult to design the optimal solution for all possible users and is not feasible for economic reasons (e.g. development and production costs and times), for reasons connected to the effective usability and ease of use of the products. An approach aimed at the actual feasibility of the products and their commercialisation on the market allows us to concentrate on making the products widely available at affordable prices, through the normal channels of consumption".[29]

The way forward is to design products that ensure the maximum level of usability for the maximum number of people possible, and that allow for the insertion of aids, accessories, and/or adjustment systems that allow the product to be adapted over time, to changing the person's physical abilities or adapting it on a case-by-case to meet the needs of different users or different conditions of use.

The problem, therefore, is the precise identification of the real and potential target audience for the product and the range of needs that the design will be able to satisfy.

In other words, it is about defining the "limits of the design response" in a conscious, responsible manner.[30]

---

[28] See, in particular: Janssen and Van der Vegt (1998).

[29] See Placencia Porrero and Ballabio (1988).

[30] See Sect. 10.4 "Limit users".

Some interesting examples are bathroom furnishing systems with different configurations, equipped with accessories that can be inserted according to your needs, thus allowing different adjustments to the system from the basic configuration without aids to movement, to the complete configuration with all the elements necessary to allow wheelchair access (see Fig. 9.3). Similar examples include furnishing systems for kitchens, which provide different configurations of low containers and tall wall-mounted containers according to different needs, from solutions for-maximum accessibility (with the space under the worktop completely free) that can be used with a wheelchair, or, in any case, from a seated position, to solutions to optimise the use of available space, with the entire surface available for use (see Fig. 9.4).

Finally, full adaptability to different needs characterises most touch technology products, specifically in terms of different visual capacities or different usage needs. The touch screens of smartphone, tablets and e-readers, as well as web pages designed with accessibility criteria in mind, allow you to adapt the font size and images to your needs, making readier easier and adaptation to your needs a typical action of our daily life.

Fig. 9.3 Bathroom for All, Studio Quadrato. www.studioquadrato.it

Fig. 9.4 Convivio. Experimental kitchen design by Giulio Iacchetti for the Japanese company Cleanup, 2014

References

Buti LB (2008) Ergonomia olistica, il progetto per la variabilità umana. FrancoAngeli, Milano
Brischetto A (2017) Nuove tecnologie e apprendimento for all. In: Rivista Italiana di Ergonomia, 14/2017
Di Bucchianico G, Kercher P (2016) Advances in design for inclusion. Springer, Switzerland
EIDD, European Institute for Design and Disability, Stockholm Declaration (2004). dfaeurope.eu/what-is-dfa/dfa-documents/the-eidd-stockholm-declaration-2004/
Elton E, Nicolle C (2015) Inclusive design and design for special population. In: Wilson JR, Sharples S (eds) Evaluation of human work. CRC Press, Boca Raton
Enciclopedia Treccani. Online www.treccani.it/enciclopedia
European Commission (2013) Implementing an action plan for design-driven innovation, EU commission staff working document, Brussels
European Commission (2013) Guide to social innovation, European Union, Brussels

Griffo G (2009) La Convenzione delle Nazioni Unite sui diritti delle persone con disabilità. In: Borgnolo G (ed) ICF e Convenzione Onu sui diritti delle persone con disabilità. Nuove prospettive per l'inclusione, Erickson, Trento

ISO 20282-1:2006, Ease of operation of everyday products—part 1: design requirements for context of use and user characteristics

ISO 20282-2:2006, Ease of operation of everyday products—part 2: test method for walk-up-and-use products

ISO 9241/210:2010, Ergonomics of human-system interaction—part 210: human-centred design for interactive systems

ISO/IEC Guide 71:2014, Guide for addressing accessibility in standards

ISTAT, Istituto nazionale di statistica, Rapporto annuale 2017, la situazione del paese, ISTAT 2017. www.istat.it/it/files/2017/05/RapportoAnnuale2017.pdf. ISTAT, Istituto nazionale di statistica, Report 2017: Il futuro demografico del paese. Previsioni della popolazione residente al 2065, ISTAT 2017. www.istat.it/it/files/2017/04/previsioni-demografiche.pdf?title=Il+futuro+demografico+del+Paese+-+26%2Fapr%2F2017+-+Testo+integrale+e+nota+metodologicapdf

Janssen HTJ, Van der Vegt H (1998) Commercial Design For All (DFA). In: Placencia Porrero I, Ballabio E (eds) Improving the quality of life for european citizen. IOS Press, Amsterdam

Leonardi M (2003) ICF. La Classificazione del Funzionamento della Disabilità e della Salute dell'Organizzazione Mondiale della Sanità. Proposte di lavoro e di discussione per l'Italia. www.silsismi.unimi.it/SILSISMI/Indirizzi/Indirizzi_doc/Sostegno/Leonardi271103.pdf

Oxford Dictionary. Online www.oxforddictionaries.com

Placencia Porrero I, Ballabio E (1988) Improving the quality of life for european citizen. IOS Press, Amsterdam

Spadolini MB (ed) (2013) Società Italiana di Ergonomia e Fattori umani. Design for better life, longevità: scenari e strategie. FrancoAngeli, Milano. www.societadiergonomia.it

Steffan IT (ed) (2014) Design for All—The project for everyone. Methods, tools, applications. Maggioli, Rimini

Tosi F (2014) Ergonomics – Design – Design for all. From the evaluation to the design: the creation of a common language, in: Steffan IT (ed) (2014) Design for All — The project for everyone. Methods, tools, applications, Maggioli, Rimini

Tosi F, Rinaldi A (2015) Il Design per l'Home Care. L'approccio Human-Centred Design nel progetto dei dispositivi medici. Didapress, Firenze

Tosi F, Brischetto A (2016) Ambienti di apprendimento 2.0: nuovi scenari progettuali per l'inclusione sociale. In: Rivista Italiana di Ergonomia, special issue 1/2016

UNI 11377-1:2010, Usabilità dei prodotti industriali—Parte 1: Principi generali, termini e definizioni

UNI 11377-2:2010, Usabilità dei prodotti industriali—Parte 2: Metodi e strumenti di intervento

Vocabolario Treccani. Online www.treccani.it/vocabolario

WHO, ICF International Classification of Functioning, Disability and Health. www.who.int/classifications/icf/en/

World Health Organisation Executive Board (2015) Multisectoral action for a life course approach to healthy ageing: draft global strategy and plan of action on ageing and health. World Health Organisation

Part II
The Components of Interaction Between People and Systems

Chapter 10
Elements of Anthropometry

10.1 Introduction

Anthropometry is the science that specifically deals with the measurable features of the human body, that is, its measurements and its physical-dimensional characteristics, through collection and processing statistical data from individuals in different population groups. The data supplied by anthropometry concerns measurements of the main physical parameters of individuals (height, width, circumference, holding and reaching distances, etc.). These are found from a sample of individuals selected to represent the variability of these measurements for a given population. The statistical processing of anthropometric data allows us to identify the minimum and maximum values of these measurements within the population being studied (e.g. the minimum and maximum height for the Italian population aged 19–65), their average value, their frequency and more. Human diversity, as it pertains to individuals or different population groups, derive from the differences in each individual's genetic heritage and the effects of the environment, nutrition, health and life habits. Similarly, anthropometric differences found in various population groups derive from micro- and macro-evolutionary processes, which are determined by genetic mutations and the natural selection of different phenotypes, according to varied natural and artificial environments.[1]

Anthropometric features, and, in particular, dynamic dimensions, are closely connected to the bio-mechanical features of posture and movement and to the characteristics and limitations of muscular exertion. These aspects, which are vital for the analysis and design of any environment and work station, product or piece of equipment intended for human use, are dealt with Chap. 12, in particular as it pertains to the risks deriving from poor posture and excessive or prolonged movements and exertion. Certain aspects pertaining to the bio-mechanical features of the skeleto-muscular system, which are closely linked to dynamic anthropometry, must be referred to in this

---

[1] See Borgognini and Masali (1987, p. 105).

© Springer Nature Switzerland AG 2020
F. Tosi, *Design for Ergonomics*, Springer Series in Design and Innovation 2,
https://doi.org/10.1007/978-3-030-33562-5_10

DIMENSIONS	MEN			WOMEN		
	5°	50°	95°	5°	50°	95°
height	1625	1740	1855	1505	1610	1710
height of the eyes (in feet)	1515	1630	1745	1405	1505	1610
height of the elbows (in feet)	1005	1090	1180	930	1005	1085
height (seated position)	850	910	965	795	850	910
height of the eyes (seated position)	735	790	845	685	740	795
height of the elbows (seated position)	195	245	295	185	235	280
width of the head	145	155	165	135	145	150
maximum holding height (in feet)	1925	2060	2190	1790	1905	2020
maximum holding height (seated position)	1145	1245	1340	1060	1150	1235

The measurements are divided based on gender, age group and the main reference percentiles, and relates to the adult population of the UK.

Fig. 10.1 The main anthropometric measurements of the adult population (19–65). Revised by Pheasant and Haslegrave (2006, p. 245)

chapter, particularly in terms of design problems and intervention criteria relating to movement within work stations (Fig. 10.1).

The data supplied by anthropometry concerns structural or static dimensions, that is, the dimensions of the human body in various static positions, and the functional or dynamic dimensions, that is, the dimensions assumed by the human body when in motion. The former includes, by way of example, dimensions relating to height, circumference of the skull, length of the arms, etc., which are measured in standard static positions (in upright and seated positions). The latter includes the dimensions of the human body when in motion, the shape assumed by the body during movements needed to carry out a given activity and the reachable areas permitted by the body's movement. Anthropometry supplies a potentially boundless amount of information for both static and dynamic dimensions. Once this information has been selected and interpreted, it allows us to define the physical-dimensional requirements of the design and can be used at any stage of an intervention. In fact, in order to define physical-dimensional requirements, it is necessary to identify the useful anthropometric parameters for the design and the target user group and, based on the data available, to select the values assumed by each parameter within that user group. The design may be targeted at a very broad user group, which is, for example, characterised by age or the type of activities performed (for example, Italian children who attend secondary school, or service sector employees in Europe), or, on the other hand, may be limited to more specific features (for example, race-car drivers or children aged 12–15 who compete in swimming meets in Palermo). Once the user group has been defined, the anthropometric data and, in particular, the distribution and frequency uncovered allow us to arrange the dimensions of environments and products that are intended for human use in accordance with users' physical features, movement abilities and the tasks they are required to perform, by eliminating or minimising the need to exert oneself or move in an incorrect manner.

10.1.1 Anthropometric Data for the Design

The use of anthropometric data, which may seem like an apparently simple process of identifying useful data for the design and translating it into design parameters (for example, measurements regarding the height or range of movement of the arms in order to define the maximum height of shelves), creates certain problems in terms of correctly selecting interpreting the available data and their use for design issues.

Availability of Data
The anthropometric data reported in the manuals derives from surveys that only coincide with the target user group for the design on very rare occasions and is typically processed on the basis of surveys performed in the military field. The data available in the manuals for designers, or used for specific address standards, are typically presented in tables. These show the measurements relating to the main anthropometric parameters, which are found within a given population (for example, measurements regarding the height or length of the arms in the Italian population aged 18–60, etc.), or in the form of graphics that display the same data regar-ding the schematic representations of the human body.

Selection of Data
Particular attention must be devoted to the selection of anthropometric data, in accordance with the specific design issue. In fact, the quantity of anthropometric data, and the extreme variability with which the measurements regarding anthropometric parameters are present within a given population group, requires us to identify the data that is useful for the design and the target population group for the design. For example, the design of a desk for an Italian elementary school requires the use of data regarding the height of elbows and eyes in a seated position and reference to the measurements uncovered in the Italian population aged 6–10.

Use of Data
Once the useful parameters have been identified, the variation of the corresponding data will supply the references needed to define the dimensional requirements for the design, in order to respond to the motion needs of the maximum number of users under consideration. The design of a door or a corridor will require the use of data regarding the height and bulkiness of the human figure and, within these, reference to the dimensions of individuals with greater height and bulk. The design of a handle will require the use of data regarding the dimensions and gripping capacity of the hand and, within these, reference to more limited dimensions and movement ability

(for example, the dimensions of children's hands or reduced gripping capacity, which may derive from joint ailments or the need to wear gloves).

In all cases, the use of anthropometric data requires both the correct setting preparation of the design problem and the subsequent selection and processing of available data, particularly:

- the identification of the necessary anthropometric data for defining the design's physical-dimensional requirements (for example, in the case of a handle, the reference parameters are the dimensions and movement capacity of the hand);
- the identification of the target user group (if the handle is intended for use in elementary schools, the user group refers to children aged 5–11);
- the selection of useful data for the design, which refers to the anthropometric features of the specific user group (the dimensions and movement capacity of children's hands aged 5–11);
- the ways in which the anthropometric data selected in this way allow us to define the dimensional requirements of the design (since the dimensions of the hands in this case are between 5 and 12 cm, the handle must have a diameter between 2 and 4 cm);
- the identification of limit users (see Sect. 10.4), that is, users characterised by special dimensional features and/or movement capacity.

The use of anthropometric data requires some knowledge of the basis for processing this data and, in particular, the criteria used to present the variability and frequency distribution of the data uncovered for a given population. The following paragraphs examine the aspects that are most closely linked to the interpretation of anthropometric data and the correct use and interpretation of the useful data for the design process.

10.2   Sources of Data

Anthropometric surveys are based on systematic measurements that are carried out on selected groups or samples of the population, using conventionally agreed-upon measurement tools and methods. The data collected is processed in a statistical manner and presented on the basis of the percentage variation found within the population in question. The collection and systematisation of data regarding the measurements of the human body requires the availability of user groups that are sufficiently broad and appropriately selected. As a result, these are complex procedures that require the involvement of expert operators and the use of ample human and economic resources. For these reasons, the majority of the available data comes from military sources and is processed on the basis of the measurements of young people beginning their military service. The reason is simple: firstly, this field was where it was first necessary to understand the specific anthropometric measurements of staff, in order to solve the problem of correct equipment; furthermore, these structures have a practically limitless sample on which to carry out research, as well as the available of significant funds to perform them.[2]

The majority of the available data, then, concerns the anthropometric features of the male population, aged approximately 18–26, and comes primarily from the military structures in Anglo-Saxon countries.

Anthropometric data that is intended to be used by designers and shown in tables and/or graphic diagrams are taken from surveys that were carried out on representative samples of user groups, which have been identified based on specific features (generally, the geographic area or age group one belongs to, etc.), or which are adapted, using the appropriate increase of decrease coefficients, starting from the data regarding groups of diverse populations, by making "adjustments" to the available data.

In the case of the female and elderly population, where sufficiently broad and systematic surveys are not available, the data is obtained by applying a decrease coefficient to the data for the male population.

We must highlight that the anthropometric data shown in manuals generally refers to the dimensions and movement capacity of undressed people, or those wearing minimal clothing, with no hats and bare feet.

Therefore, it is necessary to consider the bulk of clothing, which represents a variable factor both for the dimensions of the human body (for example, the width of the shoulders, which varies based on clothing worn, and height, since almost all of us wear shoes, which can also significantly increase out height) and for movement capacity, since the extension and agility of our movements may be diminished, for example, by the need to wear particular work clothing or, more simply, winter clothing or a pair of shoes with high heels. The movement of the hands and fingers may also be limited by the wearing of gloves.

In order to account for this variability, corrections to the available anthropometric data shall be applied: for example, Pheasant and Haslegrave (2006, p. 43) indicate a height correction—25 mm for men and 45 mm for women—that considers the type of shoes typically worn in public places for formal or semi-formal scenarios.

10.3 Presentation of Anthropometric Data

Anthropometric measurements concern the bodily dimensions of the human body in standard static positions and in motion and supplies a potentially limitless amount of dimensional data about each part of the body. For each feature considered (for example, height), the data found within a given population reveals variable values that can be shown on a graph, using a series of histograms that are made by depicting the height values on the abscissa and their frequency on the ordinate. By joining the top of the histograms with a continuous line, we obtain a distribution model for the feature; in the example of Fig. 10.2, we discover a normal distribution, that is, a curve

---

[2]See Panero and Zelnik (1983).

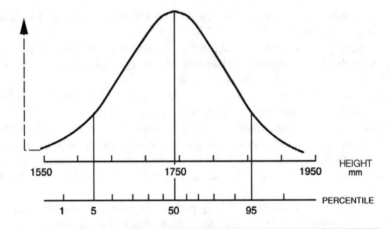

The curve obtained through approximation has a normal Gaussian distribution (the curve is symmetrical in comparison to the central value) and has a similar trend for all of the
measurements of the human body found within sufficiently broad population groups.
The data reported here refers to the adult population in the UK.

Fig. 10.2 Distribution frequency of measurements relating to the height of the adult male population. *Source* Pheasant and Haslegrave (2006, p. 19)

that has a symmetrical trend compared to the central value. The highest point of the
curve shows the average value of the feature in question, the most frequent value
found within the population in question (the trend) and, finally, the median value. The
measurements that are found can be divided into 100 parts. This corresponds to 99
averages (percentiles), which indicate the portion of the population with a parameter
value equal to, or less than, what is being considered.

The 5th percentile for height, for example, indicates that 95% of the population
is taller and that only 5% are shorter than that specific value. Conversely, the 90th
percentile indicates that 10% of the population is taller and 90% is shorter. Graphs
that represent the frequency of the data for the majority of human dimensions, which
can be found within a population with a normal distribution, assume a normal or
Gaussian trend, similar to what is shown in Fig. 10.2.

The greatest amount of data is found in the central part of the curve (in the case of
the figure, measurements of 175–180 cm are the most frequently found data for the
male population); on the other hand, the two extreme ends of the curve display a small
number of measurements, which correspond to small percentages of the population.

Observing the frequency of data related to the dimensions of the human body,
and the definition of percentiles as a percentage of data that is greater than or less
than a determined measure founding within a given population, make it clear that
it is not useful to refer to the average dimensions of the population. The average
dimension for each feature under consideration is the measurement that corresponds
to the 50th percentile, that is, the measurement where 50% of the population has
lesser dimensions and 50% of the population has greater dimensions.

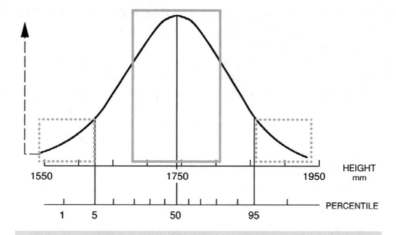

The greatest amount of data is found in the central part of the curve (in the case of hei-ght, measurements of 165–185 cm are the most frequently found data); on the other hand, the two extreme ends of the curve display a small number of measurements, which correspond to small percentages of the population.

Fig. 10.3 Distribution frequency of measurements relating to the height of the adult male population

A design based on the dimensions that correspond to the 50th percentile, therefore, exclude the majority of users (that is, the 50% with lesser dimensions and the 50% with greater dimensions).

When we consider that the greater frequency of measurements are found in the central part of the curve, and that nearly half of the population have measurements that are close to those that correspond to the 50th percentile, designing based on the average height means considering approximately 50% of users and excluding 50% of others. For example, in the case of height reported in Fig. 10.3, the height of the 50th percentile corresponds to 175 cm and nearly half of the data—from the 25th to the 75th percentile—have a height of 170–180 cm.[3]

We must also consider that the anthropometric percentiles refer to a single dimension of the human body (for example, height or the length of the arms) and that the graphic representations for men and women belonging to a certain percentile are abstractions, which do not directly correspond to real men and women.

---

[3]On a practical level, designing the width of a chair based on the width of the hips of an average user means excluding 50% of possible users, that is, all those with a body bulk greater than those in the 50th percentile, or making them uncomfortable; similarly, designing a shelf or kitchen drawers based on the height that can be reached by individuals in the 50th percentile means forcing 50% of people to stand on tip-toe or climb on a step ladder or stool, that is, all of those people with a height and range of motion for the arms that is lower than the average.

In effect, there is no case of a person who shares all of the bodily dimensions found in a single percentile, but, for each of us, individual dimensions of the body correspond to a given percentile.[4]

As a result, anthropometric data must not be used in a mechanical manner; instead, it requires the selection and interpretation of data for a specific design problem.

If used correctly, the data provided by anthropometry is a design resource that allows us to define the dimensional requirements of the design and to evaluate, based on certain references, the percentage of users who can use the environment or the product being designed in a safe and effortless manner (for example, the height of a shelf, which is defined on the basis of the height of women belonging to the 5th percentile, allows us to calculate that 95% of people—who will be taller—may comfortably use that shelf). This, in turn, allows us to respond to the needs of the maximum possible number of users.

10.4   Limit Users

With reference to the curve that represents the frequency with which the majority of the bodily dimensions are found within a given population, limit users are any individuals who possess one or more anthropometric features that are found at the extreme ends of the curve (Fig. 10.3).

Depending on the design problem, observing the frequency with which the data regarding the measurements and physical capabilities of the human body occur allows us to identify thresholds. It is impossible or impractical to respond to users' needs above or below these thresholds, that is, to establish the physical characteristics of limit users for the design on a case-by-case basis.

In the case of domestic or office furnishings—where the height of shelves and containers must be designed by considering the height of users and, in particular, the height they can reach solely by rotating their arms—the threshold will be users who are shorter, that is, beneath the 5th percentile. The number of such individuals would not justify the production costs, or the expenses that derive from not using the space available. For these cases, any many others like them, the adoption of very complex design solutions, excessive production difficulties and an excessive raise in produce costs would be required to consider the less-frequent anthropometric characteristics.

The conventionally agreed-upon compromise is not to refer to the anthropometric characteristics of all users, but those of the "maximum possible percentage of users"—that is, the maximum percentage of users that can be taken into account with the available resources—by referring to the characteristics found between the 5th and the 95th percentile of the population group or users in question.

---

[4]One common experience concerns the dimensions of young children, who can, for example, correspond to the 70th percentile for height, the 80th for the circumference of the skull, the 75th for weight and so on. In fact, the regular nature of their growth is not evaluated based on the exact correspondence of all their bodily measurements to a single reference percentile, but based on the balance between the various parameters considered.

Fig. 10.4 Italo, smart carriage. Seats provided and the space available for each seat *Source* italo-spa.italotreno.it

The reference to the "maximum possible percentage of users" and the identification of the user threshold imposed by the available resources does not only apply to the dimensional aspects of the design but, more generally, to any case where it is possible—or necessary—to refer to the percentage weight of the potential users' needs. Identifying the limit users also allows us to define the threshold within which the adopted solutions are able to ensure sufficient levels of accessibility and safety. For example, referring to the bodily dimensions of the 95th percentile allows us to define the maximum dimensions of openings and corridors, in order to prevent access to dangerous areas or potential sources of risk. Similarly, calculating the force limits required for opening doors and windows, etc. must consider the capacities of the 5th percentile and, more generally, the capabilities of the weakest users.[5]

The reference percentiles may also change based on the goals of each intervention. The most classic example—and probably the easiest—is the case of train seats in the three main service categories: first and second class on high-speed trains and high-frequency trains. In addition to the level of service offered to travellers, what varies in the three cases is the number of seats provided, in spite of the equally sized carriages, and the space available for each seat.

In high-speed train carriages, the greater amount of space available allows us to take the reference width from the 90th and 80th percentiles for men, for first and second class respectively (the same is true for the legroom provided). In second class carriages on high-frequency trains, the reference is probably the 50th percentile (Figs. 10.4, 10.5 and 10.6).

---

[5]The UNI EN 1005/3:2009 standard, "Safety of machinery. Human physical performance. Recommended Force Limits For Machinery Operation", recommends referring to the values for force capacity that correspond to the 15th percentile for the use of professional machinery (with reference to users selected based on specific skills and characteristics) and values corresponding to the 1st percentile for domestic appliances.

Fig. 10.5 Italo, first-class carriage. Seats provided and the space available for each seat *Source* italospa.italotreno.it

Fig. 10.6 Trenitalia, high-frequency train. Seats provided and the space available for each seat

10.5   Human Variability

As with all human characteristics, features related to bodily dimensions differ from person to person and vary for each individual through their life. This is the result of processes like growing and ageing, diet, activities performed, etc. Anthropometric characteristics also vary for each of us due to temporary events, such as pregnancy, a period of major weight gain or loss. In the short and long term, these can change not only the bodily dimensions of an individual, but also their capability to move and exert effort.

The variability with which anthropometric data occurs within a given population group may be linked to certain factors, which are, in order of importance, gender, age, geographical provenance, ethnicity, occupation, etc. It is also possible to note a significant variation in anthropometric data within a given population group depending on generations, that is, the generational trend (secular trend) that modifies physical characteristics: a notable example is the increase in average height that has been seen in Western countries over the last 60 years, thanks to the improvement in nutrition and living conditions.

Human variability in anthropometry is studied on a statistical basis, starting from the possibility of applying mathematical models, which are based on the probability of certain events occurring in a specific way and with a specific frequency, to human beings and data about their measurable physical features. The methods for collecting and statistically analysing anthropometrical data concerning dimensional and morphological characteristics is based on samples of individuals that represent a given population. These are identified based on one or more classification criteria used in the research being performed. In anthropometry, a statistical population is represented by a collection of subjects who may be identified based on at least one of the following characteristics[6]:

- characteristics deriving from genetic links (for example, belonging to an isolated population where endogamous marriages have occurred for generations);
- belonging to a homogeneous group through sex, age group, etc.;
- geographical provenance (for example, the population of a specific country or a given geographic area);
- cultural or religious affiliation (for example, those who practice a certain religion or who belong to a specific ethnic group, regardless of their geographical provenance, etc.);
- profession (for example, doctors, those working in the service sector, etc.);
- affiliation based on the performance of a given activity (for example, those who play a specific sport).

These classification criteria can obviously be used to define broad population groups (for example, "the residential population of Italy"), or to identify limited and/or characteristic groups (for example, "children aged 14–19, who are residents of Milan and enrolled in secondary school"). In terms of design, referring to anthropometric data allows us to define the physical-dimensional requirements of the design (that is, the height, depth, space for movement, etc.), based on the characteristics and needs of the population group—or, in this case, the user group—it is aimed at, by considering the variability and statistical distribution of the anthropometric data within said population group (Figs. 10.7 and 10.8).

Once the target population group for the design has been identified, it is then possible to identify the data, which is divided based on the sub-groupings that comprise

---

[6]See Borgognini and Masali (1987, p. 105), and the UNI EN ISO 15535:2012 standard, "General requirements for establishing anthropometric databases".

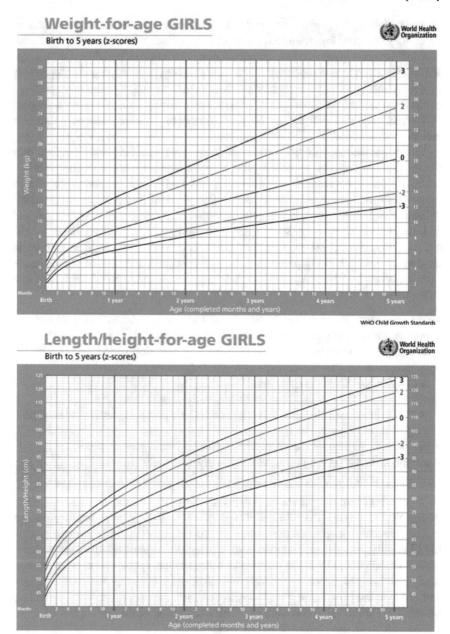

Fig. 10.7 Growth curves for weight and height in girls aged 0–5, according to standards from the World Health Organisation

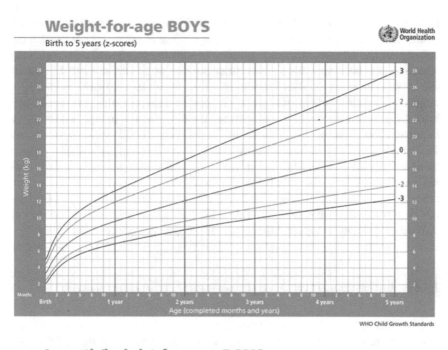

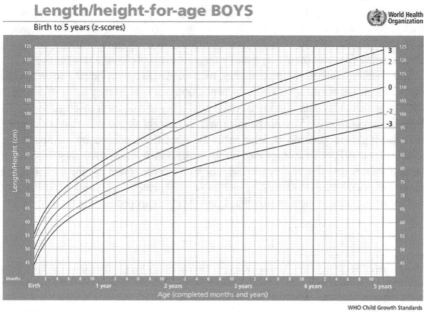

Fig. 10.8 Growth curves for weight and in boys aged 0–5, according to standards from the World Health Organisation

it. These can also be defined on the basis of gender or age, etc., based on more specific characteristics, such as playing a given sport. For example, by interpreting the anthropometric data for the adult population of a given country, and considering the measurements uncovered for a given parameter and percentile, notable differences can be seen between the data for both genders, given that men have, on average, larger bodily measurements (weight, height, muscle mass, etc.) than woman. Significant differences can also be found between the different age groups. These derive both from modifications linked to growth and ageing and generational variations. In the early years of life, bodily dimensions change more rapidly and may present more striking differences from individual to individual than would be noticed during adulthood. This is also true for proportions[7] (Figs. 10.5 and 10.6). The growth process is of course conditioned by living conditions, diet and the type of activities performed. Similarly, these same factors condition average life expectancy of different population groups and the effects of the ageing process on individuals.[8]

10.6 Anthropometric Measurements: Static Dimensions

The majority of the anthropometric measurements relating to static dimensions refer to measurements of an immobile person in two standard positions: the upright position, in which the subject is standing still, looking straight ahead with their shoulders relaxed and their arms at their side, and the seated position, in which the subject is sitting upright and still on a flat, horizontal surface, looking straight ahead, with their arms hanging freely at their side and the forearms in a horizontal position.

The postures used for anthropometric measurements are obviously conventional references, as people rarely assume these positions and, when they do, only for short periods. In any case, data relating to static postures must be used by considering the dynamic aspects and the fact that people move frequently, thus continuously changing and adapting their position in relation to the conditions offered by the space and objects they use. For all anthropometric measurements, it is also necessary to consider the corrections that derive from individual applications, such as, for example, the aforementioned increases caused by clothing or any protective elements that follow.[9] The anthropometric data is outlined in Chap. 11, "Anthropometric reference data".

---

[7]The length of the torso, which is approximately 50% of an adult's height, represents 70% of the body's entire length at birth. The growth process gradually slows down in the years that follow, finally stopping at around 18 for men and 16 for women.

[8]Ageing is different from person to person and, in addition to health conditions and possible treatments, largely depends on the lifestyle and living conditions experienced during the entire lifespan. In terms of anthropometric data, ageing is characterised by a decrease in muscle tone and resistance and a decrease in height associated with postural changes.

[9]The definitions and criteria for graphic representation are taken from: Panero and Zelnik (1983), Pheasant and Haslegrave (2006), UNI EN ISO 7250/1:2017 (2017) standard, "Basic human body measurements for technological design—Part 1: Body measurement definitions and landmarks".

Diagram 10A—Static Dimensions Upright Position

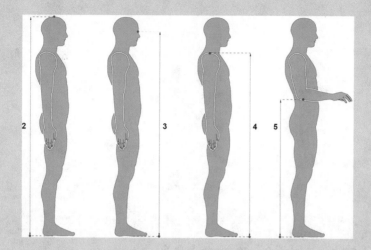

1. **Weight (body mass): total mass (weight of the body in KG). Applications: necessary to determine the weight of any element that must be borne by the person's weight.**
2. **Height (height of the body): vertical distance from the floor to the top of the head.**

    Applications: Necessary to determine the minimum height of corridors for people and any suspended elements.
    Corrections: shoes (25 mm for men, 45 mm for women), hats, helmets, boots and other protective clothing for the head and feet.

3. **Height of the eyes: vertical distance from the floor to the outer corner of the eye (canthus).[10] Applications: Necessary to define the height of the eye line.**
    Corrections: shoes. See point 2.
4. **Height of the shoulders (acromial height): vertical distance from the floor to the acromion.[11] Applications: Used to determine the height of worktops used in the upright position.**
    Corrections: shoes.
5. **Height of the elbows (radial height): vertical distance from the floor to the highest point of the radial bone (can be seen on the outer surface of the elbow); this is measured with the arm extended along the body. Applications: Used to determine the height of worktops used in the upright position.**
    Corrections: shoes. See point 2.

Diagram 10B—Static Dimensions Upright Position

6. **Width of the shoulders (bi-acromial width)**[12]**: distance between the right acromion and the left acromion.**
    Applications: Necessary to define the bulk of the individual, their size and the layout of seats.
    Corrections: 100 mm for light gear, 40 mm for heavy gear.

7. **Width of the shoulders (bi-deltoid width): distance between the external limits of the right and left deltoid muscles.**
    Applications: Necessary to define the bulk of the individual, their size and the layout of seats.
    Corrections: 100 mm for light clothing, 40 mm for heavy clothing.

8. **Width of the elbows: maximum horizontal distance between the lateral surfaces of the elbow region.**
    Applications: necessary to define the bulk of the individual, their size and the layout of seats.
    Corrections: see point 7.

---

[10]The canthus is the lateral corner of the eye. It is formed by the meeting of the upper and lower eyelids.

[11]The acromion is the prominent scapula bone that meets the clavicle.

[12]See Footnote 11.

Diagram 10C—Static Dimensions Seated Position

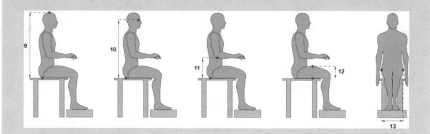

9. **Height in a seated position: vertical distance from the horizontal surface of the seat to the highest point of the body.**

   Applications: Necessary to define the minimum admissible heights above the seat surface and the minimum heights at which to position lamps and general suspended elements.
   Corrections: 10 mm for heavy gear, 25 mm for a hat and 35 mm for protective helmets.

10. Height of the eyes in a seated position: vertical distance from the horizontal surface of the seat to the external corner of the eyes (canthus).[13]
    Applications: Necessary to define the height of the eye line.
    Corrections: 10 mm for heavy gear.

11. Height of the elbows in a seated position: vertical distance from the horizontal surface of the seat to the lowest point of the elbow, bent at 90° with a horizontal arm.
    Applications: Necessary to define the height of tables, desks, surface and general work tools, with respect to the height of the seat.
    Corrections: gear (increase in height on the seat).

12. Thickness of the thigh (in a seated position): vertical distance from the horizontal surface of the seat to the highest point of the thigh.
    Applications: Necessary to define the free space beneath tables and work surfaces, with respect to the seat.
    Corrections: 100 mm for light gear, 25–50 mm for medium or heavy gear.

13. Width of the hips: width of the body, measured along the widest point of the hips when seated.
    Applications: necessary to define the bulk of the individual, their size and the layout of seats.
    Corrections: see point 12.

---

[13] See Footnote 10.

Diagram 10D—Special Static Dimensions

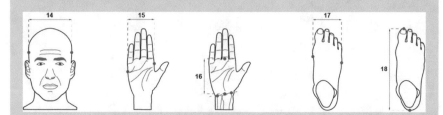

14. Width of the head: maximum width of the head above the ears. Applications: necessary to define the space needed for the passage or support of the item of clothing.
    Corrections: hats, helmets and any gear or protective element of the clothing.
15. Width of the hand: maximum width of the palm of the hand. Applications: necessary to define the space needed for inserting, supporting and gripping with the hand.
    Corrections: gloves (up to 25 mm or more) and protective gear for the hands.
16. Length of the hand: perpendicular distance between the end of the middle finger and the styloid.
    Applications: necessary to define the space needed for inserting, supporting and gripping with the hand.
    Corrections: gloves (up to 25 mm or more) and protective gear for the hands.
17. Width of the foot: maximum distance between the medial and lateral surfaces of the foot. This is measured perpendicularly along the longitudinal axis of the foot. Applications: necessary to define the space needed for inserting and supported the hand.
    Corrections: shoes and protective gear for the feet.
18. Length of the foot: maximum distance between the back of the heel and the tip of the longest toe (first or second). This is measured in parallel along the longitudinal axis of the foot.
    Applications: necessary to define the space needed for inserting and supported the hand.
    Corrections: shoes and protective gear for the feet.

Diagram 10E—Gripping Distances[14]

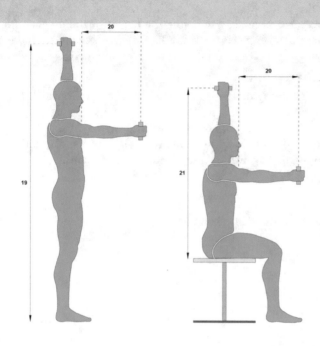

PARAMETERS	men (19–65)			women (19–65)		
	5°	50°	95°	5°	50°	95°
19. Maximum height of vertical hand grip (upright position)	1925	2060	2190	1790	1905	2020
20. Maximum height of vertical hand grip (seated position)	610	665	715	555	600	650
21. 21. Frontal gripping distance of the hand (normal gripping height)	1145	1245	1340	1060	1150	1235

19. **Maximum height of vertical hand grip (upright position): distance between the floor and the centre of a narrow cylindrical rod, which is held in the palm of the hand and carried at the maximum height above the head. Applications: necessary to define the height of the objects that the person is able to grasp with the hand when extending the arm as much as possible. Corrections: increase of 25–5 mm or more for shoes;**
20. **Maximum height of vertical hand grip (seated position): distance between the seat and the centre of a narrow cylindrical rod, which**

is held in the palm of the hand and carried at the maximum height above the head.

Applications: necessary to define the height of the objects that the person is able to grasp with the hand when extending the arm as much as possible. Corrections: seated position: increase of 10 mm for gear.

21. **Frontal gripping distance of the hand:** distance between the outer surface of the shoulders and the centre of a narrow cylindrical rod, which is held in the palm of the hand and carried at shoulder height. Applications: necessary to define the height of the objects that the person is able to grasp by merely moving their arms.

---

[14]See Pheasant and Haslegrave (2006, pp. 46 and 44) and Chengalur et al. (2004, pp. 52–53). The data refers to the English population.

Diagram 10F—Dimensions of the Hand[15]

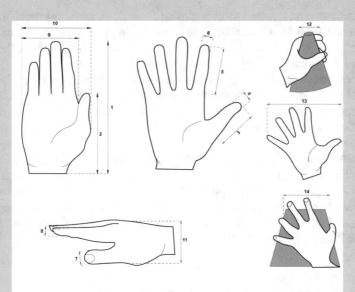

PARAMETERS	men (19-65)			women (19-65)		
	5°	50°	95°	5°	50°	95°
1. length of the hand	173	189	205	159	174	189
2. length of the palm	98	107	116	89	97	105
3. length of the thumb	44	51	58	40	47	53
4. thickness of the thumb	19	22	24	15	18	20
5. length of the index finger	64	72	79	60	67	74
6. thickness of the index finger	19	21	23	16	18	20
7. width of the thumb	20	23	26	17	19	21
8. width of the index finger	17	19	21	14	16	18
9. width of the palm at the base of the fingers	78	87	95	69	76	83
10. maximum width of the palm	97	105	114	84	92	99
11. thickness of the hand (to the thumb)	44	51	58	40	45	50
12. maximum gripping diameter	45	52	59	43	48	53
13. maximum extension of the hand	178	206	234	165	190	215
14. maximum functional extension	122	142	162	109	127	145

---

[15] See Pheasant and Haslegrave (2006, pp. 144–145) and Chengalur et al. (2004, pp. 52–53).

10.7 Anthropometric Measurements: Dynamic Dimensions

The anthropometric measurements regarding dynamic dimensions refer to measurements connected to the movement of the human body, in particular, the space needed to move the body in various positions and reachable areas, that is, the collection of distances that can be reached by moving the body and its parts.

The various positions the human body can assume and the distances that can be reached by individual parts of the body are obtained through sequences of movements that define the space for dynamic movement and reachable areas. These are depicted in graphic form by the dimensional coordinates of the space that the person occupies when moving.

While static reachability refers to the body when immobile and balanced, dynamic reachability may change based on factors that modify that balance, such as, for example, weight or unstable support. Conversely, other factors may increase the reachable areas for the subject, such as, for example, an increase in the support base for feet, and others, such as the presence of an obstacle behind the shoulders, may diminish it (Figs. 10.9 and 10.10).

10.7.1 Space for Movement and Reachable Areas

Space for movement (clearance) refers to the space needed by the human body to easily perform the movements required by a given activity. To define space for movement, it is necessary to examine the bulk of the body and the posture requirement of individual parts of the body for the movement. Dynamic reachability represents the collection of distances that can be reached by the human body through movements and may be described through the dimensional requirements of working postures, that is, the space occupied by the person during the movements needed to perform a given activity (Fig. 10.11).

Typically, reachable areas (zones of convenient reach) are areas that can be reached easily, that is, through movements requiring little effort. In this case, the range of motion for the arms also defines the areas that can be reached through movement. The reachable areas—and zones of convenient reach—are graphically represented through the arches described by the hand through the movement of the arms, and refers to the minimum and maximum dimensions of these arches, that is, the dimension for women in the 5th percentile and men in the 95th percentile.

10.8 Definition of Dimensional Requirements

The correct sizing of spaces and all of the elements that are physically employed by people depends on the functionality and efficiency with which one can carry out

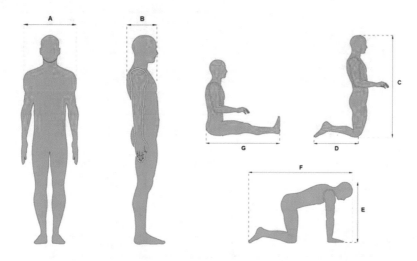

PARAMETERS	men (19-65)			women (19-65)		
	5°	50°	95°	5°	50°	95°
A. maximum width of the body	480	530	580	355	420	485
B. maximum depth of the body	255	290	330	225	275	325
C. height in a kneeling position	1210	1295	1380	1130	1205	1285
D. length of the legs and feet when kneeling	620	685	750	575	630	685
E. height of the head when hands and knees are placed on the floor	655	715	775	605	660	715
F. length of the body when hands and knees are placed on the floor	1215	1340	1465	1130	1240	1350
G. length of the legs extended on the floor when seated	985	1070	1160	875	965	1055

Fig. 10.9 Space occupied by the human body in various positions. Revised by Pheasant and Haslegrave (2006, p. 86) and Chengalur et al. (2004, pp. 52–53)

most daily activities. It also depends on the conditions of safety, well-being and discomfort in which the required physical tasks are, or will be, carried out. Ensuring the dimensional accessibility of spaces, furnishings and equipment by the maximum number of users means not only guarantee usability, but also raising the level of safety with which they can be used, by eliminating sources of risks caused by the need to behave in risky ways or to perform incorrect or excessive exertions or movements.

The dimensional requirements of environments, equipment and products are among the most widely discussed topics in both design manuals and legislation and

position	min. dimensions	
	vertical	horizontal
1	2030	760
2	2030	1020
3	1220	1170
4	460	2430

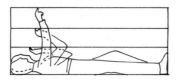

position	min. dimensions	
	vertical	horizontal
1	460	1930
2	610	1930
3	810	1930

Fig. 10.10 Space for movement in the various positions. Revised by Chengalur et al. (2003, p. 275)

standards pertaining to safety and accessibility. In particular, the dimensional requirements needed to ensure the accessibility of spaces and furnishing elements by those with physical disabilities is one of the sectors where a broad and consolidated framework of possible design solutions have been provided, albeit with some limitations. The solutions come from the provisions of law[16] and the standards and references provided by design manuals. The limitations pertain to the dimensional parameters used in legislative and standards contexts. These primarily refer to anthropometric characteristics and standard movement capacity (in this case, the bulk and space to manoeuvre of wheelchairs and the "normal" movement capacity of the arms), which lead to the exclusion of the needs of other user profiles, for whom the mechanical adoption of the prescribed or advised solutions is totally insufficient.[17]

The applicable technical standards regarding Ergonomics and the Safety of machinery supply both general ergonomic intervention principles and the parameters needed for the sizing and layout of environments and equipment intended for

---

[16]The Italian provisions regarding accessibility and elimination of architectural barriers are contained in Law. 13/1989 of Legislative Decree 236/1989, in Presidential Decree 503/1996 and the guidelines for removing architectural barriers in places of cultural interest, published in the Official Gazette no. 114 of 16 May 2008. See also Chap. 9, "Ergonomics and Design for All: Design for Inclusion".

[17]This is the case for the visually impaired, who cannot recognise, by means of a stick or tactile device, the presence of small differences in height or connecting ramps between roads and footpaths. These are, on the contrary, allowed or recommended by the legislation regarding the elimination of architectural barriers. The same is true for people who use walking sticks, crutches of similar aids, for whom the manoeuvrable space needed for wheelchairs may be insufficient or inappropriate. The design instructions and guidelines also refer to standardised dimensions and space for movement on bodily dimensions and the movement capacity of average men, without considering, in any way, the variability found within the anthropometric characteristics.

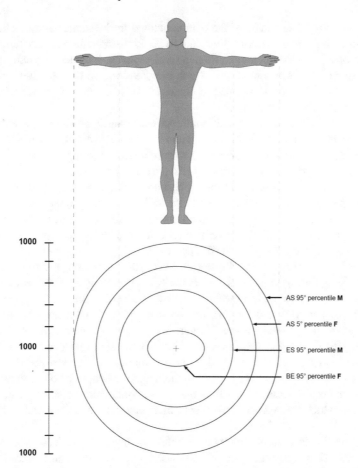

It graphically describes the space occupied by the individual. The major axis and the minor axis describe the maximum width and depth of the body. Taking the 95th percentile for men as a reference, then adding 25 mm to its bulk for clothing, the ellipse axes are 63 and 38 cm. ER (elbow room): the range of motion for the elbows. It is described by the first circle and corresponds to the horizontal extension of the elbows (102 cm diameter for the 99th percentile in men). AS (arm span): the extension of the arm. It is described by the second circle, which corresponds to the extension of the arms of woman in the 5th percentile, and the third circle, which corresponds to the extension of the arms of men in the 95th.

Fig. 10.11 Body ellipse. Revised by Pheasant and Haslegrave (2006, p. 88)

work purposes. The extent of the data contained in the standards also allows it to be applied outside of work stations, and to apply the recommended parameters and limitations to environments, equipment and products intended for daily use. The main standard recommendations, which directly concern the definition of dimensional requirements, are outlined in Chap. 15, with specific reference to the main design problems and case studies.

The definition of dimensional requirements primarily concerns space for movement (or the layout of the work station) and the reachable areas, which pose substantially similar problems but, conversely, require reference to opposing anthropometric parameters. In the case of the space needed to easily perform movements, we must use the bulk of the person with the highest anthropometric measurements as reference (typically the man who belongs to the 95th percentile). In the case of reachable areas, on the other hand, we must refer to the distances that can be reached with more reduced anthropometric and movement capacity (typically, the reference in this case is the woman who belongs to the 5th percentile).[18]

In reference to the body ellipse (see Sect. 10.7.1), the posture assumed by the human body during motion, in the source of a given activity, may be identified on the basis of anthropometric measurements that describe, for example, the range of motion for the arms and lengths, the vertical and horizontal gripping height, the range of motion of the hands and fingers, etc. Analysing the movements needed to perform the required tasks allows us to identify the space needed to allow them to be correctly executed and to identify the positioning and dimensional relationships between various elements and between them and the people who use them. In both cases, we need to examine the frequency with which various tasks are carried out and with which the relevant movement, their hierarchy and their frequency occur (Fig. 10.12).

The reachable areas also allow us to define the layout of work stations (and work areas needed to perform any activities), based on certain basic principles (Pheasant and Haslegrave 2006, p. 86):

- principle of importance: the essential components needed to carry out operations safely and efficiently must be in accessible positions;
- principle of the frequency of use: the components that are used most often must be located in accessible positions;
- principle of functions: the components with linked functions must be located close to one another;
- principle of the usage sequence: the components that are frequently used in a given sequence must be located close to one another and arranged according to the sequence of operations.

---

[18]The standard indicates the threshold that must typically be used for the conventional reference of the 5th and 95th percentile. Broader thresholds (the 1st and the 99th percentile) must be considered when health and safety is of particular importance. When the equipment (the product or environment) must be used by both men and women, we must refer to data about women in the 5th or 1st percentile for reachable areas, and data about men in the 95th or 99th percentile for space for movement. See the UNI EN 614/1:2009 (UNI EN 2009) standard, "Safety of machinery—Ergonomic design principles"—Part 1: Terminology and general principals.

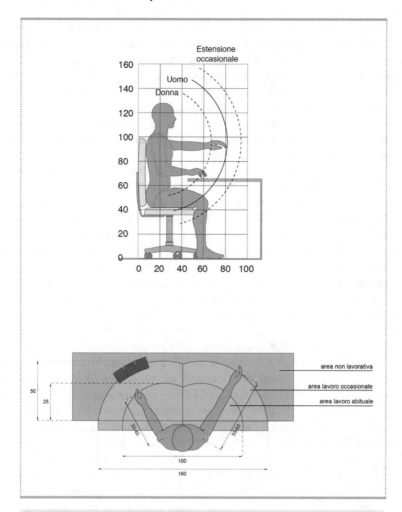

The reference for the horizontal and vertical range of motion for the arms allows us to define the work area (the reference here is the 5th percentile), within which it is possible to grab with little effort and solely by rotating the arms.

Fig. 10.12 Horizontally and vertically reachable areas related to activities that resemble office work. Revised by Pheasant and Haslegrave (2006, p. 99), Chengalur et al. (2003, p. 196)

Both the spaces for movement and the reachable zones must be taken into account, as well as the dimensional needs of wheelchair users, that is, the bulk of people seated in a wheelchair and the range of motion of the arms in a seated position.

Other variability factors must be considered on a case-by-case basis, based on the activities for which the environment or product is, or will be, used and the characteristics of the users who use, or will use, them.

The characteristics of the activities allow us to define the required tasks, the position in which they must, or can, be performed, the type and frequency of the necessary movements, etc. (see Sect. 10.9).

In many cases, problems related to space for movement and reachable areas are closely linked and can create dimensional restrictions that tend to overlap. Ensuring users can reach all of the elements in the work station with a reduced capacity to extend their arms, and ensuring, at the same time, sufficient space for movement for users with a greater physical bulk naturally creates a problem that cannot easily be resolved. Similar problems may be posed by the various needs of users and the need to use a given environment or product to carry out different activities.

Examples pertaining to the first case include the design of a bathroom that must be used by both adults and elderly people and children under 6 years old, or the design of furnishings that can be used by people in wheelchairs.

In the second case, the incompatibility of the dimensional restrictions may concern, for example, the design of a table or chair that will be used both to design and write by hand and type on a keyboard. It may also potentially be used for eating (an activity that requires a different positioning of the arms and a different spatial relationship between the table and the chair). In these cases, and many others like them, referencing anthropometric data allows us to define the threshold, below or above which it is impossible or impractical to respond to users' needs, and also allows us to define the levels and limitations within which the adopted solutions may be used and/or adapted to the various characteristics of users or different activities.[19]

Finally, we should underline that the definition of dimensional requirements not only refers to anthropometric aspects, but must also consider people's postures and movements.

---

[19] A typical case is the common adjustable office chair, which may be adjusted based on the anthropometric characteristics of the user and/or based on the type of activity they are performing. In this case, the adaptability threshold is the maximum change in height that is allowed by the chair's adjustment mechanism. By way of example, similar cases include tables with variable height and incline, which can be used for writing and drawing (or eating) and can be adapted over time to the dimensions of a growing child, kitchen furnishings with removable shelves and surfaces, which can easily be used by people in wheelchairs and, generally speaking, any environment or items that allows, and provides for, different usage possibilities and levels of adaptability.

10.9 Definition of Dimensional Requirements: Use of Anthropometric Data

The definition of dimensional requirements is based on identifying:

- the relevant dimensional parameters (that is, those that are useful for, and specific to the design case and/or intervention);
- the relevant anthropometric parameters (related to the dimensional parameters identified);
- data related to the reference user group;
- dimensional restrictions imposed by the intervention context (environmental layout, characteristics of the activities or users, etc.);
- limit users of the design and/or intervention.

The first step involves identifying the relevant dimensional parameters and relevant anthropometric parameters. The parameters must refer to both static dimensions, that is, anthropometric dimensions for people in immobile positions (for example, in the case of evaluating and/or designing an office chair: the bulk of the hips, the width of the legs, the height of the elbows in a seated position, etc.), and the dynamic dimensions, that is, the dimensions for people in motion, which are considered based on the type of activity that is predicted and/or possible. In particular, the postures of people when performing various activities, and related reachable zones, must be taken into account.

At the same time as identifying the anthropometric parameters needed to define the dimensional requirements, we must identify:

- the category(s) of people that the design is intended for and their relevant characteristics, which allows us to select the anthropometric data for the design. Users' characteristics can be determined on a relatively generic basis, such as, for example, age in the case of environments or products used primarily by children or elderly people, or based on more specific factors, such as, for example, the bodily dimensions and/or strength and movement capacity for equipment used when practising certain sports or, more simply, for "plus-size" clothing;
- the characteristics of the planned and/or possible activities, which can be used to list the physical tasks required to use the product, the position in which it must, or can, be carried out, the type and frequency of the necessary movements, etc.;
- the restrictions imposed by all of the context variables, for example, the dimensional restrictions imposed by the layout of the work station and/or environment in which it is located, the organisation of work activities, etc. Other variability factors must be considered on a case-by-case basis, based on the activities for which the environment or product is, or will be, used and the characteristics of the users who use, or will use, them. By way of example, necessary operational references for the evaluation and/or design of a table-chair system for office work or similar activities are included below. For reference anthropometric data, see Chap. 11 (Fig. 10.13).

ANTHROPOMETRIC REFERENCES FOR THE DESIGN OF A TABLE-CHAIR SYSTEM FOR OFFICE WORK

1) IDENTIFICATION OF THE DIMENSIONAL PARAMETERS AND RE-LATED ANTHROPOMETRIC PARAMETERS

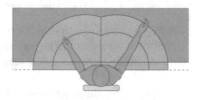

Dimensional parameters	Anthropometric parameters
1. Width of the work surface	Reachable areas related to the movement of arms and hands
2. Depth of the work surface	
3. Distance between the work surface and seat	Width of the legs
4. Distance between the seat and the foot rest	Height of the knees Length of the legs
5. Leg room: height	Leg height Space of movement necessary for the extension of the legs
width depth	Height of the legs
Space for movement needed to extend the legs	Elbow width
6. Width of the seat	Width of the hips
7. Distance between the armrests	Width of the elbows
8. Height of the armrests	Height of the elbows

Fig. 10.13 Station for office work. Reference dimensional and anthropometric parameters

References

Borgognini TS, Masali M (1987) Antropologia e antropometria. UTET, Torino

Chengalur SN, Rodgers SH, Bernard TE (2003) Kodak's ergonomic design for people at work, Wiley, New York (1st ed.: Rodgers SH (1983) Ergonomic design for people at work. Wiley, New York)

Chengalur SN, Rodgers SH, Bernard TE (2004) Kodak's ergonomic design for people at work (2nd ed.), Jhon Wiley & Sons, New York (1st ed. 1983)

Fubini E et al (1993) Le misure antropometriche della popolazione italiana nella progettazione ergonomica europea. In: Atti del Congresso della Società italiana di Ergonomia, Palermo

ISO 11226:2000 (2000) Ergonomics—evaluation of static working postures

ISO 15534-1:2000 (2000) Ergonomic design for the safety of machinery—part 1: principles for determining the dimensions required for openings for whole-body access into machinery

ISO 15534-2:2000 (2000) Ergonomic design for the safety of machinery—part 2: principles for determining the dimensions required for access openings

ISO 15534-3:2000 (2000) Ergonomic design for the safety of machinery—part 3: anthropometric data

Hall ET (1996) La dimensione nascosta, Bompiani, Milano (1st ed.: Hall ET (1966) The hidden dimension. Doubleday, Garden City)

Panero J, Zelnik M (1983) Spazi a misura d'uomo, Be-Ma, Milano (1st ed.: Panero J, Zelnik M (1979) Human dimension and interior-space. Whitney Library of Design, New York)

Pheasant S, Haslegrave CM (2006) Bodyspace: anthropometry, ergonomics and the design of work. CRC Press, Boca Raton (1st ed.: Pheasant S (1986) Body-space: anthropometry, ergonomics and design. Taylor & Francis, London and Philadelphia)

UNI EN 1005-3:2009 (2009) Sicurezza del macchinario – Prestazione fisica umana – Parte 3: Limiti di forza raccomandati per l'utilizzo del macchinario

UNI EN 1005-4:2009 (2009) Sicurezza del macchinario – Prestazione fisica umana – Parte 4: Valutazione delle posture e dei movimenti lavorativi in relazione al macchinario

UNI 349:2008 (2008) Sicurezza del macchinario – Spazi minimi per evitare lo schiacciamento di parti del corpo UNI CEN ISO/TR 7250-2:2011, Misurazioni di base del corpo umano per la progettazione tecnologica – Parte 2: Rilevazioni statistiche relative a misurazioni del corpo umano corporee provenienti da singole popolazioni ISO

UNI EN 13857:2008 (2008) Sicurezza del macchinario – Distanze di sicurezza per impedire il raggiungimento di zone pericolose con gli arti superiori e inferiori

UNI EN ISO 15535:2012 (2012) Requisiti generali per la creazione di banche dati antropometrici

UNI EN 614-1:2009 (2009) Sicurezza del macchinario – Principi ergonomici di progettazione – Parte 1: Terminologia e principi generali

UNI EN 614-2:2009 (2009) Sicurezza del macchinario – Principi ergonomici di progettazione – Parte 2: Interazioni tra la progettazione del macchinario e i compiti lavorativi

UNI EN ISO 7250-1:2017 (2017) Dimensioni del corpo umano da utilizzare la progettazione tecnologica – Parte 1: Definizioni delle dimensioni del corpo umano e dei punti di repere anatomico

Chapter 11
The Anthropometric Reference Data

Anthropometry is the most applicable, economical and non-invasive way to determine the size, proportions and composition of the human body. Moreover, since the body size at each age reflects the general state of health and well-being of individuals and populations, anthropometry can be used to estimate functionality, health status and survival (De Onis and Habicht 1996).

By the definition of anthropometry of the World Health Organization (WHO) it is clear that the detection of anthropometric data is a useful tool for the assessment of nutritional status and more generally of health, necessary both in clinical practice and for epidemiological research. The WHO, in addition to identifying anthropometric data of reference, tried to establish a guide about the practical use and interpretation of such data. The focus was mainly on children and adolescents, using anthropometry as a tool to assess the growth of children.

When anthropometric data are used for monitoring growth, it is essential to consider these values not as a comparison, but as real standards to refer to in order to define a 'normal' or at least desirable growth. This does not mean that the reference value must be considered as a fixed standard to be applied rigidly, but that the data must be interpreted; if you are dealing with individuals of different ethnic backgrounds and different socio-economic and nutritional conditions, individual differences must be taken into account. Only in this way will it be possible to monitor the individual conditions, identify any pre-existing or high-risk pathological situations, on the basis of which to decide whether or not to take corrective measures. It is essential to analyze the context in which a deviation from the standard values occurs. For example, if I have an important deviation from the reference values at the age of 6 months I could have a principle of malnutrition; but if the same deviation occurs at 6 years of life, this may reflect a previous state of malnutrition but perhaps no longer in place.

This chapter was co-authored by Francesca Fazzini, Giulio Arcangeli, Nicola Mucci and Mattia Pistolesi.

© Springer Nature Switzerland AG 2020   217
F. Tosi, *Design for Ergonomics*, Springer Series in Design and Innovation 2,
https://doi.org/10.1007/978-3-030-33562-5_11

In the same way, if a child is very different from the average, not necessarily we will have to take nutritional measures, but we will have to evaluate if child's rate of growth is adequate within his percentile.

Another factor to consider is the influence of non-pathological factors on the normal growth and state of health of the individual, simply like the sex of the individual. Each factor must be evaluated singularly and in its context and not as an absolute value.

As for children, the growth curves established by WHO are now universally accepted. To achieve them, the "Multicentre Growth Reference Study" (MGRS) was undertaken between 1997 and 2003 (De Onis et al. 2004).

The MGRS collected primary growth data and related information from about 8500 children from very different ethnic backgrounds and cultural backgrounds (Brazil, Ghana, India, Norway, Oman and USA).

The new growth curves provide a single international standard that best represents the physiological growth of all children, from birth to 5 years, and which establishes the breastfed infant as a regulatory model for growth and development.

The tables prepared at the end of the study, published in 2006, dealt with several anthropometric indicators: length/height by age, weight by age, weight by height, body mass index (BMI) by age, head circumference by age, arm circumference by age, subscapularis plica by age, tricipital plica by age, motor development phases.

For each of these indicators a distinction is made by sex and age, creating appropriate tables showing the growth curves of the normal child population. Given the considerable variability that characterizes the child's age, the indicators are not represented as an absolute value but as percentile curves. In fact, it should be kept in mind that the indicator used to evaluate growth is not the anthropometric index itself, measured once, nor the curve obtained by joining the corresponding points to subsequent measurements of the same index, but the variation in the time of the growth rate and therefore of the inclination of the curve (Figs. 11.1, 11.2, 11.3 and 11.4).

Percentile tables are the most used in pediatric clinical practice, although there is another graphical representation in terms of Z-score, which measures how many standard deviations each value deviates from the reference curve.

All the curves of the various anthropometric indicators can be consulted on the WHO[1] website. On the site there are also several publications concerning the MGRS study, a software for the analysis of individual and population data, and a training manual for the appropriate use of the new curves with exercises and practical examples.

The need to have anthropometric reference data for the adult population is evident, not only for having an index of individual health status but also for the socioeconomic conditions of individuals, as well as for the design of instruments according to ergonomic criteria. In this regard, a very important document on the anthropometric indicators of the adult population comes from the United States; it is based on the

---

[1] World Health Organization (WHO): The WHO Child Growth Standards. www.who.int/childgrowth/en/.

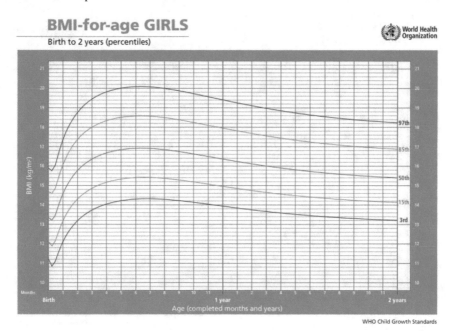

Fig. 11.1 Percentiles of BMI for age in girls from birth to 2 years

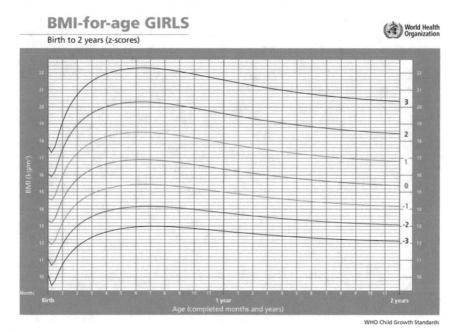

Fig. 11.2 BMI Z-score for age in girls from birth to 2 years

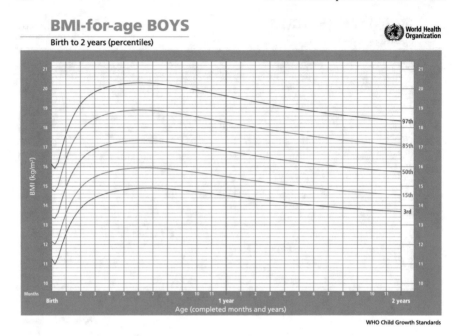

Fig. 11.3   Percentiles of BMI for age in BOYS from birth to 2 year

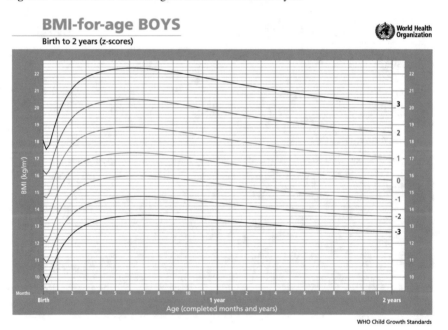

Fig. 11.4   BMI Z-score for age in children from birth to 2 years

"Nutritional Health and Nutrition Examination Survey" (NHANES), a study conducted by the National Center for Health Statistics (NCHS) with the aim of assessing the health and nutritional status of the entire American population through surveys, questionnaires, laboratory tests and physical exams. The first NHANES was conducted in 1971 and since 1999 the data are updated annually; the first report on the subject was published in 2001.[2]

From the data obtained from NAHNES, tables were created on the main anthropometric indicators (weight, height, BMI …) of the entire US population, divided by gender, age, race and ethnicity.

For each anthropometric indicator, it is indicated a mean, the standard deviation from the mean and the percentile values (Tables 11.1, 11.2, 11.3 and 11.4).

Although in this study there are also data concerning children, currently for the child age it is preferred to use the previously described WHO curves, because in the NHANES the sample was exclusively made by US breastfed children, therefore without taking into account the inevitable differences in nutrition and social conditions present at world level (Fig. 11.5).

Regarding the standardization of anthropometric measurements at the international level, reference is made to ISO standards and to those of CEN and NASA for measurement techniques. For the univocal interpretation of anthropometric data, reference is to the UNI EN ISO 15535:2013 standard. The standard specifies the general requirements for anthropometric databases and their associated studies that contain measurements taken in accordance with ISO standards. It provides the necessary information, such as user population characteristics, sampling methods, measurement elements and statistics, to enable international comparison between different population groups.[3]

The definitions and methods for detecting anthropometric data are described in the UNI EN ISO 7250/1:2017 standard, Dimensions of the human body to be used for technological design—Part 1 Definitions of the dimensions of the human body and of the anatomical landmarks.

Anthropometric data on some populations are reported in the UNI CEN ISO/TR 7250/2:2011 standard, Basic measurements of the human body for technological design—Part 2 Statistical surveys on measurement of the human body coming from individual ISO populations.

Extracts from technical standards
Summary of main anthropometric data—Italy: re-elaborated tables from UNI CEN ISO/TR 7250-2:2011 standard, Basic Measurements of the Human Body for Technological Design—Part 2: Statistical Surveys Relating to Measurements of the Human Body from Individual Populations ISO, 2011, pp. 14–17 (Table 4: Italy, statistical summary).

---

[2]Centers for Disease Control and Prevention. www.cdc.gov/nchs/nhanes/index.htm.

[3]UNI EN ISO 15535:1553. General requirements for establishing anthropometric databases. http://store.uni.com/catalogo/index.php/uni-en-iso-15535-2013.html.

Table 11.1 Weight (kg) of men over 20 in the USA between 2007 and 2010

Race and ethnicity and age	Number of examined persons	Mean	Standard error of the mean	Percentile kg								
				5th	10th	15th	25th	50th	75th	85th	90th	95th
All racial and ethnic groups[a]												
20 years and over	5651	88.7	0.45	61.5	66.5	69.7	75.0	86.1	98.9	107.2	114.4	124.1
20–29 years	894	83.4	0.85	58.4	62.6	65.3	69.5	80.1	93.7	101.6	109.0	116.8
30–39 years	948	90.5	0.75	63.3	67.8	70.8	75.4	86.7	101.1	110.1	117.8	128.0
40–49 years	933	91.0	0.96	64.4	69.5	73.5	78.5	87.9	100.6	108.6	116.2	126.3
50–59 years	934	91.3	1.10	63.8	69.0	72.9	78.2	88.6	102.9	109.8	117.6	126.6
60–69 years	933	90.5	0.76	61.7	67.9	70.9	76.4	88.5	101.2	108.3	115.1	127.4
70–79 years	649	86.5	0.36	62.6	66.8	70.4	75.1	34.7	95.1	102.9	109.4	117.9
80 years and over	360	79.3	0.81	57.6	61.5	64.1	69.0	77.9	88.2	94.3	97.3	104.5
Non-Hispanic white												
20 years and over	2738	90.4	0.42	63.7	68.9	72.2	77.3	88.0	100.6	108.7	115.6	124.1
20–39 years	796	88.3	0.77	61.5	65.9	69.4	73.8	35.3	98.5	107.9	113.5	122.7
40–59 years	832	92.9	0.71	66.4	72.1	75.0	80.4	90.0	103.1	111.9	117.7	128.7
60 years and over	1110	89.0	0.55	63.5	69.0	72.2	76.6	87.0	98.5	105.6	112.6	121.0
Non-Hispanic black												
20 years and over	1094	90.4	0.74	60.9	65.1	67.9	74.0	86.8	101.8	111.4	120.1	132.5
20–39 years	356	89.9	1.43	61.3	64.2	66.6	72.1	85.6	101.4	111.7	120.0	134.3
40–59 years	372	92.1	1.26	60.5	66.2	71.2	76.9	89.1	103.3	113.3	120.3	132.4
60 years and over	366	87.6	1.15	58.0	63.8	67.1	72.0	85.3	98.0	106.5	116.4	128.8

(continued)

Table 11.1 (continued)

Race and ethnicity and age	Number of examined persons	Mean	Standard error of the mean	Percentile										
				5th	10th	15th	25th	50th	75th	85th	90th	95th		
				kg										
Hispanic[b]														
20 years and over	1541	84.4	0.88	60.4	65.0	67.3	71.4	81.6	93.2	100.2	107.3	121.0		
20–39 years	573	84.0	1.21	58.6	63.7	66.3	70.3	79.9	92.8	100.4	109.4	123.2		
40–59 years	577	85.9	0.80	64.2	68.2	70.3	74.9	83.6	94.2	101.3	105.4	117.0		
60 years and over	391	82.0	1.13	58.6	63.0	66.5	70.2	80.4	90.9	96.6	101.5	109.4		
Mexican American														
20 years and over	991	84.1	1.04	60.4	65.0	67.3	71.3	61.4	92.8	99.2	106.9	121.4		
20–39 years	366	34.0	1.48	59.4	63.5	66.2	70.3	80.1	92.9	100.1	109.5	125.3		
40–59 years	371	85.0	0.90	64.0	68.0	70.1	73.9	82.9	93.7	99.7	104.0	115.8		
60 years and over	234	81.9	1.44	59.3	63.0	67.1	70.6	80.4	90.4	94.8	97.6	109.1		

Note [a]Persons of other races and ethnicities are included
[b]Mexican-American persons are included in the Hispanic group

Table 11.2 Weight (kg) of women over 20 in the USA between 2007 and 2010

Race and ethnicity and age	Number of examined persons	Mean	Standard error of the mean	Percentile										
				5th	10 th	15th	25thi	50th	75th	85th	90th	95th	
				kg									
All racial and ethnic groups[a]													
20 years and over	5844	75.4	0.35	50.2	53.6	56.6	61.1	71.3	65.5	95.4	102.2	113.8	
20–29 years	906	73.4	1.13	48.6	52.1	53.6	57.3	67.7	62.2	94.8	103.1	120.0	
30–39years	982	76.7	0.94	50.9	53.9	57.4	62.2	72.5	66.1	97.6	102.2	115.2	
40–49 years	1056	76.2	0.74	50.6	54.6	57.4	61.2	71.8	85.7	96.2	103.7	114.9	
50–59 years	873	77.1	0.83	51.1	55.9	58.4	62.8	73.2	87.9	96.8	104.4	115.8	
60–69 years	951	77.4	0.65	52.8	57.2	60.0	63.7	75.2	87.3	95.7	102.7	109.5	
70–79 years	679	74.8	0.69	49.9	53.5	57.0	62.1	72.3	84.9	91.5	99.1	109.2	
80 years and over	397	64.9	0.73	45.4	49.6	51.6	55.8	63.5	71.9	78.4	82.9	87.4	
Non-Hispanic white													
20 years and over	2730	75.0	0.51	50.6	53.8	56.7	61.2	71.0	85.0	95.0	101.5	112.0	
20–39 years	792	74.7	1.16	50.1	53.4	55.9	59.9	69.9	84.6	96.3	102.2	116.8	

(continued)

Table 11.2 (continued)

Race and ethnicity and age	Number of examined persons	Mean	Standard error of the mean	Percentile								
				5th	10 th	15th	25th	50th	75th	85th	90th	95th
				kg								
40–59 years	861	76.1	0.66	50.9	55.0	57.5	61.8	71.9	86.1	96.2	103.0	115.5
60 years and over	1077	73.9	0.44	50.4	53.5	57.1	61.3	71.4	84.1	91.7	90.5	106.8
Non-Hispanic black												
20 years and over	1128	85.2	0.78	53.9	58.7	61.9	68.5	80.5	97.7	107.7	115.7	130.5
20–39 years	372	84.4	1.16	51.6	57.0	60.9	65.8	79.9	96.3	109.7	117.4	133.6
40–59 years	383	88.3	1.22	56.9	60.7	65.4	71.0	83.2	100.5	110.4	116.7	131.7
60 years and over	373	80.7	1.04	52.8	57.1	61.2	67.6	77.7	90.9	100.3	106.1	114.8
Hispanic[b]												
20 years and over	1708	72.9	0.46	49.7	53.1	56.0	60.5	70.2	82.3	89.9	95.6	103.5
20–39 years	619	72.3	0.60	48.8	52.1	54.1	58.8	68.7	83.3	91.1	96.5	105.1

(continued)

Table 11.2 (continued)

Race and ethnicity and age	Number of examined persons	Mean	Standard error of the mean	Percentile								
				5th	10 th	15th	25thi	50th	75th	85th	90th	95th
				kg								
40–59 years	579	74.6	0.83	52.0	57.0	59.4	63.9	72.1	82.7	89.5	95.3	103.7
60 years and over	510	70.7	0.47	48.0	51.8	55.6	60.5	69.4	79.1	85.5	90.2	96.9
Mexican American												
20 years and over	1032	73.2	0.40	50.3	53.6	56.5	61.0	70.8	83.0	90.0	94.3	104.0
20–39 years	386	72.7	0.66	50.1	52.8	55.0	59.9	69.0	64.0	91.6	96.6	105.9
40–59 years	347	75.2	0.97	53.3	57.7	60.5	64.8	73.4	83.0	89.4	93.4	102.5
60 years and over	299	70.6	0.56	47.1	51.5	55.5	60.4	69.5	79.3	85.1	89.5	94.1

Note [a]Persons of other races and ethnicities are included
[b]Mexican-American persons are included in the Hispanic group

Table 11.3 Height (cm) of men over 20 in the USA between 2007 and 2010

Race and ethnicity and age	Number of examined persons	Mean	Standard error of the mean	Percentile								
				5th	10th	15th	25th	50th	75th	85th	90th	95th
				cm								
All racial and ethnic group[a]												
20 years and over	5647	175.9	0.20	163.2	166.0	168.0	170.9	176.1	180.9	183.6	185.4	188.2
20–29 years	895	176.3	0.33	163.6	166.4	168.8	171.2	176.3	181.7	184.3	185.8	188.3
30–39 years	948	176.4	0.36	1636	1661	168.3	171.5	176.5	181.8	184.4	186.3	188.5
40–49 years	934	176.8	0.42	164.9	167.7	169.1	171.8	176.8	181.6	184.4	186.7	188.9
50–59 years	938	176.6	0.32	163.9	167.1	168.5	172.0	176.9	181.1	183.6	185.0	189.0
60–69 years	932	174.9	025	162.2	164.6	166.9	169.9	175.3	179.7	182.6	184.5	187.0
70–79 years	646	173.2	0.34	161.7	164.0	165.7	168.5	173.1	177.7	180.2	182.6	184.7
80 years and over	354	170.7	0.38	1587	161.9	163.5	166.0	170.9	175.7	177.6	178.9	181.6
Non-Hispanic white												
20 years and over	2738	177.4	019	165.9	168 6	170.3	172.7	177.3	182.0	184.5	186.3	188.8
20–29 years	797	178.4	0.35	168.0	170.0	171.5	173.3	178.2	183.0	185.3	187.0	189.0
40–59 years	836	178.3	0.28	167.7	170.2	171.7	173.9	178.3	182.2	184.8	186.8	189.5
60 years and over	1105	174.6	0.22	163.1	165.0	166.8	169.8	175.0	179.2	182.0	183.9	186.4
Non-Hispanic black												
20 years and over	1091	176.4	0.25	165.4	167.5	169.0	171.8	176.3	180.8	183.4	185.5	188.0
20–39 years	356	176.9	0.39	166.4	168.2	170.1	172.4	176.4	181.4	183.8	186.1	187.9
40–59 years	373	176.7	0.53	165.0	168.0	169.1	171.9	176.8	181.3	183.7	185.8	188.7
60 years and over	362	174.4	0.42	163.1	164.9	166.8	169.5	175.1	178.5	181.6	183.2	185.8

(continued)

Table 11.3 (continued)

Race and ethnicity and age	Number of examined persons	Mean	Standard error of the mean	Percentile cm								
				5th	10th	15th	25th	50th	75th	85th	90th	95th
Hispanic[b]												
20 years and over	1541	170.4	0.34	159.0	161.4	162.8	165.3	170.1	175.1	178.1	180.0	183.9
20–39 years	573	171.1	0.48	159.0	161.7	163.5	165.6	170.9	176.3	179.2	181.9	185.0
40–59 years	577	170.3	0.36	160.0	161.7	162.8	165.5	170.3	174.7	177.2	179.0	181.5
60 years and over -	391	167.3	0.45	156.8	159.5	160.6	163.0	167.6	171.2	173.5	174.7	177.8
Mexican American												
20 years and over	990	169.8	0.39	158.8	160.8	162.4	165.0	169.3	174.0	177.4	179.8	183.3
20–39 yeare	386	170.5	0.61	158.7	160.8	162.5	165.2	170.0	175.4	178.6	181.7	184.9
40-59 years	371	169.5	0.35	159.7	161.2	162.8	165.3	168.9	173.3	176.2	177.8	180.8
60 years and over	233	167.2	0.54	157.2	158.9	160.5	163.0	167.4	170.6	173.1	174.0	177.3

Fryar et al. (2012, pp. 1–40)
Note [a]Persons of other races and ethnicities are included
[b]Mexican-American persons are included in the Hispanic group

Table 11.4 Height (cm) of women over 20 in the USA between 2007 and 2010

Race and ethnicity and age	Number of examined persons	Mean	Standard error of the mean	Percentile										
				5th	10th	15th	25th	50th	75th	85th	90th	95th		
				cm										
All racial and ethnic groups[a]														
20 years and over	5971	162.1	0.14	150.7	153.1	154.7	157.3	162.1	166.8	169.2	170.9	173.7		
20–29 years	980	163.1	0.24	152.0	153.9	155.7	158.1	162.9	167.6	170.2	171.8	175.1		
30–39 years	1029	163.4	0.29	151.4	154.6	156.1	158.6	163.4	167.9	170.4	172.4	174.9		
40–49 years	1060	163.1	0.22	152.0	154.4	156.1	158.6	162.7	167.5	170.0	171.7	174.5		
30–59 years	873	162.2	0.30	151.3	153.5	154.9	157.4	162.6	166.7	169.1	170.2	172.6		
60–69 years	952	161.6	0.26	150.5	152.9	154.8	157.5	162.0	165.8	168.2	169.6	171.7		
70–79 years	679	159.1	0.32	148.2	150.6	151.9	154.2	159.4	163.3	166.1	167.9	169.8		
80 years and over	398	155.9	0.36	144.6	147.0	149.3	151.9	156.1	159.8	162.1	163.3	166.3		
Non-Hispanic white														
20 years and over	2764	163.1	0.15	152.1	154.4	156.3	158.7	163.0	167.5	169.7	171.6	174.5		
20–39 years	824	164.9	0.25	153.9	156.6	158.1	160.6	164.8	168.9	171.7	173.7	176.5		
40–59 years	861	163.8	0.27	153.6	155.6	157.2	159.6	163.7	168.0	170.0	171.7	174.5		
60 years and over	1079	160.3	0.22	149.2	151.8	153.4	156.1	160.3	164.5	167.1	168.6	170.7		
Non-Hispanic black														
20 years and over	1154	163.0	0.25	152.2	154.6	156.0	158.3	162.8	167.7	169.8	171.4	173.6		
20–39 years	397	163.7	0.32	153.1	155.4	156.9	159.0	163.5	168.1	170.5	171.8	173.8		

(continued)

Table 11.4 (continued)

Race and ethnicity and age	Number of examined persons	Mean	Standard error of the mean	Percentile										
				5th	10th	15th	25th	50th	75th	85th	90th	95th	
				cm									
40–59 years-	364	163.5	0.38	152.8	155.1	156.5	158.7	163.0	168.2	170.1	171.8	173.1	
60 years and over	373	160.6	0.28	149.7	152.1	153.5	156.1	160.6	164.9	167.0	168.5	171.2	
Hispanic[b]													
20 years and over	1763	157.1	0.19	146.3	148.8	150.1	152.6	156.9	161.5	163.9	165.6	168.3	
20–39 years	673	158.2	0.23	147.7	149.7	151.5	153.8	157.9	162.2	165.0	166.6	169.5	
40–59 years	580	157.1	0.33	146.8	148.8	150.7	152.8	156.7	161.6	163.6	164.8	167.2	
60 years and over	510	153.7	0.31	143.9	146.0	147.5	149.4	153.3	157.7	159.8	161.6	164.5	
Mexican American													
20 years and over	1074	156.6	0.17	145.3	148.4	149.8	152.2	156.4	160.9	163.1	164.8	167.7	
20–39 years	427	157.5	0.27	145.9	149.5	151.2	153.3	157.5	161.4	163.9	165.4	168.4	
40–59 years	348	156.6	0.41	145.9	148.6	149.9	152.1	156.0	161.0	163.1	164.6	166.9	
60 years and over	299	153.3	0.40	143.8	145.8	147.0	149.3	153.2	157.0	159.8	161.4	163.6	

Note [a]Persons of other races and ethnicities are included
[b]Mexican-American persons are included in the Hispanic group

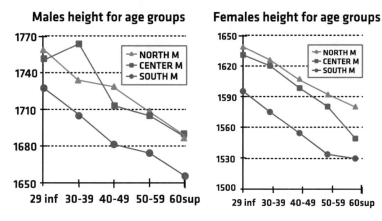

Fig. 11.5 Height for age groups in the North, Center and South Italy (Masali 2013)

Summary of main anthropometric data—Japan: tables re-elaborated from UNI CEN ISO/TR 7250-2:2011 standard, Basic Measurements of the Human Body for Technological Design—Part 2: Statistical Surveys of Measurements of the Human Body from Individual Populations ISO, 2011, pp. 20–23 (Table 6: Japan, statistical summary).

Summary of main anthropometric data—USA: tables re-elaborated from UNI CEN ISO/TR 7250-2:2011 standard, Basic Measurements of the Human Body for Technological Design—Part 2: Statistical Surveys of Measurements of the Human Body from Individual Populations ISO, 2011, pp. 48–51 (Table 16: United States, statistical summary).

References

Centers for Disease Control and Prevention, De Onis M et al (2004) WHO multicentre growth reference study (MGRS): rationale, planning and implementation. Food Nutr Bull 25(Suppl 1), S1–89

De Onis M, Habicht JP (1996) Anthropometric reference data for international use: Recommendations from a WHO Expert Committee. Am J Clin Nutr 64(4):650–658

Fryar CD et al (2012) Anthropometric reference data for children and adults: United States 2007–2010. Vital Health Stat. Data from the National Health Survey 11(252):1–48

Masali M (c - 2013). L'Italia si misura. Vent'anni di ricerca (1990–2010). Vademecum antropometrico per il design e l'ergonomia. Aracne, Roma

UNI EN ISO 15535 (2013) General requirements for establishing anthropometric databases. http://store.uni.com/catalogo/index.php/uni-en-iso-15535-2013.html

UNI CEN ISO/TR 7250-2 (2011) Misurazioni di base del corpo umano per la progettazione tecnologica – Parte 2: Rilevazioni statistiche relative a misurazioni del corpo umano corporee provenienti da singole popolazioni ISO

UNI EN ISO 7250-1 (2017) Dimensioni del corpo umano da utilizzare la progettazione tecnologica – Parte 1: Definizioni delle dimensioni del corpo umano e dei punti di repere anatomico

WHO—World Health Organization (2006). The WHO Child Growth Standards. (www.who.int/childgrowth/en/)

Chapter 12
Elements of Biomechanics of Occupational Interest

12.1 Introduction

Biomechanics is "a discipline that studies the applications of mechanics to living organisms"[1] at all macro or microscopic levels, using concepts, methods and laws of various relevant disciplines, from physics to mathematics, from physiology to biology, and anatomy.

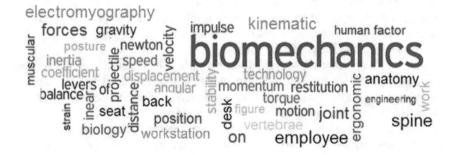

Biomechanics has been established as a scientific area[2] and independent research field but not as a subject of independent study; it acts in fact between the area of medical-biological research and the physical-technical one, therefore it is configured as an interdisciplinary field that combines many professional figures interested in its application among which it is possible to mention those who study locomotion, improvement of sports performance, living organisms in their complex mechanical interactions such as in the air or in the aquatic medium, human-machine interaction,

---

This chapter was co-authored by Marco Petranelli, Nicola Mucci, (Riccardo Baldassini), Francesca Fazzini and Giulio Arcangeli.

[1] Hatze (1974, pp. 89–90).

[2] Knudson (2007).

F. Tosi, *Design for Ergonomics*, Springer Series in Design and Innovation 2,
https://doi.org/10.1007/978-3-030-33562-5_12

the creation of prostheses and orthoses, robotics, the occupational field, even the theatrical teaching.[3]

Each applied sector has developed models and tools appropriate to its field of intervention, however the physical and biological principles used are common and superimposable.

Thus, while in the biomechanics applied to sport or to the control of human movement, the field of study is the movement of man in its many expressions and components, in the biomechanics applied to robotics the mechanics of the human body and of living beings are applied in order to project robotic devices and machines, which act in physical contact and in synergy with the man, while—and only to cite some significant examples—in occupational biomechanics the field of interest is that of the analysis of posture and of the finalized movement to the quantitative description of the modalities of execution of the task and of the load that intervenes during the manual work by acting on the musculoskeletal structures of man.

As can be seen from the examples shown, while the context and the aims of the approach are necessarily differentiated, the contents are widely common and concern, on one hand, the physical-mathematical laws that study the balance of the body, the conditions to maintain it and the forces that allow and regulate the movement; on the other, the anatomical and physiological structures that underlie postures and movements.

The path that we present here starts from an excursus on the development of biomechanics up to the present day, starting from when, in the seventeenth century, with the establishment of the first scientific disciplines, such as physics, began to assume the role and characteristics of reference point of a myriad of studies and applied technical and scientific disciplines.

Among the different sciences that form the basis of biomechanics lie Newtonian mechanics and biology. In the central part of the study we examined the fundamental concepts that link these two fields to biomechanics, one through the scientific method of approach, the other through the field of study and application.

In the third and last part, the functional components of the biomechanics (functional biomechanics for some) have been analyzed, that is the relationships and interactions that the different parts of the organism, of the human body in particular, have with each other and with the environment in operation of the various global or analytical actions and activities that can be developed.

The presentation ends with a brief example of biomechanical evaluation applied to a common gesture such as walking, transferable—starting from the approach method—to most of the areas of application of the biomechanical discipline.

---

[3]Trubočkin (2011, pp. 10–15).

12.2 From the First Systematic Studies On Human Movement to Robotics

Although the study and description of human movement began in very ancient times, there is a broad agreement in considering Giovanni Alfonso Borelli (1608–1679) the father of modern biomechanics, a mathematical scientist and Neapolitan physiologist, a student of Galileo Galilei, the first, as he described in his main work the "De motu animalium", to apply the principles and laws of geometry, mathematics and mechanics to movement, in particular to that of man. Borelli studied and described complex motor activities such as walking, jumping, running, swimming, etc. conducting an in-depth analysis of the nature and role of muscles and levers. To him, or in any case to the results of his work, we owe the very name of the discipline, which right in the seventeenth century because of the fact that it consisted to apply the mechanics to biological systems began to be called Bio-Mechanics.

Only during the nineteenth century, however, biomechanics has taken the characteristics of the scientific discipline that we know today, and it is due to scholars like the French physiologist Jules Marey (1830–1904) who developed tools for the detection of dynamic quantities for the calculation of the mechanical work associated with a given movement, or to Christian Wilhelm Braune and Otto Fischer, of the German School, who conceived, the one anatomist, the other physicist and mathematician, an innovative experimental method that allowed a study in 3D of the body segments and of the inertial forces acting on the motion of the barycentre during locomotion, thus anticipating the so called "gait analysis" which is now applied in a very wide range of sectors.

Remaining in the first half of the nineteenth century, it is right to point out Guillaume Benjamin Duchenne[4] (1806–1875) who first developed a solution that will become extremely useful and productive in the years to come until today, that of the classification of the functions of the individual muscles in relation to the movements they are able to generate, that is what is called Functional Anatomy, a discipline strongly linked to the biomechanical evaluation of human movement.

After recalling the work of Eadweard Muybridge (1830–1904), famous photographer who developed through his photographic plates a revolutionary system able to fix, as never before, a gesture frame by frame, thus allowing a very detailed analysis of the "kinematics" of the main gestures of everyday life, we arrive in the twentieth century, which saw the strongest development of biomechanics with the many applications we know today.

From the early years of the new century, with the contribution of numerous neurophysiologists like Sir Charles Sherringhton (1857–1952) on the role of sensory nerves in muscles (proprioception) and their involvement in movement and its control, the biomechanics has been enriched by the contribution of neurosciences that have profoundly widened its horizons and fields of application. In the first part of the twentieth century, two schools in particular provided, independently of each other because of

---

[4]Duchenne (1867).

the particular socio-political conditions of that period, the largest contribution to the development of biomechanics.

Firstly, the contribution of the Anglo-American school led by Nobel Prize-winner physiologist Archibald Hill (1886–1977) who—together with the pupil Andrew F. Huxley and followed by the American William Fenn, Herbert Helftman and others—has tackled the problems related to production of skeletal muscle tension—the so-called "Hill model",[5] namely the force-length and force-speed ratio (isometric and isotonic contraction) that occurs in the different types of muscle contraction.

A significant contribution was provided by Russian neurophysiologists and biomechanics headed by Nobel Prize winner Ivan Petrovic Pavlov (1849–1936), founder of the School of Physiology of the Russian Academy, and Nicolaj Bernstein (1896–1966), a great scientist able to produce fundamental works in the biomechanical, physiological, psychological and bio-cybernetic fields, which focused his research on the field of the neurosciences of the movement combined with the biomechanics. These contributions have allowed biomechanics to be applied to many fields, starting from sports performance and human movement regulation, to the production of prostheses and orthoses for amputees, from the optimization and protection of manual labor in factories up to that of gender differences in the transport and handling of workloads, giving life to occupational biomechanics.[6]

In the second half of the last century, thanks to the increasingly massive use of electronic-information technologies and the advance of cybernetics (a discipline that brings the systems of control of the behavior of man, machine, economy or social organization closer together), there is a radical change and a reconsideration of the relationship between man, environment, machine and society that inevitably ends up influencing even the human sciences that become increasingly specialized and interconnected, as well as increasingly related to new technologies, especially digital and information. The interest in static mechanics (posture) is reduced, while interest for dynamic analysis increases everywhere, supported by the use of digital imaging devices combined with electromyographs and piezoelectric pressure sensors that can provide very accurate information on all forms of movement, human or animal, with a strong impact on all fields affected by biomechanics, from sports to work, from rehabilitation and disability, to that of industrial design, biomedical engineering, the relationship between man and environment, up to the most recent applications in the field of robotics that sum up and make the most of the contributions examined up to now.

The science and technology of biomechanics and robotics promise to be one of the most influential research and application direction of the 21st century. In fact, biomechanics and robotics go beyond the single areas of biomechanics, robotics, biomedical engineering, mechatronics, biocommunication and biologically inspired

---

[5]From the graph it can be seen that the speed is maximum at zero load and the maximum force at zero speed with the maximum power (L/Dt) between 30 and 40% of the F max.

[6]Bernstein (1967).

robotics, to create models and products that are increasingly replacing human manual activities, thus opening the way to an anthropological revolution that can be considered also scientific, technological and industrial.

12.3 The Theoretical Bases of Biomechanics

In order to better understand biomechanical science, it is essential to refer to the development of physics and the biological sciences that make up the theoretical basis of reference, as we saw in the historical exursus; in particular we have to refer to the classical mechanics that before and more than others gave it a solid theoretical basis.

12.4 Mechanics and Biomechanics[7]

The three principles of the dynamics of Isaac Newton, "axioms or laws on motion" as he himself called them, together with the law on universal gravitation, and even before the observations on the fall of the grave of Galileo Galilei, or Galilean relativity, constitute the foundations of classical mechanics, without a deep understanding and application of which the development of mechanics and biomechanics in many disciplinary fields would have been extremely difficult if not quite unthinkable.

12.4.1 Mass, Weight and Inertia

The mass of a body is the measure of the quantity of matter in the body; the greater the quantity of matter in a body, the greater will be its "weight", weight expressed in arithmetic terms, that has not to be confused with "weight force". While in fact the mass of a body represents its intrinsic property, independent of the position in space and of any other physical quantity, the weight is the effect produced on this mass by the presence of a gravitational field, therefore, while the mass of a body it is constant and does not change during its motion, its weight varies according to the place in which it is measured.

A body with a mass of 1 kg has a weight, expressed in Newton, of 9.81 m/sec^2, but this happens, as we know, on the Earth and in any case under certain conditions (at sea level, etc.); mass, on the other hand, is an invariant and enormous quantities of energy would be needed to reduce the amount of matter. An object on Earth or on Mars would have the same mass, but its weight, possessing Mars a mass much

---

[7]Bell (1998).

smaller than that of the Earth, would be much smaller than that measured on our planet.

On the basis of the Second principle of dynamics, the force applied on a body is a quantity directly proportional to its acceleration with which it shares the direction and verse, with a constant of proportionality given by the mass of the body itself, that means, with equal force agent, that the acceleration results inversely proportional to the mass of the body. The mass therefore represents the resistance that the body opposes when it applies a force that tends to modify its motion, or the relationship between the applied force and the acceleration impressed. This internal resistance to motion which possesses any body is called inertia. However, the mass presents another characteristic, identified by Newton with the law of universal gravitation: the characteristic of attracting other bodies in a way directly proportional to their mass and inversely to the square of the distance that separates them. Inertial mass and gravitational mass are two different characteristics, conceptually at least, of bodies and this difference, has for a long time divided scientists until Einstein established with the principle of equivalence, at the base of the law on relativity, that the gravitational field and the inertial field are the same in spite of the laws on dynamics based on different principles.

Mass, strength, inertial and gravitational properties of a body are the basic concepts of classical mechanics, of extreme interest and importance for applications in the field of biomechanics, as we will see.

12.4.2   Center of Mass and Center of Gravity

The center of mass (CoM) is the point where the distribution of mass, of matter, is the same, converges, in all directions and does not depend on the gravitational field. The center of gravity (CoG) is instead the point where weight distribution is the same, converges, in all directions and depends on the gravitational field.

Center of mass (and center of gravity do not indicate the same thing however, if the gravitational field in which the object is placed is uniform, they end up coinciding and in most cases this is true, just think that even at the top of Everest (8848 m) the gravitational field strength is still 99.6% of its standard value. In conclusion, since the mass in a uniform gravitational field is subjected to the same acceleration (g) of the weight (9.81 m/sec^2), the two values coincide (Fig. 12.1).

When the laws of mechanics are applied to biological systems, the problem of evaluating these parameters arises. Biological systems, including the human body system, have such characteristics that it is difficult to apply the simplifications that are often assumed in the classic approach of mechanics.

The CoM of the human body is not a fixed anatomical point, it does not have a fixed position with respect to the anatomical body points, in some cases it can also be outside the body, its position varies according to how the masses associated with each body segment.

Fig. 12.1 Center of mass
and center of gravity

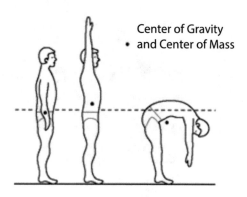

Center of Gravity
• and Center of Mass

The position of the CoM of a moving individual cannot be measured directly as it changes its position as the position of the body segments changes. To develop a biomechanical model useful for measuring, for example, the CoM of a body, we must then divide the human body into a chain of bone segments with known and relative characteristics (segmentation) and, starting from the one with appropriate regression equations and the use of anthropometric tables, we can get to calculate, although not without a certain approximation, the CoM which, as we have seen, coincides with its CoG.

12.4.3 The Static

Static is the branch of mechanics that deals with the analysis of the forces (loads) that weigh on the physical systems in static equilibrium, i.e. in a state in which the relative positions of the subsystems do not vary over time (acceleration = 0), or where components and structures are at constant speed. The static studies the conditions necessary for a body initially at rest to remain in balance even after the intervention of external actions, called forces.

When in a static equilibrium condition, a system is said to be in a quiet or resting state, or its center of mass or gravity is moving at a constant rate (Fig. 12.2).

Posture analysis is performed starting from the mutual position of the three body axes, transversal, sagittal and longitudinal, axes that allow the human body to be divided into three spatial planes, the FRONT one (intersection of the longitudinal axis with the transversal axis, forward and behind), the SAGITTAL one (intersection of the sagittal axis with the longitudinal one, right and left) and the HORIZONTAL one (intersection of the transversal axis with the sagittal axis, the upper part and the lower part of the body).

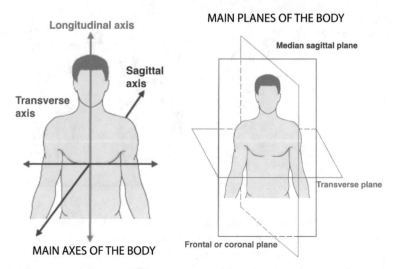

Fig. 12.2   Main axes and planes of the body

12.4.4   The Kinematics, the Scalar Quantities and the Vector Ones

The kinematics is applied in the description of the movement independently of the forces that produce it and includes a spatial reference system—linear displacements, angular or rotary displacements—and a space-time reference system—speed and acceleration.

A movement is called linear when all the parts that make up an object travel the same distance at the same time. The center of gravity of the object can travel either a straight line or a curved line, provided the basic conditions are met. When a curved line is evident and continuous we will talk about curvilinear motion; on the contrary, when the linear pattern is evident, we will talk about rectilinear movement.

A movement is called angular or rotary when an object rotates around an axis, called rotation axis, which can be represented by a fixed point or by the center of gravity of the object; in this last case the parts of the object closest to the rotation axis make a shorter path or trajectory than the more distant ones, so the latter will move at a higher speed than the nearest ones.

All kinematics variables of linear or angular displacement are vectors, elements that are inside a space called vector space, which can be added to or multiplied by numbers called scalars ($\pm$). Vectors are commonly used in physics to indicate quantities that are defined only when a magnitude is specified (the length of the same vector indicating its intensity), a direction and a versor relative to another vector or a system of vectors. The quantities that can be described in this way are called vector quantities, forces such as velocity, acceleration, displacement, capable of modifying the motion of a body or producing a deformation. On the other hand, when the measure is a simple (and only) measure of size (how large, how fast, how long, etc.) is defined as scalar quantity or quantity. In the aglo-saxon language there are two different terms to define the velocity: one, speed, refers to a scalar quantity in which only one quantity is given (the derivative of a space traveled with respect to a determined time); the other, velocity, is instead the physical term used to define in cinematic a scalar quantity that must be accompanied by the indication of the direction and represents a delta, that is a variation of speed in relation to a variation of time, therefore not a single value medium as it is done using a scalar quantity (speed).

The instantaneous speed represents a particular case in which the reference to time tends to the value 0, i.e. it is extremely small.

Both with speed and with velocity we have the indication of a constant speed; if, on the other hand, we intend to record a speed variation with respect to a certain time, it is necessary to resort to another, equally vectorial quantity, the acceleration, which represents the derivative with respect to the time of the velocity vector. The average acceleration coincides with the instantaneous acceleration when the latter is constant over time, and we refer to this condition as uniformly accelerated motion.

12.4.5 Moment of Force and Inertial Force

The moment of force, or torque (T) in the Anglo-Saxon field is a particular type of vector, and it is considered a pseudo-vector because unlike the classical vectors is not defined in relation to a reference system which could be the Cartesian axes. The moment of force indicates "the ability of a force to give a rotation to a rigid body" around a point, if 2D, or to an axis in space, if 3D, when it is not applied on its center of gravity or center of mass (otherwise there would be a linear translation) (Fig. 12.3).

Among the vector quantities it is useful to remember the impulse, a type of vector that indicates the change of the momentum of a particular body in a time interval, which makes us understand that the variation of speed of a given object does not depend only from the force that is applied to him, but also from the duration of this operation.

12.4.6 Kinetic (or Dynamic Tout Court)

The kinetic or dynamic, with a term now acquired, is that branch of classical mechanics that studies the forces that produce or modify the movement of a body, internal or external to the body itself. Examples of internal forces are the muscular activity, the action of the ligaments or the friction that comes from the muscular system or from the tendinous and ligamentous system. Examples instead of external forces may be the soil, external agents of any kind such as gravity, the activity of other interacting bodies or passive sources such as the action of the wind (Fig. 12.4).

Fig. 12.3 Example of
moment of force (torque)

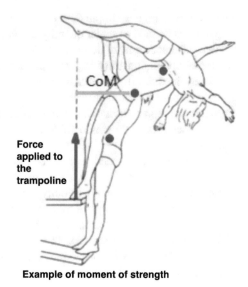

Example of moment of strength

Fig. 12.4 Categories
applied to the analysis of the
movement used in
biomechanics

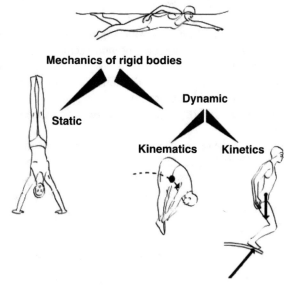

Categories applied to movement analysis used in Biomechanics

   In general, the dynamic problem is more difficult to solve than the kinematic
one because it involves the forces acting on the mechanism and the inertial characteristics (mass and tensor of inertia) of each of the elements that compose it and

also implies the (numerical) resolution of complex equation systems. A wide variety of dynamic analysis can however be carried out, and thanks to those and the great support provided by digital technologies we can achieve a thorough evaluation and understanding of the mechanisms and strategies involved in the production and control of movements.

12.5 The Biological and Anatomo-physiological Component

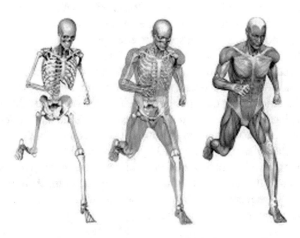

From the beginning, biomechanics has been applied to the study of the structural (anatomical) and functional (physiological) components of the human body and of the animal, addressing a particular albeit not exclusive, attention to the locomotor system and those tissues and organs that refer to that such as musculoskeletal system and the nervous system.

12.5.1 The Musculoskeletal System in Humans

The musculoskeletal system[8] is composed of a combination of bones, joints, muscles and connective tissue (tendons, ligaments and cartilage)[9] which, in addition to the

---

[8] An apparatus is defined as a set of tissues and organs different in structure, function and origin, but which cooperate with each other for the performance of the same functions (respiratory, digestive, locomotor, etc.). A system, on the other hand, is a group of organs and tissues that perform a common function, but which are not necessarily connected to each other from the anatomical point of view (nervous, muscular, skeletal, etc.).

[9] We can count around 320 pairs of muscles in the adult human body (depending on the type of classification), 213 bones (excluding sesamoid bones) and about 360 joints (from rigid to semi-rigid to mobile ones).

functions of form, protection and support, allows, with the contribution of the system nervous, the realization of all types of movement, volunteers, automatics and reflexes.

It is beyond the scope of this work to examine the morphological-structural component of the musculoskeletal system; we refer for this examination to the many texts of anatomy, especially the functional one,[10] which allows us not only to examine the origins and insertions or connections of the main body segments, be they bone, muscle or tendon-ligaments, but above all to understand how these work together, synergistically, according to the task.

Biomechanics, as we have seen, is interested in the forces and external perturbations that modify the state of quiet or movement of a body (the mechanics), and also deals with the forces, or rather, the internal loads acting on the living organisms, on that of man in particular, modifying its state and condition.

Among the vertebrates the musculoskeletal system represents the supporting structure: while the bones bear the gravitational forces and the internal forces produced by the contractions of the skeletal muscles in order to maintain the posture, the ligaments hold the bone segments together through the joints to which they are connected. The tendons for their part play the role of connecting the muscle to the bone, transmitting the forces exerted by the muscle to the skeletal system to produce the movements. As consequence of the forces exerted on the body, the movements produced, the elastic energy transmitted by the contractions or the thermal energy released by them, the adjacent tissues, ligaments, tendons, connective tissue, muscles, vertebral discs and the nerves are constantly subjected to mechanical and thermal overloads with consequent inevitable wear and tear, if not just exposure to trauma.

In-depth information on each system follows.

12.5.2   Ligaments and Connective Tissue

Ligaments reduce the degrees of freedom of the joints by limiting non-physiological movements. In particular, their main function is to "guide" the movement of the joint, to maintain the congruence of the joint and to act as a deformation sensor for the joint. When a tendon is subjected to traction overload, initially it opposes a low resistance to elongation, however, as the trajectory increases (stress-tension curve) and the maximum amplitude or elongation range is approached, it quickly begins to oppose a strong resistance. This means that the stability of any joint in the low-tension area of the stress-tension curve must be exercised by muscle contraction and that the ligaments help to support the mechanical load on the affected body joints only to oppose extreme movement.

---

[10]Kapandji (2011), Kendall and Mc-Creary (2005).

12.5.3 Tendons

Tendons are composed of about 80% of collagen tissue, which forms a link between muscle and bone; they are essentially fibrous "cords" that bind the muscles to the bones and ensure mechanical continuity; they transmit the tensile load from the muscle to the bone and thus assure the movement of the joint. The tendons bear a tensile load higher than that of the ligaments, but have a stretching and deformation capacity reduced by about half compared to those, so they are very resistant but also relatively not very elastic.

12.5.4 The Muscles

Skeletal muscles provide movement and maintenance of posture by transferring tension to the skeletal system through the tendons (myotendinous junctions), which develops through active contractions or passive stretching of contractile units.

The musculoskeletal system uses a mechanism as simple as levers to produce more or less large angular changes in adjacent body segments. As a result, the amount of muscle force (power) necessary to produce a desired effort or movement depends on the characteristics of the external or resistant force (resistance) and the distance from the fulcrum—the point which remains stationary with respect to both power and resistance—at the point of application of the agent force.

Thus, we can distinguish 3 different species or types of levers, more or less advantageous for the purposes of the muscular strength required to move the resistant one relative to the support point or fulcrum. A greater length of the arm (distance from the fulcrum) of the power compared to that of the resistance leads to an advantage of the lever, i.e. the power applied may be less than the resistance in order to win it. Vice versa, if the resistance arm is greater than the power arm, a great power must be applied to move a small resistance: the lever is said, in this case, disadvantageous (Fig. 12.5).

A lever is of the first type if the fulcrum is between the acting force (power) and the resisting force (resistance). It is advantageous if the power (Fm) is more distant from the fulcrum compared to the resistance (Fr), and vice versa when the resistance is closer to the fulcrum of power. Example: in the human body we have a fairly small

Fig. 12.5 Levers of first type

number of levers of this kind, an example is represented by the flexion-extension of the head by the occipital muscles having as its fulcrum the first cervical vertebra (atlas) and the occiput and as resistance the weight of the head. In this case the power arm is less than that of the resistance; to bend the head or to keep it steady, we have to use a greater amount of force than that represented by the weight of the head; it will therefore be a disadvantageous lever. On the contrary, extending the arm having the elbow (fulcrum) raised holding a counterweight (resistance) using the posterior musculature of the arm itself (triceps), is an example of an advantageous first-rate lever (Fig. 12.6).

A lever is said to be of the second type if the resistance (Fr) is on the same side of the power (Fm) and the latter is more distant from the fulcrum than the resistance. The second type levers are always advantageous because the distance of the power from the fulcrum is always greater than that of the resistance. Example: raise on the forefoot by the calf. In this case, the fulcrum is represented by the forefoot, the resistance is represented by the weight of the body that weighs on the foot and the power is represented by the strength of the muscles of the legs (calves). The lever is advantageous, the effort we make for walking is reduced compared to the weight represented by our body to be lifted (Fig. 12.7).

Finally, a lever is of the third type if the fulcrum is on the same side of the muscular force (Fm), at the same time it must be closer to the fulcrum than the resistance (Fr). The third type levers are always disadvantageous because the resistance arm will always be greater than that of the power. Example: lift an object with your hand by closing the elbow by the brachii biceps. In this case the fulcrum is represented

Fig. 12.6 Levers of second type

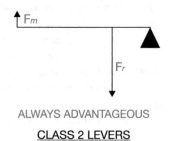

Fig. 12.7 Levers of third type

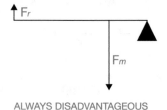

by the elbow, the power is represented by the front musculature of the arm and the resistance is represented by the object placed on the hand.

Levels of the second type, advantageous, are very poorly represented in the human body; most of those present in fact, in the arts above all, are of the third type and therefore disadvantageous from the point of view of the force required to overcome the resistance. This may seem contradictory, in fact the function of the muscles involved in locomotion or manipulation is not that of moving large loads or overcoming large resistances, but rather allowing fast and wide movements, and therefore the disadvantage with respect to the force to be exerted is largely compensated by a greater angular velocity and a greater capacity of articular excursion.

Each muscle works associated with the antagonist muscle, that is to say, to each muscle it corresponds to another muscle with opposite functions: when one contracts, the other relaxes and vice versa. Most body movements are made possible thanks to the coordinated action of the agonist and antagonist muscles.

12.5.5 Muscle Contraction

The muscular contraction is generated by a stimulus that can come from outside (load) or from inside (movement); this gives rise to a contraction of the muscle after a brief period of latency due to the complex (invisible) processes of receiving, processing and conducting the stimulus. This is followed by the muscle contraction phase in which the energy produced by chemical bonds (ATP) is converted into mechanical energy with the shortening action of the contractile units (sarcomeres with the related actin and myosin filaments) arranged in series on the muscle fibers or myofibrils.

Once the contraction has been made, the muscle fiber relaxes and returns to the original condition through the action of particular enzymes which block the prolongation of the contraction phase, making the muscle return to the initial relaxation state. This process corresponds to a single "muscular shock" that involves only a minimal part of the muscle, the one related to a myofibril and the series of sarcomeres placed longitudinally next to each other. The individual shocks are different from fiber to fiber for what concerns the speed with which they develop voltage (duration of the ascending phase of the curve), for the maximum voltage that they reach (height of the curve) and for the duration of the shock itself (amplitude of the curve).

12.5.6 The Different Types of Muscle Contraction

In relation to the parameters of the muscle shock listed above, speed, width and duration, we have two different types of muscle contraction, one of a static type, the isometric contraction, and one of a dynamic type, to which two different types of contraction, isotonic, the most common, and isokinetic, (isocinetic = constant speed) obtainable with particular equipment.

Isometric contractions are static contractions that occur at constant muscle length and are obtained when the muscle shortening is prevented by a load equal to the muscle tension, or when a load is supported in a fixed position by the muscle tension. The isometric contraction occurs when the muscle contracts without changing its length, without moving a load, even if often the force applied, or the recruitment of muscle fibers, is very high. The picture is different in the dynamic contractions that occur when the muscular length varies, producing a work, that is a movement, of a part of the body.

An isotonic contraction occurs when a muscle is shortened by shifting a load that remains constant for the entire duration of the shortening period. There are two types of isotonic contraction: the concentric contraction and the eccentric contraction.

The concentric contraction (positive work = the applied muscular force acts in the same direction as the angular velocity of the interested joint) is that which causes the muscle to shorten when it contracts. It is the most common form of muscle contraction and occurs frequently in daily activities.

The eccentric contraction (negative work = the applied muscular force acts in the opposite direction to that of the angular velocity of the affected joint) is instead that which occurs when the muscle stretches while it remains under tension due to the pressure imposed by an external resistance, that is, when the force applied to the muscles is greater than the force exerted by the muscle. An eccentric contraction, mainly antigravity, requires the consumption of a surplus of energy to make the action of stretching of the muscle by the external load (gravity, weight, resistance, etc.) less traumatic to slow down.

There is a second type of dynamic contraction, that is isokinetic, which occurs when the muscle develops the maximum effort for the whole width of the movement by shortening at a constant speed, contravening the so-called Hill law (see Fig. 12.8), according to which the speed is inversely proportional to force; in fact this is true in the common contractions and in order to carry out an isokinetic work it is necessary to use special equipment.

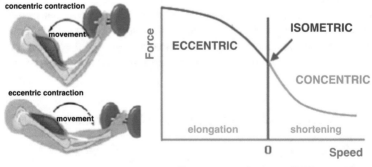

Force-speed curve (Hill graph)

Fig. 12.8 Hill chart and Force-Velocity curve

According to Hill's law, we can say that at maximum speed the force is equal to 0, while at speed 0 the force is very high; in other words, the expressed force is maximum during eccentric contractions, it is reduced in the isometric ones and even more in the concentric ones.

Thus, a great speed of muscle shortening can only be achieved with small forces, while high forces are only accompanied by small speeds.

12.5.7 The Efficiency

The term and concept of efficiency is probably the most abused and the most misunderstood one of everything concerning human movement. Confusion and errors come from an imprecise definition of both the terms that are included in the efficiency equation with the numerator, the mechanical work output and, in the denominator, the metabolic cost or consumption in input; that is, efficiency is inversely proportional to the metabolic cost and directly to the mechanical product.

Mechanical inefficiency derives mainly from two causes: from the conversion of metabolic energy into mechanical energy and from the inefficiency of the nervous system in the control of this energy, as the result of the combined work produced by each muscle, tendon, ligament and joint involved in movement and effectiveness of the control action of the nervous system and its ability to produce the response with the maximum speed, strength and precision required by the situation.

Since it is obviously not possible to directly measure the metabolic energy delivered, and indirect method such as comparing the oxygen (O_2) consumed with the exhaled carbon dioxide (CO_2) should be used.

12.5.8 Performance, Work, Energy in Human Performance

How good is the performance of muscle work[11] in the different types of human activity and how can it be calculated?

In classical mechanics the performance of a given machine is the relationship between the work produced and the energy[12] spent to make it, therefore the mechanical efficiency can never assume an absolute value higher than one, which represents the ideal performance; all the developed energy is used/transferred completely without losses. This is hardly true in reality, because internal forces that arise from the components of the machine, for example frictions, or external ones, which arise from

---

[11] In physics, work is a quantity that indicates a change in energy following the change in the point of application of a force for moving it (Work = force × displacement).

[12] Energy, whatever the form in which it may occur—electric energy, solar energy, thermal energy, chemical energy, nuclear energy, gravitational energy, etc.—represents the ability of a system to do work.

The yield is   YIELD   Output energy
expressed by the   ⟶   (in percentage)   =   ————————————   x 100
formula: Incoming energy

Fig. 12.9 Formula for calculating the efficiency (mechanical)

the action of forces or fields of forces external to the machine itself, such as weight or inertial forces, cause loss of a significant quantity of energy as in the case of the presence of friction or the production/dissipation of thermal energy (heat).

In applications to human performance, considering that the energy produced remains mechanical energy expressed generally in joules (1 j = 0.230 calories), or in Kg m, kilograms-meter (1 Cal = 426 kg m) or even, and often preferably, in calories (1 Cal = 4186 j), the metabolic energy spent is expressed in liters of oxygen consumed per minute (lO_2 min) to be converted for the determination of the efficiency in equivalent value in kilocalories (1 lO_2 = 5 kcal) (Fig. 12.9).

By comparing the output value (work) with the input value (consumption) we have the percentage value that indicates the efficiency.

As an example, suppose that a person is using a metabolic amount of 15 kcal/min for the mechanical work done and is producing an external (mechanical) work equal to an energetic equivalent of 3 kcal/min. In this case the efficiency is 20%—(3 kcal/min/15 kcal/min) × 100—this means that 12 kcal/min or 80% of the energy consumption has been dissipated.

Work is done by spending energy. When you do a greater amount of work with the same energy expenditure or even when a certain amount of work is performed at a lower energy cost, the efficiency is greater.

Other factors that must be taken into consideration when evaluating performance are the duration of work, the speed of execution, the load and not least the quality of work performed. Therefore, the calculation of the performance of a muscular work is a complex operation that involves several variables that are not all easily and exactly quantifiable.

The efficiency of the muscle cell is quite high, about 25%, comparable to that of a dynamo, higher than that of an internal combustion engine (20–25%). However, muscle performance during the execution of a complex movement such as march and running is surprisingly higher, reaching 40–50%. This is achieved by a combined action between muscle and tendon in the particular condition in which the muscle stretches during contraction (eccentric). During a gait, the fixation of the supporting leg is achieved with a knee flexion and a contraction of the quadriceps. The same leg in support will then be the one that provides the thrust that is realized with the extension of the leg caused by the contraction of the quadriceps. Therefore the quadriceps muscle remains in contraction in the support phase (contraction-elongation) and in the subsequent extension phase (contraction-shortening): from the mechanical point of view, in the contraction-elongation phase elastic energy is stored which is released in the following extension phase. This mechanism allows a considerable energy saving as the force for the extension of the leg derives from a recovery of elastic energy and not from metabolic activity. The physiological consequence of this mechanism is the low energy cost of march and running: about 1 kcal per kg of mass per km traveled.

If the efficiency of march and running is, as we have seen, very high, this is not the case for other human activities in which, due to the dissipation of energy, internal (frictions, heat) or external (load displacement, conditions environmental, etc.), the efficiency is much lower. For cycling, for example, the mechanical efficiency produced represents about 15–25% of that supplied, with a range justified by the different qualification of the subjects; in this case the loss of efficiency is mainly due to the heat dissipation connected with the muscular work and also to a significant extent to the forces of external friction that, in this case, correspond to the aerodynamic resistance and grow with the square of the speed.

In swimming, taking into account that the technique influences to a much greater extent than in other activities, the coefficient of performance is much lower getting around 5–15%, and this is due to the internal energies spent to allow the floatation, as for the strong resistance opposed by the fluid, which is calculated in the cube of speed, which however is much lower—and therefore less expensive in terms of energy—in the aquatic environment compared to what happens in other environments.

The comparison between different subjects regarding their energy expenditure and performance must be made taking into account the same amount of work; in this case it is practically identical for units of body weight and gender.

Very slow pace of locomotion are not favorable in terms of efficiency when a certain distance has to be traveled. Walking in the plains at 3.7 km/h requires a cost of 56 kcal per km while it requires 52 at a speed of 5.6 km/h (<51%), so also the walk in the plain at 1.6 km/h with a load of 20 kg requires a cost of 131 kcal per km, while doing it at a pace of 3.2 km/h requires one of 84 kcal per km (<36%). The reason for the low economic performance of slow gait is that a large part of the energy used during work serves to maintain body functions that do not play a role to support the execution of the work. When the same distance is traveled in a shorter time, the energy cost of maintaining vital functions is proportionally reduced.

12.5.9 Performance and Potency

An important factor for human performance is the time taken to perform the work: the longer the work period the lower the performance. In order to achieve maximum efficiency, the work must be performed with the maximum power[13] reachable, within the limits given by the individual work possibilities.

---

[13] In physics, power is defined as the work accomplished/ transferred in the unit of time, or as its temporal derivative ($P = dL/dt$). In the international system of units of measurement, the power is measured in watts as a ratio between units of energy or work in joules and units of time in seconds. $1 \, w = 1 \, j * s$—or even—$1 \, w = 859.85 \, cal/h$.

When the distance to travel is large, the speed must be reduced so that the subject does not have to be exhausted from the job. To perform a long-term job it is necessary to sacrifice the efficiency to the resistance. If the distance is short, the speed must be increased or the loads will be increased if it is to be carried out with maximum efficiency.

12.6   The Functional Components: Posture, Movement and Strength

Biomechanics allows you to examine and evaluate gestures or movements in the framework of the body's global activities, regardless of their nature or purpose. Understanding its relationships and intertwining is the key to understanding and describing complexity. Referring to the human body, the main functions, supported by the cardiovascular system, are regulated by the nervous system, the musculoskeletal system that are functionally linked to each other and therefore indicated by the synthetic term of the neuro-musculoskeletal system (NMS). The biomechanical understanding of the interactions of the NMS requires the knowledge of the functioning of each of the three subsystems and above all of their combination within the environment in which it operates, in order to understand and use the biomechanics in the different fields of application, from that sporting to rehabilitation, from engineering to employment, etc.

12.6.1   Posture

Posture is defined as the relative orientation of the different body segments in space, aimed at maintaining and consolidating equilibrium.

In order to maintain this objective even for prolonged periods of time, from a biomechanical point of view it is necessary:

- to keep the head and the trunk aligned, resisting the force of gravity;
- to maintain the center of gravity inside the support surface;
- to stabilize the slower, capillarized (red fibers) tonic-postural musculature, subject to wear and rigidity, as a function of stabilization and support to the phasic-motor musculatur, more rapid (white fibers), fatiguing, dedicated to the dynamic and finalized gesture.

Consider a subject who is standing upright forward without a support (see Fig. 12.10). In this condition the mechanical load on the ischiocrural muscles (hip extensors) and sacrospinal muscles (extensors of the spine) is proportional to the distance (d) of the hips from the center of gravity (CoG) of the upper body, which increases with the removal of shoulders from the point of fall of the pelvis. Muscle tissue, like other soft tissues involved, do not respond positively to this type of load

Fig. 12.10 Left, subject
inclined forward, right
posture incorrect (angles
should have been more
correctly 130–140°)

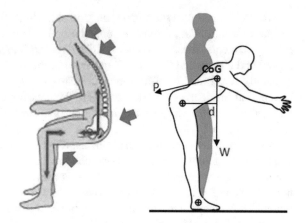

called "static work" (P), especially if prolonged. In this case the blood supply and consequently of oxygen are reduced, so the biochemical balance of the organism is disrupted with the formation of catabolites (lactic acid) which accelerate the appearance of fatigue, first, in the form of soreness, therefore in the form of real pain (alarm reaction) that forces the subject to change the position to find relief and avoid worse problems such as muscle trauma or loss of balance with the risk of falling.

If the uncomfortable or unbalanced posture is not changed - which can happen if repeated in a systematic way as it happens, for example, in office workstations—after the first warning signs begin to appear more acute and increasingly pains that open the way to the next phase in which it is possible to establish frameworks that are not easily reversible, such true musculoskeletal, neurosensory or even, in some cases, circulatory pathologies.

We will then talk about overuse syndrome due to static overload, protracted repetitive movements, excessive stress or even a combination of these. Note that psychological factors may also be involved since it is shown that psychological stress is associated with muscle tension.

In conclusion we can arrive at a definition of the posture that is wider than the one made in the opening, not just a "sum of reflexes and a complex multi-sensorial and polysegmentary interaction", but the result of the adaptation of each individual to the physical, psychic and emotional environment. whose purpose is to realize a position, to maintain an attitude, to guide and reinforce the movement and also to communicate sensations and moods.

12.6.2   Posture Control

It represents a person's ability to maintain the stability of the body and of the different segments in response to the forces that disturb the structural balance of the body itself.

Posture has mainly a stabilizing function of the body balance. To support posture, the postural musculature must support the body against gravity, stabilize the supporting elements (muscles, tendons, ligaments and joints) when other elements are moved and ensure that the body is balanced in the vertical projection of the center of gravity (CoG) and/or of the center of mass (CoM), which tends to remain or to be reported inside the support base where the pressure center is located (CoP), the manifestation of the forces acting on the basis of support in the maintenance of the standing station.

Posture is maintained through a constant re-elaboration of the parameters of muscular activity (length and tension), which is indispensable for maintaining equilibrium. The center of gravity is in continuous movement both for the action on the body of external forces and for the movements caused by the voluntary and in part, minimal movement, even for the small adjustments that the body makes when the static posture is prolonged over time.

In conditions of rest, for example standing, most physiological systems demonstrate very irregular and complex dynamics that represent interactive regulation processes that operate on multiple time scales. These processes require the organism to have an adaptive response ready and able to react to sudden stresses, for example, in the case of micro postural adjustments to maintain a certain position in time, which may require no movement and therefore respond to eminently proprioceptive control mechanisms that occur without the help of the locomotor apparatus, or, if the center of gravity is brought out of the support base, which requires more articulated responses that not only involve processes of a sensory nature. perceptive. Finally, there is a third condition that arises when postural control takes place within the framework of targeted voluntary movements, of lesser interest for our treatment, distinguishing when the voluntary movement does not cause the loss of equilibrium from when it provokes it. If the voluntary movement does not jeopardize the state of equilibrium, the consequent reactions have only a compensatory effect on the shift of the center of gravity, they are weak and not closely linked, in a temporal sense, to the execution of the voluntary movement. If the voluntary movement instead jeopardizes the equilibrium state, then the anticipatory postural reactions, the so called feedforward reactions, are observed.

In the postural control in the erected station, the postural system of man is able to respond to the oscillations of the barycentre given by minimal forces, with a suitable and contrary force exercised in a timely manner, which allows the maintenance of the upright position. This postural response is partly linked to the intrinsic mechanical properties of the muscle, in part to a set of nervous-type reflex mechanisms that, by gathering information from different sensory channels, are able to modulate the muscle tone (postural) in order to allow the keeping upright position.

When the organism is disrupted, for example when the center of gravity tends to come out suddenly from the base of support and the postural tone is no longer able to maintain the equilibrium position on its own, the control mechanism mostly uses automatic responses, closed-circuit feedback that operate on relatively short time scales to restore the body's balance.

Maintaining the center of gravity inside the support base is ensured by a complex system of sensory afferences that carry information about the deviations from the orthogonal position of the body with respect to the ground, in particular from the cutaneous receptors under the soles of the feet, from the muscular proprioceptors and articulations sensitive to changes in muscle tension and to changes in the position of the articular segments and from the vestibular receptors which, on the basis of the movements of the head, indicate the inclination of the body. Finally, the visual afferences that transmit information on the movement of the visual field, that is to say on the surrounding spatial context, are added to those we've seen before.

12.6.3 The Vertebral Column

The human vertebral column consists of twenty-four movable and 9 fixed vertebrae separated by deformable fibrocartilaginous disks, the intervertebral disks. The column is surmounted by the skull and rests on the sacrum, which is firmly bound to the hip bones through the sacro-iliac articulation. The vertebrae can be subdivided into 4–5 groups: seven cervical, twelve thoracic and five lumbar, followed by 5 sacral vertebrae fused in the sacrum and 4–5 coccygeas fused in the coccyx. The spine is a flexible structure and is held together by numerous muscles and ligaments.

In the upright position, the human spine normally presents a series of curves when viewed on the sagittal plane (laterally): the concave cervical region, the convex thoracic-dorsal region and the concave lumbar region; the concavities are known as lordosis and convexities as kyphosis.

In the upright position, the pelvis-at the base-is more or less vertical and between the first lumbar vertebra and the sacrum there is an angle of about 25–30° with respect to the horizontal plane. When sitting on a chair of normal height like a dining chair, we bend our knees at 90° and open another similar angle between the thighs and the trunk, in which case most of our weight is discharged on the ischial tuberosities. Note that part of the angle between the thighs and the pelvis is obtained by the hip joint which in that case rotates the pelvis forward by about 25–30°, compensated by an equivalent flexion of the lumbar spine when it remains in a vertical position (40–50° if you lean forward, 10–20° if you lean back), then, while sitting in a vertical position, you tend to flatten the normal lordotic curve of the lumbar region.

The pivotal element of the vertebral column is the vertebral body. This in the cervical tract has a smaller size, is convex superiorly and concave in the lower part and has a quadrangular shaped surface, in the thoracic tract it becomes larger as we move away from the cervical one with a circular shape, finally in the lumbar tract it has a wedge shape and a considerable volume obviously designed to bear the greater load of the overlying part. The load capacity of the vertebral body depends a lot on its content of water and nutrients. With the increase of the age a rarefaction of the structures of the spongy bone of which the vertebra is composed is produced and together a less hydration of the disks with a decrease of the load capacity. The load capacity of the vertebral body increases from top to bottom and is increased by a physical training that foresees axial loads.

The spine is particularly prone to postural stress. The long muscles of the spine called spine erectors are important in lifting loads and supporting the weight of the upper trunk when it is tilted forward. Since they run parallel to the vertebral column, each tension of these muscles exerts a compression of the same amount on the spine.

Each pair of vertebrae is connected to an intervertebral disc that acts as a hydraulic pad; the discs have a composite structure consisting of a central gelatinous body, the nucleus pulposus, rich in proteoglycans that retain high amounts of water and an outer part rich in collagen, the fibrous ring. The nucleus pulposus for its composition is able to withstand strong compressions, expanding towards the fibrous ring that can sustain, if in good condition and thanks to the particular arrangement of its fibers, strong stresses produced by the transmitted hydrostatic pressure.

A famous study (see Fig. 12.11) produced years ago by prof. Alf Nachemson, director of the Department of Orthopedics at the University of Goteborg (SW), high-lighted the effects of different positions on the intervertebral discs of the lumbar spine, i.e. that part of the spine that is more stressed and that more easily it goes to wear and consequent pathologies. Using a standing posture as a reference point at 100% of the pressure, sitting without assistance with good posture, the pressure increases to 140% probably for the deactivation of the muscles of the central region (core). Always sitting but increasing the pressure on the lumbar region and moving the CoG forward, the pressure goes to 185%, which is reduced by 50–80% if you lean on a backrest. The positions most at risk are those in front free flexion, espe-cially in an upright position: they cause an increase in the pressure from 50 to 175% depending on the weight lifted anteriorly. The study makes us understand how some positions that are taken daily (at work, in the car, in housework, etc.) can be danger-ous for the column, although it should be noted that a healthy rachis is easily able to withstand the stresses described; problems can arise with the prolonged maintenance and repetition of these positions in relation to the age and also to the state of health and physical condition of the interested parties.

Fig. 12.11 Effects on intervertebral discs of different positions (Alf Nachemson, University of Goteborg, SW)

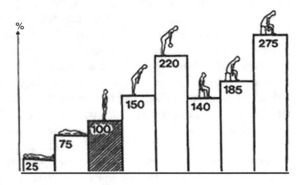

12.6.4 Strength, Tone and Posture

The muscular force[14] is that motor capacity that allows to overcome an external resistance or to oppose it by the development of tension by the musculature. It should be noted that this definition is suitable for the type of force commonly used in everyday life, called general force, which is expressed through contractions, isometric or isotonic, normally-but not necessarily-of the sub-maximal type, mostly short or very short duration, 3–5 s. Other types of force, defined as special, refer to the integration of the musculoskeletal system with other systems of the human organism—the nervous one for the development of the rapid or dynamic force (elastic-explosive-reflex) or the cardio-circulatory system for the development of resistant force (endurance), both of which are not significant for the purposes of our study and in any case far more difficult to measure than for general or static force. Especially in order to control the center of gravity and counteract the action of external forces avoiding imbalances, rather than actions determined by muscle strength we speak of control of the postural tone, a state of normal, light and continuous contraction of the muscles, especially of those extensors, called, indeed, stabilizers, which are opposed to gravity, regulated by a complex system that connects the upper nerve centers with sensorimotor receptors placed inside the voluntary muscles and with numerous peripheral receptors (vestibular, podalic, ocular) (Fig. 12.12).

Muscular strength and postural tone, while being related to functioning mechanisms and different activation conditions, are however closely linked and their bond is particularly evident in the face of alterations or central or peripheral pathologies

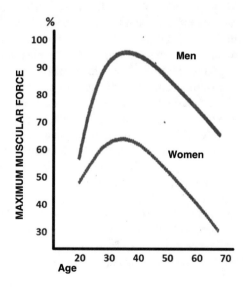

Fig. 12.12 Maximum values of strength of men and women in relation to age

---

[14]Hettinger and Thurlwell (2012).

affecting the neuro-directional-motor system. A good muscle tone, which corresponds to a right "tension" of the muscle at rest and the speed with which it manages to contract, facilitates a good condition of the muscular system which in turn acts positively on maintaining a good posture; an excessive muscular tension, an hypertonic, can be a sign of serious CNS pathologies which makes the motor response and the maintenance of a correct posture less fluid and precise. Finally, the presence of a low level of muscular tone, a hypotonous, makes the voluntary response less ready and effective, as well as strongly affects the maintenance of a correct posture. "The strength that a subject is able to produce in every kind of physical activity is almost always strongly conditioned by the posture he adopts, and problems of strength are almost always also problems of posture. Studies in which force is measured in a different number of positions, generally show that it affects posture to a greater extent than individual differences (age, sex, etc.)".[15]

12.6.5 Trend of Postural Control and Dynamic Strength in Relation to Age and Gender

Multiple studies have shown that with aging, both sensory inputs and the ability to provide appropriate motor responses are reduced by making changes to the strategies used for postural control. With aging, both static (maximal) and dynamic muscle strength is significantly reduced, and the mechanisms of coordinated activation or co-activation of the agonist and antagonist muscles are also significantly worsened. These processes of fall of functional abilities linked to postural control and dynamic movements, have highlighted strong gender differences and with respect to age, but the two abilities, one more related to control, the other to dynamic aspects, have highlighted a trend not coincident in relation to the two variables mentioned above.

Hettinger (1961)[16] studied the maximum values of strength developed by both men and women when the three groups of muscles which for the author are particularly significant for the evaluation of human muscular power (quadriceps femoris, forearm and muscle erector spinae) are used.

As can be seen, under normal conditions, the peak of maximum force performance is reached between 20 and 30 years with significant differences between males and females only partially attenuated by the difference in muscle mass (in the composition as in the amount of the fibers) between the two sexes. Subsequently the strength remains relatively stable or decreases slightly for the next 20 years. After 60 years there is a very marked decrease (up to 75–80%) of the strength in both men and women, more significant in the latter, a decrease that however does not change the difference in values of force already observed in the age of maturity, which remains fairly stable lower than about 1/3 for women compared to men.

---

[15]Pheasant (1986, pp. 154–156).

[16]Hettinger (1961).

	Maschi			Femmine			In combinazione		
	90^0	60^0	30^0	90^0	60^0	30^0	90^0	60^0	30^0
KE (%)	14.28 (10.37)	11.42 (7.37)	10.06 (5.18)	35.20 (43.49)	20.56 (17.69)	23.26 (11.79)	25.16 (33.30)	16.17 (14.25)	16.93 (11.27)
KF (%)	5.01 (2.55)	8.68 (7.67)	9.23 (13.06)	14.89 (15.77)	15.72 (14.48)	18.22 (19.58)	10.14 (12.36)	12.34 (12.03)	13.90 (17.05)
	$+20^0$	Neutro	-20^0	$+20^0$	Neutro	-20^0	$+20^0$	Neutro	-20^0
AP (%)	7.94 (4.87)	6.78 (4.30)	8.26 (7.87)	17.13 (11.22)	14.13 (8.40)	15.67 (14.99)	12.72 (9.79)	10.60 (7.60)	12.11 (12.45)
AD (%)	8.94 (7.85)	9.02 (7.32)	7.88 (6.09)	20.65 (10.34)	19.62 (10.81)	23.06 (13.63)	15.03 (10.83)	14.53 (10.59)	15.78 (13.03)

Legenda: KE = knee exstensors (quadriceps f.) - KF = knee flexors (hamstrings) - AD ankle dorsiflexors (tibialis ant., extensors digitorum) - AP = ankle plantarflexors (soleus, gastrocnemius, tibialis post.)

Fig. 12.13 Level (%) of co-activation of the antagonist muscles during a maximal voluntary contraction (M/F and in combination). Crowley-McHattan (2013)

Another question of great interest for our study is to understand if the marked differences in gender and age detected with respect to the maximum (static) force are confirmed also in relation to other muscle parameters, above all qualitative, as, for example, the co-activation of the antagonist muscles that we know to be indicative for the purpose of fluidity and coordination of movements, a quality that physiologists use to define intramuscular coordination.

This quality, referring to the whole of the neuromuscular function, or better, to the integration of the nervous system with the muscular system, would refer both to "central functions such as the activation of cortical and sub-cortical areas, and to the transmission of the signal through the motoneurons as well as peripheral factors such as the excitability of the effector organ"[17].

In the study reported here, carried out in 2013 by an American researcher, women[18] showed a higher level of coactivation compared to the male group and this would suggest that not only the two functions, the postural (perceptual) control and the maximum force, as we have seen, they refer to different functional mechanisms, but that this difference also follows a trend of gender as well as of age (Fig. 12.13).

Finally, we report the results of a research[19] carried out years ago, which had a great diffusion and success in the field of biomechanics applied to ergonomics, concerning the combined effects of the optimal angle of application of the arm, application of force and effect stabilization allowed by a different height of support on a backrest on the maximal force (Fig. 12.14).

The test consisted of pushing with one hand a handle connected to a height-adjustable dynamometer, seated, at a fixed distance of 80 cm, acting with elbow angles ranging from 25 to 160° and leaning on a high backrest 20 to 80 cm or

---

[17]Clark (1974).

[18]Crowley-McHattan (2013).

[19]Caldwell (1947).

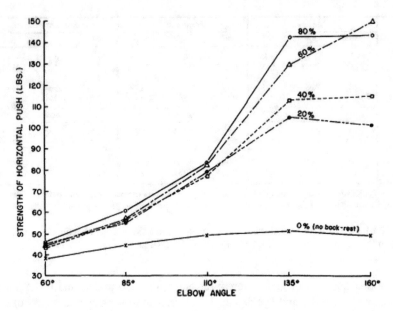

Fig. 12.14 Maximum thrust force expressed by the arm against an 80 cm handle with 5 different angles of the elbow and with 5 different height conditions of the support backrest

even without any support. The study had two purposes, that of measuring the effect of different degrees of stabilization of the body on the maximum force expressed through an effort (work) produced by the hand and that of measuring the effect of the different angles of the joint on the expression of strength by evaluating how much the stabilization effect provided by the back support can be a predictor of the test results. The result was that up to an angle of 135° elbow the backrest had no effect on the production of muscle strain; when, however, the angle of the elbow has arrived or has exceeded 135°, this has instead proved to be decisive for the result: the force is practically tripled compared to the initial levels with a reduced angle and without or with a low back support. The result of this test was very significant to affirm the principle of the importance of both the angle of application and the stabilization (postural) in the expression and production of muscle strength.

12.7   Biomechanical Evaluation of a Common Gesture: Walking

12.7.1   Biomechanical Evaluation

The analysis of a gesture is usually evaluated by evaluating the kinematic and dynamic behavior of individually assumed variables or parameters such as angles, velocities or accelerations, muscle or joint strengths or forces, direction (s), CoG or CoM trend, reaction forces with ground reaction force (GRF), etc. The technological development has allowed to create tools and equipment able to acquire with precision and accuracy an increasing amount of data related to the movement in order to better understand its organizational complexity and to verify and quantify the existence of a greater or less internal variability of gestures belonging to the same individual. this identifies two different levels of bimeccanic analysis and evaluation of the movement, different in terms of organization, purpose and result. One, more traditional, refers above all to the evaluation of single kinematic, dynamic or physiological variables, the other linked to the availability of advanced technologies (photogrammetric, dynamometric, electromyographic, etc.). This identifies two different levels of bimeccanic analysis and evaluation of the movement, different in terms of organization, purpose and result. One, more traditional, refers above all to the evaluation of single kinematic, dynamic or physiological variables, the other linked to the availability of advanced technologies (photogrammetric, dynamometric, electromyographic, etc.). Unlike the traditional approach, the second approach is able to collect and process information relating to several variables at the same time, providing answers on how the joints and segments are organized each time a specific movement is produced and where and when variability or compensation is found in executive strategies compared to reference models or performances.

12.7.2 The Gesture of Walking

Walking is certainly one of the most common gestures in human life but it is also one of the most complex movements to study and to describe in absolute terms.

A correct approach to this gesture, the most studied for the importance it has in human life, able to set in motion, between primary and secondary effects, with different functions, more than half of the muscles of the human body, is that of starting from the highest levels of organization, that of the control nerve centers, to involve therefore the whole of the sub systems and secondary functions (anatomical, kinematics, biomechanics, etc.).

From the functional point of view we must first take into account the role of gravity and the consequent reaction on the ground (GRF) for the search for balance, implemented in all phases of the step, from the preparatory one to the next pendulum, until the first leg completes its movement by resting the heel on the ground again with a slightly braked movement and cushioned by the different joints of the foot and leg.

The CoG that varies according to people (age, sex, general conditions) as in relation to external or environmental conditions, is in line below the shoulders and above the hips and tends by nature to remain stable; in fact, it represents the point where the potential energy of the body is equivalent to that of a single particle acting on that point, exactly the point where the body mass is able to maintain its equilibrium without external forces acting on it.

Biomechanics of the walking

Fig. 12.15 Variations in hip and CoG oscillation in walking

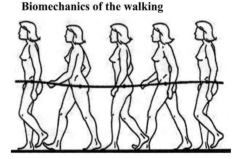

For the constraints generated by the joints the linear movement, which projects the body towards an imaginary point placed forward on the sagittal axis, represents the result of a series of rotations, so from the purely cinematic point of view the action of walking represents the result of several roto-translational actions able to make the action very economical from the energetic point of view and strongly aerodynamic (Fig. 12.15).

12.7.3   Biomechanics of the Walking

The action of walking involves a succession of alternate rhythmic movements of the lower limbs, pelvis, trunk, upper limbs and head which, causing a forward displacement of the center of gravity, produce, through a series of rototranslations of all the articular segments involved, the progression of the body forward.

The walking body barycentre has a sinusoidal shape on the sagittal plane reaching the lowest point in the double (bipodalic) support and the maximum height in monopodalic support, with a 4–5 cm excursion. Thanks to the high positioning of the body's center of gravity, the acceleration of our body is essentially of a gravity origin (potential energy that is transformed into kinetic energy).

Only modest accelerating muscle contractions come into play and this is the reason why it is possible for people to prolong the walking for a long time. In fact, it can be said that in walking, muscular work is required only in the periodic ascent of the center of gravity.

The biomechanical analysis of the path made it possible to distinguish two main phases: the stance phase, single support phase equal to 60–62% of the entire movement, and the swing phase, oscillation phase equal to 35–40%; the two phases define the "step" or the set of movements included between two successive moments of contact of the same foot with the ground. The difference of the walking with respect to the running is that in walking there is never a phase in which both feet are detached from the ground (in flight); this implies that at the moment of contact with the ground it touches the heel to absorb the impact, then moving on to the next phase, that of the metatarsal support, the task of exploiting the elastic energy released through the extensor muscles (leg stiffness) from contact with the ground.

From a functional point of view, the walking is a cyclic movement in which the position of the head must remain relatively stable to ensure control of the whole system and the surrounding environment, the arms and legs move synchronously in opposition, the pelvis and the shoulders rotate forward-backward-up and down around the vertebral column.

It is a movement resulting from the interactions between internal and external forces directed by a complex postural control system and the balance that regulates moment by moment, through the muscles, the relationships between the forces. Most of the lower limb muscle groups are active during walking (the lower limb has 29° of freedom of movement which corresponds to 48 muscles).

12.7.4 Computerized Biomechanical Evaluation of the Gait or Gait Analysis

The study of the biomechanics of the gait allows to know the movement and the position of the different body segments involved to obtain information useful for understanding how the gesture is organized through the neuro-musculoskeletal system of the subject, comparing it with the range of normality widely available in the anatomo-physiological literature and in the technical-professional practice not only clinical.

Gait analysis, or computerized analysis of locomotion, makes it possible to carry out, through the use of specific equipment integrated with each other, a description, a quantification and an evaluation of the motor patterns of a given subject.

Thanks to some of its peculiar properties such as non-invasiveness, the repeatability of the examination in a short time, the quantitative character and the three-dimensionality of the results provided, this method based on standardized protocols, is a fundamental instrument of investigation in the analysis biomechanics of human movement.

Following the integrated processing of the acquired data, the computerized tool provides a stick representation (segments) or three-dimensional representation (skeleton, parallelepiped) of the subject's movement with the obtained data compared with the normal ranges or with similar tests carried out previously.

In Fig. 12.16, we can see the vertical components related to the reaction forces on the ground during a step cycle and the comparison between the path of a normal subject (a) and that of a Parkinsonian (b). The support phase, or stance, begins with the support of the heel (HS) and ends with the detachment of the heel itself (60%). The swing or swing phase, takes place from 60 to 100% of the cycle.

A gait analysis exam provides information in 3 different areas: a cinematic area, angles and accelerations of the joints examined; a dynamic area, with moments and powers related to the joints of the pelvis, knee and ankle evaluated during the walking phase; a third area, the electromyographic one, with the activation and deactivation data of the examined groups.

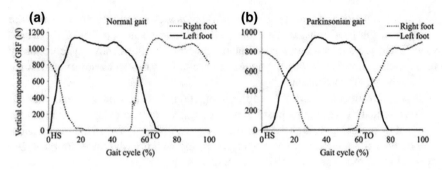

Fig. 12.16   Computerized analysis of the gait

It should be specified that today such tests do not necessarily have to be carried out in clinical laboratories but, thanks to the availability of miniaturized and portable instrumentation and wireless communication methods, where required, it has been possible to carry out "field" examinations that despite the presence of environmental factors that may influence the examination itself would also have the significant advantage of proposing closer examination conditions, for example in sports or in the workplace, to the actual operating conditions of the interested parties.

References

Bell F (1998) Principles of mechanics and biomechanics. Nelson Thornes Ltd, Londra

Bernstein NA (1967) The coordination and regulation of movements. Perfamon Press, Oxford

Caldwell LS (1947) Body stabilization and strength of arm extension. US Army Research Laboratory, Fort Knox

Clark WE (1974) The anatomy of work. In: Floyd WF, Welford AT (eds) Symposium on human factors in equipment design. H.K. Lewis, Londra

Crowley-McHattan ZJ (2013) Effects of ageing and exercise training on postural stability in relation to strength ratios of the lower limb muscles. Southern Cross University, Lismore

Duchenne GB (1867) Physiologie des mouvements démontrée à l'aide de l'ex-perimentation électrique et de l'observation clinique: et applicable à l'étude des paralysies et des déformations, Baillière, Parigi

Hatze H (1974) The meaning of the term biomechanics. J Biomech 7(2):189–190

Hettinger T (1961) Physiology of strength. Charles C Thomas, Springfield Ill

Hettinger T, Thurlwell MH (2012) Physiology of strength. Literary licensing, Miami

Kapandji IA (2011) Anatomia Funzionale. Monduzzi, Milano

Kendall F, Mc-Creary EK (2005) I muscoli, funzioni e test con postura e dolore. Verduci, Roma

Knudson D (2007) Fundamentals of biomechanics. Springer Science & Business Media, Canada

Pheasant S (1986) Bodyspace: anthropometry, ergonomics and design. Taylor & Francis, Londra e Philadelphia

Trubočkin D (2011) La biomeccanica nella didattica teatrale del GITIS di Mosca. Acting Archive Review

Chapter 13
Cognitive Aspects in User Experience Design: From Perception to Action

13.1 Introduction

For human beings, as for every other living being, the knowledge of objects and events in the surrounding environment is essential for survival. The ability to identify what is around us is related to the possibility of judging everything as good or bad, useful or superfluous, dangerous, animated, attractive, enjoyable, suited to our goals. But these qualities are not just characteristics of the things or of the systems with which we interact, they are also in our eyes and in our mind. Trying to understand how to get to know the world through the senses does not mean just trying to understand how it presents itself to us, but also how we process it.

Perception starts from physical stimuli, but it is not just a photographic record of events. Rather, perceptual activity produces representations and organizations of reality, structures of the world according to the point of view of the observer. The contexts that we process, distributed in space and time, have a value and a sense for the individual because we attribute meanings to them, our meanings. What we already know about the world, ineluctably, interacts with the perception of the present and allows us to construct mental representations that in turn guide the interactions with the contexts of use.

We are in a circular interaction with the world that on the basis of the understanding of objects, of the identification of what they are and of the environments in which they are located, leads human beings to formulate intentions that guide their actions. But intentions do not derive exclusively and directly from what we perceive; on the contrary, it can be exactly the opposite. Our needs, our desires, our skills, can lead us to perceive some things instead of others, to illuminate the presence of some objects around us and to completely hide others. In the interaction with an environment we assume personal interpretive perspectives on the basis of some factors such as our skills, our previous knowledge, our goals, the fact that we must work alone or in collaboration with another person or with an artifact. In doing so, the individual

---

This chapter was co-authored by Oronzo Parlangeli and Maria Cristina Caratozzolo.

© Springer Nature Switzerland AG 2020
F. Tosi, *Design for Ergonomics*, Springer Series in Design and Innovation 2,
https://doi.org/10.1007/978-3-030-33562-5_13

selects, among all those available, only the information that is recognized as useful for achieving his goals, that is, information that can be perceived, understood and able to create meaning. This set of complex processes is aimed at the development of mental models that represent reality and make it possible to carry out our activities in the best way possible. Our interaction with the world is mediated by the mental models that we structure to understand reality.

The world around us is designed in a human-centered perspective when it facilitates the elaboration of efficient mental models, when it makes available clues that are perceptible, interpretable, meaningful. The stimuli in the environment must have the characteristics to meet the subjective needs and solicit new possibilities for interaction. The way in which objects and environments are designed ultimately influences the formulation of intentions, can simplify the implementation of plans of actions and qualify user experiences.

13.2 The Structuring of the World in Perceptual Units: The Importance of Vision

For a long time it was thought that each sense led to specific perceptual experiences. And indeed the knowledge of the world can be derived from different physical properties that can be considered in isolation. We can think of the stimuli we perceive as electromagnetic radiations, sound waves, more or less volatile molecules, and in reference to these stimuli we can distinguish different perceptive modes: *visual*, *auditory*, *tactile*, *olfactory* and *gustatory*. And for each of these modes we have specific receptors. Today, however, we tend to think that the various senses cooperate in leading to relational experiences with the world that we can define as unitary. And indeed it is not difficult to describe a sound as high or low, a color as hot or cold, a pain as dull, a fragrance as enveloping, a taste as pungent. A perception, therefore, often involves complex sensory experiences that go beyond the sum of the single received stimulations (Palmer et al. 2013).

However, each of the perceptual modalities is important in its own way to guide our behavior, even though, perhaps, they are not all equally relevant. Research has shown that in cases where visual information is inconsistent with that coming from other senses, we then tend to trust the former more easily (Reason 1990). Therefore, vision is probably the most important source of information about the environment for humans. It not only provides us with the awareness of what objects are present in the world and where they are placed. More than all the other senses, vision produces a knowledge of the very structure of the space in which we are immersed, allowing us to orient ourselves and to move from one point to another. To try to understand how to get to know the world through the senses means, therefore, to explore the acquisition of information coming mainly, though not exclusively, through vision.

13.2.1   Stages and Problems in Visual Perception

Visual perception begins with the elaboration of images that are formed on our retinas. We can think of the retina as a surface on which the world is projected, but in the passage from 3 to 2 dimensions the information on the distance between things is basically lost. Moreover, the projections of parts of distinct and separate objects can be found next to each other in the retinal images, so that even just identifying the correct spatial relationships becomes problematic. In other words, therefore, neither the objects present in the world nor their distance are unambiguously given in the images that are processed at a first stage of vision.

To solve these difficult problems and provide us with the knowledge we need, the visual system exploits the previous knowledge that has been developed both in the course of individual development and in that of the species. Thanks to this knowledge it is possible to use the spatial and temporal relations between the points of the images as *clues* to reconstruct the state of the world that most probably could have given rise to it. These inferences are unconscious, and the processes at the neural level realize them as a series of *elaboration stages* (Marr 1982) that render explicit—in distinct mental representations—aspects of the world around us that are progressively more objective. In the early stages, the small oriented discontinuities of light present in the retinal images are analyzed, arriving then at identifying the margins between the regions of the images potentially corresponding to different objects in the world. The output of this stage is then further elaborated, up to the construction of a representation of the shape of the surfaces, of their spatial orientation and their depth relative to the observer and to that of the other surfaces present. This representation can then be used to finally identify the three-dimensional organizations of surfaces as specific objects, and access all the knowledge that we have stored of them in our memory.

In the transition from one stage to the next, we often have information that is lacking, ambiguous or incongruous. The visual system, therefore, in order to reach highly probable conclusions on the state of the world, uses particular clues and uses its own rules that are the result of a very long cognitive evolution (Guidi et al. 2011).

13.2.2   Perceptual Organization and Its Laws

When we look at the world, stimuli reach our eyes and are subjected to a series of elaborations. Considering what happens in the fundus of our eyes, that is, on the organized set of cells that we call retina, the stimulus configurations are qualifiable as two-dimensional organizations of points of different luminance and color. Starting from these two-dimensional organizations, one for each eye, how do we then understand which parts of an image correspond to the same object and which to another? How do we distinguish an object from the background on which it is placed? What makes a figure, a unitary whole, part of the visual space? These seem like little questions, because we are usually very efficient in answering these questions and because

Fig. 13.1 Emerging figure. In the figure are points that, due to their greater density, give rise to the perception of a diamond that is placed in front of a background made of sparser points

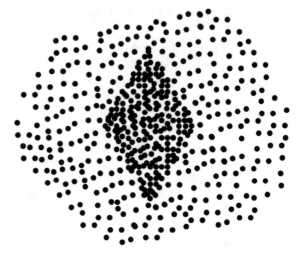

they are processes that occur unconsciously. For example, we are all immediately able to say that in Fig. 13.1 something like a diamond is represented. But, in fact, we should only say that in the figure there are only areas of more or less dense points. What brings us therefore to identify margins, unitary elements, geometric figures?

To solve these issues we use strategies that have been described as laws of perceptual organization that serve to analyze local and global relationships between the points and regions in the visual field even if, in many cases, they can lead to results that are not exactly optimal and that can be qualified as optical illusions (Parlangeli O and Roncato 2010; Roncato et al. 2016). These laws were identified for the first time at the beginning of the last century by a group of German psychologists belonging to a current of thought known as *Gestalt* psychology, and can be seen as *heuristics*, that is, optimal interpretative strategies, even if not always correct, based on assumptions about the typical structure of the world (Koffka 1935; Wertheimer 1923).

Gestalt laws can be considered universal processes through which we structure the experience with the world by identifying that which is an image, an object, something that makes sense to us.

Several laws aimed at visual grouping have been identified. The main ones are "*proximity*", "*resemblance*", "*good continuation*", "*closure*", "*common movement*", and "*simplicity*" (Kanizsa 1979). According to the law of proximity, elements of images that are close to each other tend to be grouped together and perceived as unity. This is exemplified in Fig. 13.2a, in which we see an organization of "two" groups of squares and not a set of twelve square figures, each in isolation.

The law of resemblance, on the other hand, states that the visual system tends to group together the elements present in the images that have similar characteristics of shape, color, size. Thus, in Fig. 13.2b the larger rectangles positioned at the center seem to be "together", perhaps forming a figure placed on top of a carpet of smaller rectangles.

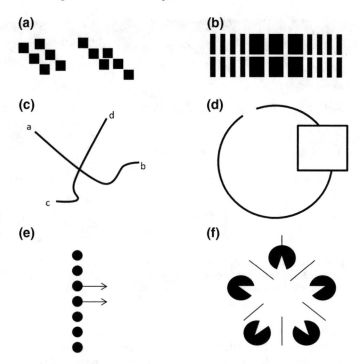

Fig. 13.2 Laws of perceptual organization. **a** *Proximity*: the squares of the nearest image tend to be seen as unity. **b** *Similarity* due to size causes the central rectangles to be seen as unity. **c** *Continuity* of direction causes two lines, a–b and c–d, to be seen intersected, instead of discontinuous lines such as a–c and b–d. **d** The tendency to *Closure* leads to seeing a circle even if at least in two points the circumference is interrupted. The *Common Movement* (**e**) of the two disks according to the arrows would lead them to be perceived as a unique configuration. **f** *Simplicity* leads to interpreting the various graphic elements in a result that foresees the presence of a white star that overlaps with black circles and intersecting segments

The law of good continuation (Fig. 13.2c) refers to the fact that we tend to interpret lines that intersect in order to preserve their continuity. So a–b is a line and c–d another. The perception of an a–c or d–b line is very unlikely.

The law of closure (Fig. 13.2d), on the other hand, states that lines that form closed figures tend to be perceived as a figural unit; thus it seems irrelevant that the line that forms the circle is interrupted at one point and, even, preserves its unity when there is a square that interrupts it and that, in this case, is seen as occluding.

In Fig. 13.2e a row of small disks is seen, but if two of them were moving in the direction indicated by the arrows, they would be perceived as a single separate unit.

In Fig. 13.2f graphic elements are present, each separated from the other but which, together, give rise to global percepts, thanks to meaningful, harmonious solutions, in which a five-pointed white star is placed on five black disks and on three segments that intersect each other below it.

Fig. 13.3 The interface of a site for online sales. The information categories and the possibilities of action are organized in coherent spatial areas to facilitate perceptual processes and consequently the activity of the user in interaction with this site for the online sale of clothing

Knowledge of these laws has proved particularly important for the design of artifacts and interfaces. These laws, in fact, correspond to the strategies with which the visual system groups the elements present in the two-dimensional stimuli and can be exploited to predict and control the interpretations that are most likely elaborated with reference to the reality represented.

As can be seen in Fig. 13.3, assuming that you have to interact with an online clothing sales site, the concepts and operational possibilities are matched with different spatial areas, in order to correctly guide the user's interaction with the system. Sometimes clickable areas are simply identified by verbal labels; in other cases a different chromatism is added to these, and the perception of geometric figures, such as the ellipse or the rectangle, is also exploited.

13.3   Perception and Design

The user, thanks to what has been implemented by the designers, should be able to establish a relationship with the environment, with the systems with which it interacts, which can be defined as adequate from a functional, cognitive and affective point of view. Taking advantage of the structuring principles of perception according to the Gestalt laws, the designer has at his disposal different strategies for organizing the information in such a way as to make the user's activity flow in the right direction. Among these, the most studied and exploited are *mapping* and *affordance* (Norman 1988).

13.3.1 Mapping

Mapping refers to the spatial organization of the elements of a system which must correspond to the mental representation necessary to interact profitably with them.

If we press a switch, or slide a cursor, we expect the consequences to be consistent with the action, so what is above must correspond to interactive elements placed at the top, and what is below must represent possibilities of use towards the bottom, as in the case of the arrows represented on the screen of our computer and which allow us to scroll the pages up or down.

Mapping can concern different aspects. Among these, for example, the correspondence between the position of the figural elements with what we know of the systems with which we interact. In Fig. 13.4 on the left we see a warning light placed on the dashboard of a car to signal (a) the amount of fuel in the tank, and (b) when it is necessary to go to the gas station. But if the emptying of the tank is represented by a progressive reduction of the notches from top to bottom, the moment of the need for the refueling action is positioned at the top, that is, when the tank is still full. Therefore, the icon of the gas pump correctly indicates that the rectangular warning lights refer to the fuel and its progressive reduction, but does not respect the mapping related to the supply action.

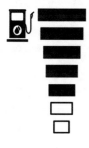
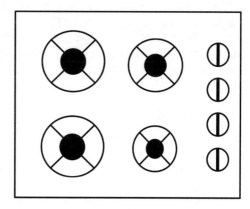

Fig. 13.4 Examples of violations of natural mapping. On the left the warning lights on a car dashboard that indicate the amount of fuel in the tank, but the need to go to the gas station is signaled in the part where the tank is still full. On the right, instead, a hob in which it is impossible to establish a relationship between knobs and burners

Colors can also have a role due to a more or less correct mapping. It has now become a universal convention that makes red the sign of ban, arrest, alarm. Just as there can be a difference between that which represents full, on the color level, and that which represents empty. In this case, the fuel gauge in Fig. 13.4, correctly shows the notches of the tank towards the bottom with two empty rectangles to indicate the need to refuel.

Another aspect of mapping can be related to the spatial correspondence between elements that are connected to each other. Figure 13.4 on the right shows a hob with burners and knobs for switching on that are arranged vertically on the side. This arrangement makes it practically impossible to establish with certainty which knob controls which burner.

13.3.2   Affordance

Affordance concerns the relationship between the possibilities of use offered by the environment and the characteristics of people interacting with that environment: depending on who perceives a given thing, different possibilities of use can be seized. In practice, this is the possibility that the environment has to indicate directly to the user, without the need for further cognitive processing, its multiplicity of use. A glass indicates that it is made to be filled, a bench that something can be placed upon it, a mirror that you can get a reflection of your own image. Depending on our ability to act, the environment is revealed in the opportunities it provides for these actions to be realized.

Over the years the concept of affordance has been enriched with different connotations according to the authors who have considered it in the context of their projects and theoretical proposals (Table 13.1).

It was introduced initially by Gibson (1979) who adopted the term affordance to describe the ability of the environment to directly suggest to the observer the actions that are allowed within a specific context of reference.

Table 13.1 Comparison between affordance, action and change (revised by Galvao and Sato 2005)

	Relationship between affordance and action	Relationship between affordance and change
Gibson (1979)	Dependent on possible actions	Does not change, is in the environment
Warren (1984)	Dependent on individual abilities	Can change according to the user's abilities
Norman (1988)	Dependent on possible actions, individual abilities, and cultural characteristics	Can change according to user experience
Maier and Fadel (2003)	Dependent on perception of utility	Can change according to the design objectives

According to Gibson (1979) affordance:

- is dependent on our perception;
- exists in relation to our actions;
- does not change as our goals change.

Warren (1984), subsequently, used the same term by considering the properties of human beings in addition to the properties of the environment. Thus affordance becomes a relationship between the possibilities offered by the environment and those given by the characteristics of the people who interact with that environment: depending on who perceives a given thing, the suggested affordances can change. For a human being a ladder invites you to climb, while for hens it has the value of a sequence of perches.

Norman (1988), on the other hand, is the author to whom this concept is usually referred. In the elaboration of his theory he has dealt with the affordances of the objects of daily life, like doors, telephones, but above all computers, referring to the perceived properties of the object that determine how that object could be used. According to Norman, even the concepts of past knowledge and experience, which help people to elaborate a mental representation of the use of a given object, are part of the concept of affordance. For example, the establishment of some affordances by citrus juicers of a very innovative design depends almost entirely on the experience of use, since the function of these objects, for users who have never seen them, is completely incomprehensible.

Finally, a more recent theory (Maier and Fadel 2003) has introduced into the concept of affordance the possibility of positive and negative boundaries, so as to emphasize what the artifact or system should invite to do and what instead should invite not to do. Thus, a knife presents not only the sense of being grasped by the part of the handle, but also that of not being picked up by gripping the blade.

It is possible to identify various types of affordances (Hartson 2003) in consideration of the different functions they support:

- physical affordances, which help the user in his physical activities; for example, the size of a button invites to the fact that this can be pressed and not to other actions;
- cognitive affordances, which help the user in his cognitive activities; as metaphorical icons or language labels that facilitate the understanding of the meaning of a button in an interface;
- sensory affordances, which help the user in the perception actions; for example, the presence of a word in italics within a text takes on particular importance and can contribute to increasing readability;
- functional affordances, related to the use and utility of a system, an artifact, an environment; as the possibility of organizing the documents in a folder by date or by name.

The route of the user, from perception to knowledge to action, which can be done in interaction with the environment, shows how much each type of affordance is involved both in the ease of learning and in the ease of use of interactive environments.

13.4 Designing Mental Models

Mental representations are the cognitive tool that allows human beings to establish relationships with reality. The knowledge of a given system, in fact, is not a homologous representation of the same, but rather a personal re-elaboration of a set of stimuli which, however, are filtered from the user's point of view. During the interaction with an artifact, personal perspectives are assumed, determined by certain factors such as our skills, our previous knowledge, our goals. In doing so, we select, among all those available, only useful information to achieve our goals, we develop *mental models* suitable for the optimal performance of our actions (Johnson-Laird 1983). And this will have as a positive consequence a lightening, for example, of the cognitive workload, or the possibility of rendering interpretable and *almost known* the set of contextual data that do not easily integrate with our previous knowledge. On the other hand, however, the mental models thus elaborated, inevitably put the user, his abilities, his functions, in an egocentric relationship with the environment and, in the event unforeseen situations should arise during the design phase, these themselves could transform into an obstacle (Parlangeli et al. 2012).

Avoiding these possible negative consequences implies the need to analyze the way in which people interact with the environment during its design, to understand their needs and identify the actions performed to achieve objectives. It is therefore necessary, during the course of the design process, to structure situations in which users interact directly with the environment of reference, or even only with prototype versions of the same (Gamberini et al. 2012).

In this analytical attempt there are many factors that must be considered as intervening variables. Among these, for example, the level of previous experience is absolutely relevant, but it cannot be considered a static element. The human beings are active systems in continuous evolution that are composed of a set of physiological/cognitive factors that direct their development and functioning, but which also allow them to play an active role in determining the course of their own life. More interesting and informative, therefore, is the analysis aimed at establishing the cognitive pathways that lead to the development of the mature mental models of the environment, to understand the difficulties that are encountered, if and how they are overcome and which are the stimuli that push to the exploration of the system and consequently to its apprehension (Bannon 1991).

At the end of this analytical journey (Fig. 13.5), an ideal situation should be reached in which the designer has an effective user model for an adequate design of the environment.

Sometimes this relationship can be entirely based on the attribution to the system of characteristics that are very distant from those possessed. Naturally, in fact, we tend to interpret reality as moved by intentions, motivations, thoughts. Thus it is not infrequent that one interacts with mechanical or electronic systems by imagining them endowed with their own real mind (Parlangeli et al. 2013).

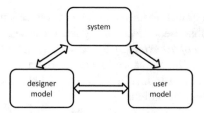

Fig. 13.5 Complex interaction between user models and those of the project team. The interaction that a user establishes with a system depends on the mental models that he is able to process with reference to its structure and functioning. This process is influenced by the representation of the system that the project team has developed in reference to the user-system interaction

As Dennett has explained (1987), we can assume three different explanatory levels in the interaction with reality. At a first level of abstraction—*physical stance*—the explanations are essentially physical, that is to say that we use the knowledge of physics, chemistry, the material properties of things to interpret and predict the behavior of the systems with which we interact. We are at this level of abstraction when we say that a car is proceeding too fast to stop in time in front of an obstacle. At a second level we assume a design perspective—*design stance*—that is, we interpret the behavior of the systems with reference to the intentions that have guided the project. Do planes fly? It is because they have been designed with wings made in a certain way and have engines made in other ways that allow them to take off.

The last level of abstraction—*intentional stance*—concerns the attribution of mental characteristics to the systems. Our computer, but also our car or our watch, are entities able to decide on their own, to have intention, also to comply with us or to spite us (Parlangeli et al. 2012).

Dennett (1987) has claimed that we naturally tend towards an intentional perspective and therefore, the project should take into account our social approach with reality, as if it were able to establish with us a relationship of mutual understanding and collaboration. The user, thanks to what has been realized by the designer, should be able to establish a relationship with the environment from a functional, cognitive, relational and emotional point of view. Finally, the environment should result improved by the designer's activity and continually redefined by the actions of the user interacting with it, establishing a mutually evolutionary relationship.

For the construction of an adequate conceptual and relational model of the system, the designer can focus on a few aspects of maximum relevance. Besides taking into account natural mapping and affordances, one can refer to the following principles (Norman 1988).

1. *visibility*: render visible parts of the system or important information, as, for example, the log-in and log-out buttons to access or exit a website;
2. *feedback*: provide return information that highlights the result that was obtained following an action such as, after sending an e-mail message, finding the message amongst the sent mail;

3. *minimize the memory workload*: such as promoting recognition rather than memory, and ensuring that user expectations guide the access to information;
4. *proper management of the error*: try to reduce the possibility of error and, where this is not possible, allow the cancelling of the effect that has caused that action, introduce functional constraints in order to avoid actions that lead to error, ask confirmation for risky operations;
5. *create meaningful structures*: facilitate the creation of categories by visually grouping the elements of the same category; structure the information following an organization of the contents in order to facilitate memory.

13.5   From Perception to Action

Almost all actions are performed after an intention has been formulated, that is, we have a previous goal that leads us to act. In reference to this goal we intentionally elaborate an action plan to achieve it and, during the execution of these actions, we gradually evaluate the results we are achieving.

Often the execution of a plan of actions implies the use of one or more instruments, artifacts, objects created by man for man. To explain what happens in these circumstances, that is, when we execute a plan of action involving the use of tools, a model of intentional action has been formulated that is usually cited as the *Norman's model of action*, (Hutchins et al. 1985). This model has been developed to provide a conceptual structure able to represent human behavior in the use of tools, but also to account for the difficulties encountered in pursuing the objectives set by using these tools. And, therefore, the action model makes it possible to identify the cognitive distances that could interfere with the achievement of the objective during the interaction with any artifact.

The action is represented on two paths: that of the execution and that of the evaluation, that is, the first is relative to the moment in which an action is performed and the second is that concerning the verification aimed at determining if the effects of said action have achieved the desired goals (Fig. 13.6).

Norman (1988) describes seven stages of the action, one for the goal, three for the execution (intention, action, execution), three for the evaluation (perception, interpretation, evaluation).

1. Development of the goal: this happens when we establish an objective, an aim, a goal to pursue.
   Path relative to the execution: from the planning to the execution of an action.
2. Formulation of the intention: we establish whether, given a certain goal, we intend to take action.
3. Specification of the action: in consideration of the goal and the relative intention we decide what plan of action to take.
4. Execution of an action: we begin the formulated plan of action.

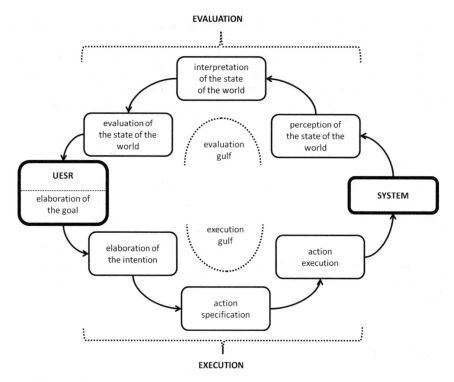

Fig. 13.6 The Norman's model of action. The two paths of execution and evaluation of the actions in interaction with a system are shown. In the passages from one stage to the next, depending on the different cognitive activities, physical, cognitive, sensory and functional affordances come into play

Evaluation path, that is, the comparison between what happens and the results we wanted to achieve.

5. Perception of the state of the world: we gather in the environment the stimuli we have produced as a result of our actions.
6. Interpretation of the state of the world: we give meaning to the stimuli selected within an interpretive framework of the world.
7. Evaluation of the state of the world: we establish whether the state of the world produced as a result of our actions coincides with the objectives we had set ourselves.

The difficulties inherent in the paths of execution and evaluation are conceptualized as distances between the beginning and the conclusion of the relative processes and are called, respectively, the *gulf of the execution* and the *gulf of the evaluation*.

If one imagines a situation in which a child tries to ride a bicycle for the first time, one can understand how, in some cases, the distance between intentions and the implementation of actions can be almost unbridgeable. In the same example, the gulf of the evaluation can be particularly wide, when the child can not correctly perceive the function of the brake handles on the handlebars, tightening them when starting to peddle, and not understanding why the bike does not move.

This model, however, does not take into account the possibility that a goal can be changed during the activity. It is necessary, instead, to take into account the possibility in which the user can not reach the prefixed objective and that a shift toward a new objective can occur. The distance between two objectives, mediated by an action or by the evaluation of a result, is defined distance of the scenario (Rizzo et al. 1997). This type of distance refers to the effort that the user needs to understand that a given objective can not be achieved and that it is necessary, therefore, to shift his attention to an objective which, at that moment, is more suitable.

The model of the action is a reference point for design processes in that it places the user's point of view under the designer's attention, and it emphasizes the need for action and interpretive relationship with reality. What is the purpose of an object that I can not understand? What good is a system that does not seem to fit my ability or does not match any of my goals? The *Design Thinking* (Weinschenk 2011; Kahneman 2011) must, in fact, direct the project towards hypotheses that achieve a harmonious relationship between user and system, perhaps, at times, imagining that the bicycle can be equipped with two small side support wheels and thinking of coaster brakes.

References

Bannon L (1991) From human factors to human actors. The role of psychology and human-computer interaction studies in systems design. In: Greenbaum J, Kyng M (eds) Design at work: cooperative design of computer systems. Lawrence Erlbaum Associates, Hillsdale

Dennett DC (1987) The intentional stance. MIT Press, Cambridge

Galvao AB, Sato K (2005) Affordances in product architecture: linking technical functions and users' tasks. In: ASME 2005 international design engineering technical conferences and computers and information in engineering conference. American Society of Mechanical Engineers, Canada

Gamberini L et al (eds) (2012) Human-computer interaction. Pearson, Milano

Gibson JJ (1979) The ecological approach to visual perception. Houghton, Mifflin and Company, Boston

Guidi S, Parlangeli O, Bettella S, Roncato S (2011) Features of the selectivity for contrast polarity in contour integration revealed by a novel tilt illusion. Perception 40:1357–1375

Hartson HR (2003) Cognitive, physical, sensory and functional affordances in interaction design. In: Behaviour and information technology, vol 22, pp 315–338

Hutchins EL et al (1985) Direct manipulation interfaces. Human Comput Interact 1(4):311–338

Johnson-Laird PN (1983) Mental models: towards a cognitive science of language, inference and consciousness. Harvard University Press, Cambridge

Kahneman D (2011) Thinking, fast and slow. Farrar, Straus and Giroux, New York

Kanizsa G (1979) Organization in vision: essays on Gestalt perception. Praeger Publishers

Koffka K (1935) Principles of gestalt psychology. Lund Hunphries, Londra

Maier JRA, Fadel GM (2003) Affordances-based methods for design. In: Proceedings of ASME DETC03/DTM 48673, vol 3, pp 785–794

Marr D (1982) Vision: a computational investigation into the human representation and processing of visual information. W.H. Freeman and Company, New York

Norman DA (1988) La caffettiera del masochista. Giunti, Firenze (1st ed.: Norman DA (1988) The psychology of everyday things. Basic Books, New York)

Palmer SE et al (2013) Music-color associations are mediated by emotion. PNAS Natl Acad Sci 110(22):8836–8841

Parlangeli O, Chiantini T, Guidi S (2012) A mind in a disk: the attribution of mental state to technological systems. Work 41:1118–1123

Parlangeli O et al (2013) The attribution of mental states to technological systems. Paper presented at IV joint workshop Rutgers-Siena on cognitive sciences, RUCCS, Rutgers University

Parlangeli O, Roncato S (2010) Draughtsmen at work. Perception 39:255–259

Reason J (1994) L'errore umano. Il Mulino, Bologna (ed. originale: Reason J (1990) Human error. Cambridge University Press, Cambridge)

Rizzo A et al (1997) The AVANTI project: prototyping and evaluation with a cognitive walkthrough based on the Norman's model of action. In: Proceedings of the designing interactive systems (DIS '97) conference proceedings. ACM Press, Amsterdam, pp 305–309

Roncato S, Guidi S, Parlangeli O, Battaglini L (23 June 2016) Illusory streaks from corners and their perceptual integration. 7(959)

Warren WH (1984) Perceiving affordances: visual guidance of stair climbing. J Exp Psychol Hum Percept Perform 10:683–703

Weinschenk S (2011) 100 things every designer needs to know about people. New Riders, Berkeley

Wertheimer M (1923) Laws of organization in perceptual forms. Psycologische Forsch 4:301–350

Chapter 14
Cognitive Aspects in User Experience Design: From Perception to Emotions

14.1 Introduction

As we have seen, a correct perception of objects and their modes of functioning is at the basis of the decisions and behaviors of human beings. However, this is not sufficient in itself in the implementation of consequent choices of use: the driving force of every behavior is in fact the intention.

The formulation of an intention is an essential step in any decision making and behavior; it is the result of a process of preventive evaluation on the action itself: that is, on our general attitude towards a particular situation or object, on the estimation of the consequences that the action may have and also on the importance we give to the expectations of others on our behavior or, in any case, on elements relating to the cultural and social context of reference.

A further element that intervenes in our decision to adopt a certain behavior is the evaluation that each subject makes of his own ability in controlling the behavior itself, in that particular situation.

Regarding in particular the use of objects, our perception and evaluation of them—and consequently our intention to use them—will therefore be conditioned by the general knowledge we have about the object in question, by our evaluation of the possible outcomes of its use and also on the feeling of being able to use it correctly.

Clearly every attitude of ours towards an object (of the behavior regarding its use), while still remaining in some way anchored to the previous assessments mentioned above, will change and update in relation to the specific experiences made of and through that object, being therefore influenced not only by the outcomes of the actions (achievement or otherwise of the objectives that have led us to use it), but also by the emotional value of the experience itself. In other words, a user experience that entails the achievement of one's goals in a pleasant way, and gives the subject a sense of self-efficacy, will positively influence the possibility that the object will be used again.

This chapter was co-authored by Maria Cristina Caratozzolo and Oronzo Parlangeli.

© Springer Nature Switzerland AG 2020
F. Tosi, *Design for Ergonomics*, Springer Series in Design and Innovation 2,
https://doi.org/10.1007/978-3-030-33562-5_14

The design of objects, therefore, is able to influence both the correct perception and understanding of the objects themselves and, as will be seen, the pleasure of their use and the motivation to continue using them. It is no coincidence that we speak—for technological objects, but not only—of design of User Experience Design.

14.2 From Perception of Use to Motivation to Use

The motivation and the consequent intention to perform a behaviour are predicative factors on the effective implementation of the same. This intention is known as behavioral intention and derives from a belief that the result that follows the behavior will be positive or negative for the subject, that is, from her/his evaluation (Fishbein and Ajzen 1975). The decision of an individual to engage in a particular behavior is therefore based on the results that the individual expects following the execution of the behaviour.

Many of our evaluations are based on a procedure of heuristic anchoring and subsequent accommodation (Tversky and Kahneman 1974), that is, we rely on general and previous knowledge about a situation or object and then adjust it in our opinion as we gradually acquire specific information. In particular, according to the Theory of Reasoned Action–TRA, Theory of Reasoned Action (Fishbein and Ajzen 1975)—behavior is influenced by the attitude that the subject has towards a particular action and the outcome that follows; the attitude is in turn defined as the predisposition towards an object, a person or a situation formed on the basis of general knowledge and previous experiences. For example, starting from the general information at his disposal, a subject might think that "the use of a smartphone is supportive in many work and leisure activities".

According to the theory, people then experience the influence of subjective norms linked, that is, to the social expectations of that particular behavior, or to the representation of how to implement a certain behavior to be considered acceptable by other persons who are important to the subject and by the motivation of this to please them—for example: "my colleagues and friends expect me to own and use a smartphone" and also "it is important for me to do what my friends expect, because they are important to me" (see Fig. 14.1).

The model is very general and does not refer to specific types of behavior and, for this reason, it is applicable to different areas, for which it is possible from time to time to identify specific attitudes and subjective norms.

To sum up, the TRA assumes that people behave in a rational way, based on intentions reasoned on the basis of a more or less accurate estimate of the potential effects of their behavior on the world and on others. The central idea is therefore that the most proximate cause of behavior is the subjective intention of wanting to do it.

However, the two factors foreseen by the model—subjective attitudes and norms—are not sufficient to effectively predict the behavior of human beings in all situations. In a more analytical version of this, more specific variables are introduced that constitute the two main determinants: as regards attitude, beliefs are considered,

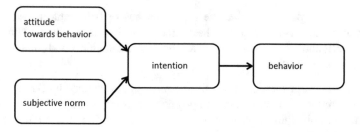

Fig. 14.1 The theory of reasoned action. Behavior is influenced by the attitude that the subject has towards that particular action and towards the result that follows

that is the possibility that a behavior can generate certain results and produce specific outcomes, and the evaluation that the single subject attributes to the specific outcome. Regarding, conversely, subjective norms, on the one hand the normative convictions (the beliefs about what certain referents expect with respect to a specific behaviour) are taken into consideration and, on the other, the willingness that the individual has in adapting his behavior to expectations of his own contacts.

The Theory of Planned Behavior—TPB (Ajzen 1991)—completes the previous theoretical framework in a more organic way, adding the perception of behavioral control among the determining factors, that is, the perception that a subject has to be able to put into effect the desired behavior (see Fig. 14.2). It is not a matter of the actual control of the individual on behavior but of his subjective, situational perception, therefore linked to the context of individual behavior (for example: "I believe I have the skills and knowledge necessary to use a smartphone").

The author talks of *control beliefs* that influence the perception of behavioral control; these beliefs concern the possible obstacles that a subject believes she/he can meet in the specific situation:

- in which the behavior would be implemented, defined in terms of resources and necessary skills;

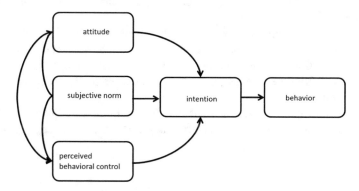

Fig. 14.2 The theory of planned behavior. In the theory of planned behavior, to the previous variables is added that relative to behavioral control

- the subjective perception that the individual has of being able to successfully overcome the previous obstacles, that is, a self-assessment regarding the possession of the skills and resources necessary for the implementation—and completion—of the actions in question.

This last concept corresponds to that of *self-efficacy*, defined by Bandura (2000) as the perception we have of our competence regarding a task and the consequent expectation of obtaining a positive outcome. The dimension of self-efficacy is also conditioned by the extent in which a given task is salient or not for us. It therefore becomes clear how this concept is linked to that of motivation—which completes the theoretical framework regarding the process of decision and action of the subjects.

Self-efficacy assumes a key role in the self-regulation of motivations through the influence exerted on the cognitive representation of objectives, on causal attribution and on the expectations of the result. In particular, it influences the origin of the motivation itself, which can be internal or external to the subject; the motivational drive is not in fact innate in the individual, nor does it depend on the structure of his personality: it is arranged in relation to a given situation.

Motivational drive is defined as the "*complex process of the forces that activate, direct and sustain behavior over time*" (Avallone 1994), characterized by three dimensions: the direction, which concerns the orientation of strategies and actions implemented functionally to the achievement of one's own purposes; the intensity, that is, the entity of the force, how much energy at that moment can be produced to sustain one's own behavior. Finally the persistence, that is, the durability of energy, the ability to sustain behavior over time. Although not objectively measurable, motivation can therefore be quantified according to these three parameters and is defined in a contextualised way in relation to the perception of the subjects—and to their consequent evaluations—in relation to objects, environments and situations.

A relevant aspect of the motivation concerns its origin, internal or external to the individual; in fact, the drive to implement behavior is defined as intrinsic for the pleasure of doing it, to do it well, in a way that is satisfactory for oneself. On the contrary, external solicitations and rewards which induce subjects to perform tasks, in order to obtain benefits or avoid negative circumstances, identify an extrinsic motivation.

The substantial difference between these two types of motivation is therefore in the search for a type of gratification (or reward) that can be defined in terms of self-esteem, gratification, control or, instead, of success, social recognition, utility—that respond to self-fulfillment needs on one side and esteem on the other, as shown in Fig. 14.3 (Maslow 1954).

It is clear how these two declinations of behavioral thrust are linked in different ways to conditioning factors such as control, subjective norms, attitude and emotions.

Fig. 14.3 Maslow's pyramid of needs. The "*Hierarchy of Needs*" conceived by Maslow in 1954 is a scale of needs subdivided into five different levels, from the most basic (necessary for the survival of the individual) to the most complex (of a social nature); the realization of the individual passes progressively through the various stages

14.3   The Theories of the Technology Acceptance Model (TAM)

Every human behavior is therefore linked to an anchoring and adjustment process, defined by individual and situational variables such as control, motivation and emotion/gratification.

The particular type of behavior that is the behavior of use is also, consequently, conditioned by the same determinants and particular attention has been turned, in the area of research on Human-Computer Interaction, to the technologies or, rather, to the factors that determine the disposition or not of the subjects to use the technologies. Therefore, various theoretical models have been developed that attempt to interpret the attitude of the individual towards the use of information technology, and all are derivations and evolutions of the aforementioned TRAs and TPBs.

The Technology Acceptance Model (TAM) of Davis (1989) and Venkatesh takes up the constructs of the TPB and reinterprets them suggesting two measures of technological acceptance and disposition of the subjects to use technologies, with particular reference to the professional sphere.

Specifically, in the proposed model, the following are considered direct predictors of behavior towards the use of technological objects:

- perceived utility: "*the degree in which a person believes that the use of a particular system can improve his work performance*";
- perceived ease of use: "*the degree in which a person believes that the use of a particular system would be effortless*" (Davis 1989).

A certain combination of these two evaluation parameters would lead the subject to conceive the intention to use an object (see Fig. 14.4).

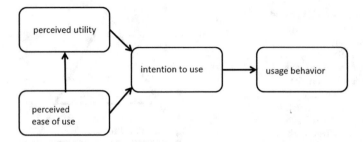

Fig. 14.4 TAM model. The model shows how the intentions of use relative to technological objects are determined by the utility and by the subject's perceived ease of use in relation to the object in question

The validity of the model, which has been repeatedly re-elaborated and extended over the years by the same authors, has been proven by numerous successive studies which have provided empirical evidence on the relationships between utility, ease of use and effective use of the system, showing also how the attitudes had only partial influence on the real intentions and how the subjective norms did not have any (Davis et al. 1989). These results suggest the possibility of simple but powerful models of the determinants of acceptance on the part of the user but are strongly linked to organizational and work contexts in which the social norms and intrinsic motivations of the subjects are not very relevant.

In particular, it emerges that the most relevant belief in determining the attitude towards technological objects is the perceived utility (Davis 1993); in other words, the subjects are willing to use a complex and difficult-to-use technological tool, provided that its utility in achieving its objectives is seen as high. An easy-to-use tool that was perceived as not very useful would instead be accepted to a lesser extent. To cite the previously-used example, a subject will be more willing to use a smartphone, even if it might be complex, if he considers it useful for his work (or to keep in touch with friends).

On the other hand, considering these parameters in a longitudinal sense over time, that is no longer only in relation to the first use of a technological object but as predictive tools for a repetition of the behavior itself (that is the readiness to use again the technology in question), it was stressed how ease of use influences the perception of utility (as illustrated in Fig. 14.4). People tend to attribute a greater utility to objects that have proved usable, compared to others with the same functions but more complicated. Our smartphone, that is, will appear to us to be even more useful when, after having used it for a while, we will have ascertained that it is easy to use.

These considerations call into question the construct of behavioral control: a study by Venkatesh and Davis (1996) investigated the possible links between the level of self-efficacy towards technologies and the usability of software on the one hand and perception of ease of use on the other, considering both the cases of first approach to the technological object and those of repeated, and therefore expert, use.

Experimental results have shown that beliefs about the ease of use of the system are related to the perception of the subjects about their own ability to use the computer (computer self-efficacy), regardless of whether they have actually used the system or not. The usability of the system becomes relevant only after the subjects have started using the new technology.

Both the factors that the original TAM model defines as determining behavioral intention are, in turn, influenced by external variables, such as training in the use of the technology (Parlangeli et al. 2011), knowledge more or less general about it, or even any eventual support tools that the subject can benefit from. In other words, these are the evaluative elements constituting the attitude and the subjective rules already explained by the TRA and the TPB.

The TAM model in its original formulation, thanks to its good predictive value and to the solid theoretical and empirical basis, represents a valid starting point for studies on the propensity to use technological objects; its main limitation in this version—as has emerged above—is represented by the fact that the analysis focuses almost exclusively on the acceptability of an instrument and is mainly applied to workplaces, to artefacts whose motivation of use is strongly linked to functionality and work performance.

14.4 The Role of the Determining Factors: Evolutions of the TAM Model

According to what emerged from the first empirical studies conducted within the TAM model, perceived utility is the fundamental factor in influencing people's intentions to use; moreover, it has been found that subjective rules have no significant effect on intentions beyond the perception of utility and ease of use (Davis et al. 1989), so much so that this factor is missing in the original model.

However, the same authors of the model recognized the need for further research to "*investigate the conditions and mechanisms that govern the impact of social influences on behavioral use*" (ibidem). Alongside these types of influence it was also important to understand the determinants of the construct of perceived utility and the way in which their influence changes over time as the experience in the use of the system increases.

An extension of the model is TAM2 (Venkatesh and Davis 2000), in which the external variables are grouped so as to show their relationship with the perceived utility; this version of the model demonstrates how the acceptance of technology is also influenced by variables related to social processes such as the subjective norms and the image, as well as instrumental cognitive processes.

The experimental study demonstrates that subjective norms significantly influence the perceived utility through internalization on the one hand, namely the tendency to incorporate social influences into one's own perceptions of utility, and identification on the other, that is, the use of a system to obtain status and influence within the

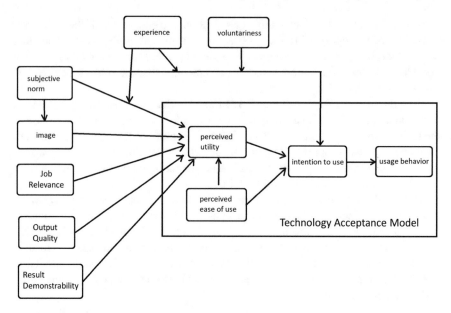

Fig. 14.5 Technology acceptance model 2. This version of the TAM model integrates factors of social influence and instrumental cognitive determinants of perceived utility

working group, thus improving one's own work performance. To use a by-now known example, the use of a smartphone is incentivized both by the fact that the people of my social group use it (and therefore expect me to do so) and by the ambition to earn—through its possession and use—a status of greater prestige (or improved efficiency) within the group itself.

In addition to these two indirect effects (that is, mediated by the construct of perceived utility), the subjective norms have produced a direct effect on the intentions in mandatory use contexts, that is, those in which the user has no discretion on the use or otherwise of the technological object, but not on the voluntary ones (see Fig. 14.5).

Moreover, it emerges from the study that corresponding to the acquisition on the part of the subjects of direct experience with a system is a lesser influence of social information on the definition of perceived utility and intention, but rather the tendency to judge the utility of a system on the basis of the potential benefits deriving from the use in terms of status (therefore of image).

In addition to the processes of social influence on the perceived utility and the intention of use discussed above, four instrumental cognitive determinants of perceived utility are theorized in the model: they demonstrate how people form—at least in part—judgements of perceived utility, comparing cognitively what a system is capable of doing with what they need to be done in order to perform their task.

The determinants in question are defined as:

- job relevance: the degree of perception regarding the utility in the context of one's own work on the part of the user;

- output quality: the degree in which the user believes that the technology improves the result of his work performance; beyond the considerations on what tasks a system is able to perform and the degree in which these tasks correspond to the objectives of the subject (that is, job relevance), people will consider the way in which the system performs those tasks;
- results demonstrability is the quantitative version of quality, that is, the degree of tangibility of the results related to the use of the technological artefact.

In essence, the model maintains that the opinions regarding the utility of a system are influenced by the individual cognitive correspondence of their work objectives with the consequences of use of the system (job relevance), and that the quality of the output assumes a proportionally greater importance to this latter. My opinion on the utility of the smartphone that I use will be, that is, determined on the basis of how (and how well) the use of this object will allow me to achieve the objectives that I have set myself.

Unlike the processes of social influence, the effects of cognitive instrumental processes remain significant even with the consolidation of the experience.

These results suggest that the determinants of perceived ease of use evolve from the early stages of experience with the system during the stages of consolidation of the same; it therefore appears necessary to consider other determining factors, such as constructs related to control, to intrinsic motivation and to emotions as general anchors for the definition of perceived ease of use compared to a new system and— aspect this of extreme interest—factors of adjustment such as objective usability and satisfaction with actual use (Venkatesh 2000).

Specifically, as shown in Fig. 14.6, computer self-efficacy and the perception of external control (that is, the facilitating conditions) are considered as anchor factors related to control; the intrinsic motivation is instead conceptualized as pleasure in the use of the computer (computer playfulness). The inclusion of this factor is particularly interesting because, if extrinsic motivation had been captured by the construct of perceived utility, the intrinsic motivation was only now considered as a determining factor (and independent of the system) of the perceived ease of use—and therefore of intention of use of the technology. Higher levels of pleasure in use will in fact also lead to a lesser perception of effort.

Emotions are also considered determining factors in the formulation of behavioral intentions and, in particular, constituents of the perceived ease of use. Specifically, the model talks of "computer anxiety", defined as the apprehension of being faced with the possibility of using a technological artifact. While self-efficacy refers to judgments about ability and playfulness to spontaneity in the individual's interaction with a computer, computer anxiety is an emotional reaction that exerts a negative influence on the perceived ease of use of a new system. The consequences of a high level of anxiety about the behavior and performance include a negative impact on cognitive responses, in particular on the expectations.

Summing up then, computer self-efficacy, facilitating conditions, computer play-fulness and computer anxiety are anchoring constructs independent of the system that play a fundamental role in shaping the perceived ease of use of a new system,

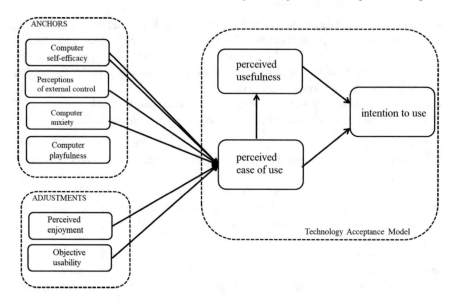

Fig. 14.6 Extended TAM. This extended version of the model considers how the determinants of perceived ease of use evolve from the early phases of experience with the system during the stages of consolidation of the same. It, in fact, inserts determinants related to the control, intrinsic motivation and emotions as general anchors for the definition of perceived ease of use. Furthermore, objective usability and satisfaction with the effective usability of the system are considered adjustment factors

particularly in the early stages of the user experience with an artifact. As experience increases, the objective usability and the enjoyment perceived by the use of the system represent specific accommodation factors—deriving, that is, from user-system interaction—that will have a further influence on the perceived ease of use.

People's choices are ultimately influenced by beliefs regarding their own efficacy in managing objects and situations, from the fear of executing wrong, embarrassing or even merely unpleasant actions. The design of the artefacts must therefore take into account the different aspects of the experience that the use of the same will produce: the characteristics of the system in terms of performance and functionality—therefore its usability—the specific emotions and perceptions of each individual user and the "symbolic" dimension concerning the collocation of the object in a given cultural and social context.

14.5 Hedonic and Utilitarian Aspects in User Interaction and Experience

For that particular type of experience that is experience of use we can affirm that it concerns both the cognitive and perceptive spheres of each subject and the emotional sphere, which includes personal tastes and conceptions of pleasure but also considerations related to the achievement of the prefixed goals that motivated the experience itself. The experience of use has in fact origin in an intention that, as we have seen, derives from the combination of pragmatic/utilitarian conditioning (we foresee that it allows us to reach our goal which is, in other words, efficient and effective) but also of a hedonic nature (we want the fruition to be pleasant, flowing and simple).

Finally, to complete the picture, contextual factors cannot be neglected: not only in relation to the physical space in which the interaction takes place but also in relation to the cultural and symbolic scenario, to the system of social rules and conditioning within which the user is situated. In interacting with a (technological) object we inevitably perform an act that has a social value because the identity and the image of a product on the market also condition the formulation of the intention to use and, therefore, the ultimate user experience.

That of User Experience (UX) is therefore configured as a complex and multifaceted concept, articulated in at least three directions (Pasquini et al. 2018):

- the characteristics of the system in terms of performance and functionality and therefore its usability—understood as a property resulting from the interaction of the system itself with the person;
- the individual sphere, which regards emotions, preferences, opinions, perceptions and behaviors of each individual user;
- the "symbolic" dimension that regards the collocation of the product/service in a specific context—cultural and of market—the presentation of one's own identity, the strength of the brand, the image.

These result quite clearly attributable to the determinant factors seen above and presented within theoretical frameworks related to behavior, with particular reference to the use of technological artefacts.

Speaking of emotional design, Norman (2004) explains how at a visceral level people become "hooked" on the emotional level by a series of physical and symbolic aspects of an object, which are able to condition, at the reflective level, their overall opinion of the experience of use. In other words, an object that offers us an experience that is also emotionally positive induces a positive conditioning on the formulation of our future intentions of use.

An aesthetics dimension, but not only that, is then added to the utilitarian dimension of the interaction with artifacts, linked to the concept of usability; the current technological objects, no longer relegated to work environments and now part of everyday life, are used in variable, individual and social contexts, in which the relationship between user and object is no longer simply limited to use, but becomes involvement and influences the process of defining the identity of individuals, their lifestyle, their system of values in the cultural context of reference.

In other words, in addition to responding to a practical need and to an aesthetic one, a digital product represents an instrument of enunciation of one's own personality, of adherence to a cultural and social system. Therefore its fruition is a moment of realization and self-determination (Ryan's Self Determination Theory: Deci and Ryan 2000) in individual, social and symbolic/cultural terms.

Given the pervasiveness of technological artefacts and the frequency of interactions that humans have with them, the activities related to their use can be assimilated to consumer activities, and as such analyzed according to an experiential approach (Holbrook and Hirschman 1982). The fact that "*consumption is of a utilitarian nature when the relevance of characteristics serves as a channel through which the consumer benefits from the functionality through use of the product. Consumption is of a hedonic nature when the subjective response of the consumer is greater than the weight of objective characteristics*" (Holbrook and Hirschman 1982), is totally in line with the distinction above between extrinsic and intrinsic motivation to action, and—exactly like this—is of difficult identification in many types of behavior.

In fact, "*some products may have the same or similar weighting of the objective characteristics and subjective response*", just as some behaviors may be dictated by a combination of intrinsic and extrinsic motivations. Moreover, they could influence each other, conditioning each other reciprocally.

In effect, the utilitarian dimension versus the hedonic can be considered a continuum along which the positioning of each type of object (and therefore of interaction) depends on the relative weights of the objective characteristics (of the product/service) and of the subjective response of the consumer (see Fig. 14.7). But is it the goal of the act of consumption (the nature of the motivation that triggers our decision to use it) or the product that is consumed that which determines the prevailing hedonism or the prevailing instrumentality of the consumption itself? According to the definition given above, reference should be made to the "*purpose of consumption*" in order to define it as hedonistic or utilitarian and not so much to the intrinsic characteristics of the product/service that is the object of the consumption.

Fig. 14.7 Hedonic and utilitarian products (modified by Addis and Holbrook 2001). The utilitarian dimension versus the hedonic can be considered a continuum along which the positioning of each type of object depends on the relative weights of the objective characteristics and the subjective response of the consumer

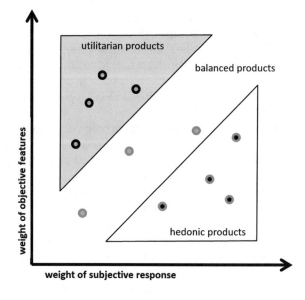

Often hedonic consumption is associated with products/services that are hedonistic for their nature, vice versa utilitarian consumption is associated to products/services that are mainly functional in nature. As the experience is subjective, it is the individual who creates his own categories of hedonistic and utilitarian products/services. A product/service normally considered functional/utilitarian may not be considered as such by some consumers, or possibly even by the same consumer in a different way.

From an experiential point of view, the consequence of the consumption is the pleasure that comes from the use of the product, and this can derive both from functional factors (the product satisfies me because it performs its functions and allows me to achieve my goals), emotional factors (using the product gives me pleasure because it is aesthetically pleasing, because it makes me feel capable, etc.).

Ultimately then all these aspects are a matter of design interest, in an approach that evolves from the principle "form follows function"—in which the function refers to the concept of efficiency in terms of performance—passing from the now too simple one of *user friendliness* (efficiency and effectiveness in the performance of one's own function), to the search for a qualitative attribute that appears more and more as essential: satisfaction. Think, for instance of risky behavior while driving (Parlangeli et al. 2018).

The appearance of the pleasantness, of the aesthetic/emotional response of a subject to the use of a technological system assumes an ever increasing weight in the contemporary nature of *ubiquitous computing*, so much so that it has emerged as necessary to recognize the semiotic and symbolic relevance of digital fruition, accompanying the study of the emotional component in the human-digital artifact relationship to *design thinking*.

References

Addis M, Holbrook MB (2001) On the conceptual link between mass customisation and experiential consumption: an explosion of subjectivity. J Consum Behav Int Res Rev 1(1):50–66

Ajzen I (1991) The theory of planned behaviour. Organ Behav Human Decis Process 50:179–211

Avallone F (1994) Psicologia del lavoro. Storia, modelli, applicazioni. Carocci, Roma

Bandura A (2000) Autoefficacia. Teoria e Applicazioni. Erikson, Trento

Davis FD (1989) Perceived usefulness, perceived ease of use, and user acceptance of information technology. MIS Quart 13(3):319–340

Davis FD (1993) User acceptance of information technology: system characteristics, user perceptions and behavioral impacts. Int J Man Mach Stud 38(3):475–487

Davis FD et al (1989) User acceptance of computer technology: a comparison of two theoretical models. Manag Sci 35(8)

Deci EL, Ryan RM (2000) The "what" and "why" of goal pursuits: human needs and the self-determination of behaviour. Psychol Inq 11:227–268

Fishbein M, Ajzen I (1975) Belief, attitude, intention, and behavior: an introduction to theory and research. Addison-Wesley, Reading

Holbrook M, Hirschman E (1982) The experiential aspects of consumption: consumer fantasies, feelings, and fun. J Consum Res 9(2):132–140

Maslow AH (1954) Motivation and personality. Harper & Row Publishers, New York, NY.

Norman DA (2004) Emotional design. Apogeo, Milano (1st ed.: Basic Books, Cambridge)

Parlangeli O, Mengoni G, Guidi S (2011) The effect of system usability and multitasking activities in distance learning. In: Proceedings of the CHItaly conference, 13–16 september, Alghero, ACM Library, pp 59–64

Parlangeli O, Bracci M, Guidi S, Marchigiani E, Duguid AM (2018) Risk perception and emotions regulation strategies in driving behaviour: an analysis of the self-reported data of adolescents and young adults. Int J Hum Factor Ergon 5(2):166–187

Pasquini J et al (2018) #UX designer. Dalla user experience alla digital experience. FrancoAngeli, Milano

Tversky A, Kahneman D (1974) Judgment under uncertainty: heuristics and biases. Science 185:1124–1131

Venkatesh V (2000) Determinants of perceived ease of use: integrating control, Intrinsic motivation and emotion into the technology acceptance model. Inf Syst Res 11(4):342–365

Venkatesh V, Davis FD (1996) A model of the antecedents of perceived ease of use: development and test. Decis Sci 27(3):451–481

Venkatesh V, Davis FD (2000) A theoretical extension of the technology acceptance model: four longitudinal field studies. Manag Sci 46(2):186–204

Part III
Project Problems and Intervention Criteria

Chapter 15
Furniture, Work Stations, Hand Tools

15.1 Introduction

The main design problems related to the layout, sizing and organisation of the environment, the sizing of products and equipment and the organisation of the required activities, require the joint consideration of aspects relating to physical tasks and cognitive tasks, as well as the emotional aspects related to the subjective situation of each one.

A similar approach applies to the assessment of potential risks, which may derive from the partial or total incompatibility of the physical, sensory or cognitive characteristics and capabilities of the person, and the levels of performance required (and/or the constraints imposed) by the physical, social and organisational context in which they operate.

Please refer to Chaps. 6 and 7 for the description and analysis of the evaluation methods for Human-Centred Design.

Operationally, this involves identification of:

- the people to whom the design is addressed (who use the product[1] to be evaluated);
- the main activities for which the product/system to be evaluated and/or designed are, or may be, used;
- the physical and mental tasks required;

This chapter was co-authored by Francesca Tosi (Sects. 15.1, 15.2 and 15.6) and Mattia Pistolesi (Sects. 15.3, 15.4 and 15.5).

[1] In this chapter, as in the rest of the volume, they are used for brevity:

• the term "product" in its literal meaning of "result of human activity" and, in the case of industrial products, the result of a design and production process, aimed at responding to a specific need. Products are therefore the physical environments, the objects of use, the services, be they physical or virtual. This meaning also extends to the equipment and machinery used in work activities;

• the term system defined as "any object of study that, despite being composed of different elements mutually interconnected and interacting with each other or with the external environment, reacts or evolves as a whole". See Treccani Online Vocabulary, www.treccani.it/vocabolario (consulted in December 2017). See par. 1.1 and notes 3, 4 and 5, Chap. 1.

© Springer Nature Switzerland AG 2020

F. Tosi, *Design for Ergonomics*, Springer Series in Design and Innovation 2,
https://doi.org/10.1007/978-3-030-33562-5_15

- the anthropometric reference parameters (for example the height, the height of the eyes, the size of the hand, etc.) and the data relating to the group of users to whom they are addressed;
- the physical and mental abilities (of movement and physical and cognitive effort) of the group of users under consideration, and the relative acceptance thresholds;
- users' limits.

You must also define:

- the limits of the design solution;
- the acceptability thresholds and the related dimensional and functional constraints;
- the dimensional requirements for the accessibility and dimensional usability of the environments, the products and their components, and in particular the spaces for movement and accessible areas;
- the requirements relating to the characteristics, duration and/or intensity of the postures, movements and efforts required by each expected physical and mental task.

The steps necessary for the correct definition of the dimensional constraints, and the limits of intervention of the project, can be further broken down and follow a succession different from the one just listed.

15.2 Analysis of Physical and Mental Tasks

The standards identify the two preliminary phases for the selection of the parameters and the data necessary for the planning and/or evaluation of the environment and/or the product and the definition of the dimensional and functional requirements in the analysis of the tasks and in the analysis of the users.

The task is defined as the activity or set of activities that are required to achieve a given goal.[2]

The tasks are also defined as *"the minimum descriptive unit of a work activity and as the goal to be achieved under certain conditions or with certain resources"*.[3]

The conditions are represented by whatever may influence the activity that must be carried out.

The conditions are called "constraints" when it is felt that they may have negative effects. The tasks are defined as "prescribed" when the objectives and conditions are set by a hierarchical and external authority (the company, the administration, etc.) and can be set with different levels of precision. In complex work, as is typical of the productive or nuclear sector, the objectives and conditions are determined, in a meticulous way, by the execution procedures. At the other extreme, the tasks are

[2] See ISO 9241-11/1998, Ergonomic requirements for office work with visual display terminals (VDTs)—Part 11: Guidance on usability; UNI EN 614-1:2009, Safety of machinery—Ergonomic design principles—Part 1: Terminology and general principles.
[3] See Di Martino and Corlett (1999, p. 33).

made explicit simply by their goals. In this case we speak of a mission, and it is the operator who defines his tasks, their priorities and their succession, as in the case, for example, of design activities.[4]

The first phase in designing the tasks provides their description through[5]:

- observation and recording of existing tasks (or tasks similar to those to be designed);
- the definition of the functions necessary to successfully achieve the objectives;
- the description of what must be done at the level of individual tasks to achieve these objectives (the description does not say how the task is currently performed but how it should be carried out under ideal conditions).

The tasks are then defined on the basis of the critical characteristics—such as sequence, completion time or time constraints, criticality, complexity and interrelations, cognitive load required—and based on the technical, relational and environmental constraints on the full execution of the task (for example: the physical environment, the organisation of work, the social environment, etc.). Task analysis requires you to:

- identify the characteristics of current or potential users who per-form, or can perform, the task in the context considered;
- verify the correspondence between the task's characteristics (for example, movements and efforts required) and those of the users.

The data processed in this way forms the basis for the subsequent decisions relating to the definition of the dimensional and functional constraints, the placement and distribution of the functions among the operators, the layout of the equipment and machinery, the location of the command interfaces, etc.

According to UNI EN 614 of 2009,[6] the design of the tasks can be described as a process that requires the definition of the project objectives, the analysis of the required functions and their allocation, the specification of the work activities and the assignment of the work tasks among the various operators.

The standards provide an analysis model for the tasks that are carried out through the following steps (see Table 15.1), basing decisions on relevant information and evaluating the consequences that design decisions may have on users' activities.

Once what needs to be done to achieve the objectives has been defined, the tasks must be evaluated based on the characteristics, capabilities and needs of those who will have to carry them out.

The methods and criteria for this evaluation depend on the complexity of the process and, in any case, must be outlined.

Among the methods that can be used are: group discussions, interviews, questionnaires, checklists, direct observations, critical analysis of accidents and psychometric evaluations using standardised scales.

---

[4]See Leplat (2000, p. 355).

[5]See Di Martino and Corlett (1999, p. 34).

[6]See UNI EN 614-2:2009, Safety of machinery. Ergonomic design principles. Interactions between the design of machinery and work tasks, p. 9.

Table 15.1 Reworking of the analyses description for the tasks, taken from UNI EN 614-2/2009

DESCRIPTION OF THE WORK TASK DESIGN PROCESS

N.	DESIGN STAGE	DESCRIPTION OF STAGES
1	ESTABLISHING THE DESIGN OBJECTIVES	• gather information on comparable existing machinery; • work out the general design objectives and design specifications; • establish general performance requirements and evaluation criteria.
2	FUNCTION ANALYSIS	• identify fuctions and sub-functions and specify them in their hierarchy and functional relationships; • specify functions together with their performance criteria; • evaluate the specified functions against the design specifications.
3	FUNCTION ALLOCATION	• allocate functions and sub-functions to the operator or the machine or, where appropriate, to both; • evaluate the suitability of the functions as human activity or machine operation; • Outline alternative design solutions and analyse the benefits and drawbacks of them.
4	WORK TASK SPECIFICATION	• gather information on comparable existing tasks; • specify the operator tasks in detail; • evaluate the workload each task imposes on the operator
5	ASSIGNMENT OF WORK TASKS TO OPERATORS	• specify the number of operators required; • assign the tasks to the operators; • evaluate the total operator workload and fulfilment of the characteristics of well-designed operator work tasks.

The evaluation and design of physical tasks must therefore ensure that these can be implemented with the minimum effort possible, taking into account the physical capabilities of operators and/or users.

The analysis of the tasks allows us to define the dimensional requirements of the design and the physical-dimensional parameters necessary to evaluate the acceptability of the design solutions (or existing solutions), based on the characteristics of the user.

From a design point of view, the most important aspects concern the identification and evaluation of the discomforts and/or pathologies that can derive from the constraints posed by the physical environment and by the structure and the succession of physical and cognitive tasks required by a given set of activities.

A work station should be designed by considering the variability of the characteristics of the operators and/or users. The latter should include[7]:

- body size: the work station should be designed with consideration for the body size of the "expected or predicted" population of the operators;
- posture: the work station should not force the user (or operator) to perform repetitive (identical) movements that cause damage, illness or injury;
- body movements: the work station should be designed with reference to the work process, to allow the body to move according to the rhythm and naturally. In particular, the operator should not be required to perform very frequent movements involving extreme angles, for a prolonged period of time;
- physical strength: actions requiring the application of high force can cause fatigue to the musculoskeletal system;
- mental skills[8]: the work station and its associated components (displays, signals, control actuators, instructions, etc.) should be de-signed with consideration for the mental abilities of the "expected or predicted" population of the operators, during machine and/or equipment control from work, and during the performance of tasks.

In particular work stations, the elements that must be considered are:

- The position of the operator[9]:

  - distance and orientation of the work surface;
  - visual alignment;
  - work surface height (standing position);
  - relationship between the seat surface and the work surface (seated position).

---

[7]See UNI EN 614-1:2009, Safety of machinery—Ergonomic design principles—Part 1: Terminology and general principles.

[8]The UNI EN ISO 10075-1/2003 standard, Ergonomic principles related to mental workload—Part 1: General issues and concepts, terms and definitions, defines terms in the field of mental workload, including mental stress and mental strain (solicitation).
Mental stress: The collection of external influences on a human being to the point of mentally conditioning it.
Strain (mental solicitation): The immediate (non-long-term) effect of mental stress on the individual who suffers from the typical and current preconditions, including the personal styles adopted to deal with it.

[9]See UNI EN ISO 14738/2009, Safety of machinery—Anthropometric requirements for the design of workstation at machinery.

- Spaces for movement and accessible zones[10]:

  - spaces for access and exit;
  - spaces for adaptation;
  - spaces for movement;
  - constraints posed by the layout of the work station (layout of furnishings and equipment, arrangement of controls etc.).

- Body size[11]:

  - body size of adult users and users with special needs;
  - variability in body size and joint movements;
  - safe distance;
  - dimensions for access (for use, installation, maintenance, registration, cleaning, repair and transport).

- Posture:

  - characteristics and duration of posture during working hours (sight line, observation distance, visual discrimination, duration and frequency of the activity and any limitation for groups of special users);
  - positions required for use of controls and displays;
  - constraints caused by wearing PPE and the layout of the work station or work (arrangement of the furnishings and equipment, position of the controls, etc.).

- Body movements:

  - freedom of movement to avoid a static posture;
  - extension, duration and repetition of the requested movements.

- Applications of force:

  - extent, duration and repetition of the required efforts;
  - forces that can be exercised in positions required by physical tasks;
  - load handling;
  - aids and equipment for handling loads.

- Mental skills:

  - cognitive effort needed by the operator to perform a certain series of activities;
  - controllability (the operator must have control of the machine and its components)
  - compliance (the machine and its components should be compatible with the operator's expectations);
  - autonomy (the machines should guarantee an adequate autonomy for the operator, with respect to decisions regarding priorities and procedures);

---

[10]See Di Martino and Corlett (1999).

[11]See UNI EN 614-1/2009, Safety of machinery—Ergonomic design principles—Part 1: Terminology and general principles.

- ability to understand (the machine and its components should be adequately understandable, in relation to variations in the cognitive abilities of the population of operators);
- flexibility (the machine and its components should be appropriately flexible to the varied skills of the population of operators and, if necessary, as needed).

15.3 Work Stations for Office Work

The office work position, which has been covered in detail in many anthropometry texts, and ergonomics texts relating to all physical, cognitive and organisational aspects, represents one of the most common intervention cases, with the work stations used also suitable for assimilation outside the workplace.

In this case, the activities to which we can refer are reading, writing, word processing and drawing (both manual or digital).

The table and chair system, with all the possible pieces of furniture and equipment needed for this type of activity, can be used in work environments, domestic environments, schools, etc., and users, who can obviously be very diverse, must be identified on a case-by-case basis.

The anthropometric reference parameters are those that can be used for identifying the correct dimensional relationship between the elements that must be used and the dimensions of the human body in a static and moving position.

The parameters for postures, movements and forces required by physical tasks must take into account the type of activity mainly required of the operator and/or user and, in particular, the positions and movements of the head, torso and arms.

The constraints that must be considered, in particular, are:

- the accessible areas needed to perform the required physical tasks;
- the space for movement required by the tasks and entry and exit to and from the work station;
- the risk factors deriving from the positions and movements of the head, torso and arms, which are required by the use of the machinery or equipment to be used;
- the risk factors that derive from the duration and frequency of the postures assumed and the extension or intensity of movements and forces.

The standards[12] provide some parameters that we need for the evaluation of postures and movements required by work activities and recommendations regarding their acceptability, depending on the visual field required by the different work activities. As has already been highlighted, the recommendations can also be used for the evaluation of things other than work stations (Figs. 15.1, 15.2 and 15.3).

---

[12]See UNI EN ISO 14738/2009, Safety of machinery—Anthropometric Requirements For The Design Of Workstations At Machinery.

Work station for office work

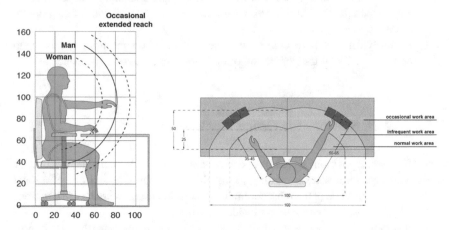

Fig. 15.1 Work station for office. Accessible zones (picture on the left) related to activities similar to office work (picture on the right). The reference to the horizontal and vertical range of movement of the arms allows us to define the work area (referred to here at the 5th percentile) within which the gripping action can take place without effort and with the sole rotation of the arms. Reworked by Pheasant and Haslegrave (2006, p. 99) and Chengalur et al. (2004, p. 196)

15.3.1   Tables for Office Work

The main dimensional constraints concern: the width and depth of the work surface, the distance between it and the seat surface, the space for the legs, the height of the seat surface (or the distance between it and the foot-support surface), the size and shape of the back of the chair and the angle between it and the seat surface. The dimensional constraints can be defined by identifying the most significant anthropometric parameters and, for each parameter, analysing the dimensional data relating to the group of users considered (a schematic indication of the constraints and anthropometric parameters is contained in Table 15.1).

The zone defined by the movements of the person during the activities carried out at the work table must be defined based on the breakdown of the tasks and must, in all cases, take into account the anthropometric dimensions and the most limited movement abilities.

In this case, the dimensional constraints concern the space necessary to sit and stand up, to move in the chair, to reach all the elements necessary for the user's work (shelves, objects, drawers etc. that must be used from the sitting position) (Fig. 15.4).

Additional constraints relate to the position of objects and equipment and the postures and movements necessary to use them. The most relevant case concerns the layout of the screen and the keyboard of the monitor, which must take into account: the position and the distance between the arms and the work surface (while typing, the arms are kept raised and not resting on the work surface) and the need to continually raise and rotate the head to control the keyboard and the screen.

Work station for office work: Size constraints and anthropometric parameters	
Size constraints	**Anthropometric parameters**
1. Width of the work surface	Reachability zones related to the movement of the arms and hands
2. Depth of the work surface	
3. Distance between work surface and seat	Width of the legs
4. Distance between seat and foot support surface	Height of the knees
	Width of the legs
5. Legroom:	Height of the knees
height	Space for movement necessary for the extension of the legs
width	
depth	
6. Width of the seat surface	Width of the hips
	Width of the elbows
7. Height of the armrests	Height of the elbows
8. Distance between seat surface and foot support surface	Length of the legs
9. Width of the back support	Width of the shoulders

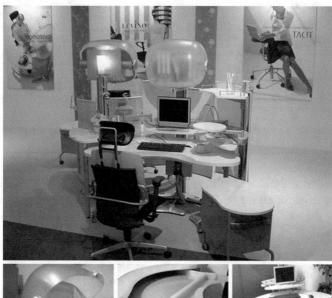

Fig. 15.2 Work-chair office work station. Solutions for reaching the work surface (Itoki table: Momotaro, Isao Hosoe Design)

Fig. 15.3 Work-chair office work station. Solutions for reaching the work station (from the Herman Miller catalogue)

Fig. 15.4 Layout of the furnishing and the movements required to use them, with a common office work station

The position and distance of the eyes from the objects to be observed and used are also essential for all the required actions, as is the precision of the movements that require the user to look frequently across several objects in the work area.

15.3.2 Chairs for Office Work

The design of the chairs used for office work, or similar activities, poses problems relating to both anthropometric aspects (seat width and height, height of the armrests and possible footrests, dimensional relationship with the work surface etc.) and to postural aspects (Fig. 15.5).

The latter pertain, in particular, to the position and inclination of the back with respect to the horizontal plane and with respect to the seat surface. The hypothetical activities presuppose, in most cases, a long stay in a seated position and the assumption of the postures necessary to write manually or using the monitor.

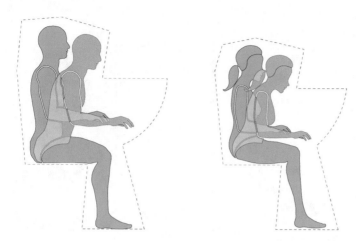

Fig. 15.5 Representation of the movement space in a sitting position. The movement space refers to the 5th and 95th percentile. Reworked from UNI EN ISO 14738:2009

The risks, therefore, concern the possible overload of the vertebrae discs, which derives from the long stay in positions that require, for example, an excessive inclination of the head and/or the vertebral column, an excessive or insufficient distance of the elbows from the table surface while typing, excessive or insufficient distance between the seat and the floor, etc.

Grandjean and Kroemer (1997, p. 76) report the results of studies conducted on the influence of the inclination of back rests on the pressure of the vertebrae discs, from which some useful indications for the design can be drawn:

- the pressure inside the vertebrae discs is minimal when the torso is folded and relaxed;
- the load on the muscles decreases as the inclination of the backrest increases;
- pressure on the vertebrae discs and muscular activity are minor with a backrest endowed with lumbar support;
- in terms of the pressure of the vertebrae discs and muscular activity, optimal conditions are guaranteed by a backrest and a seat inclined by 120° and 14° respectively and by the presence of lumbar support (Figs. 15.6 and 15.7).

The chairs used in activities like office work must have:

- the seat plane slightly tilted backwards, so that the legs and buttocks cannot slide (14–24° with respect to the horizontal plane);
- inclination of the backrest to 105–110° with respect to the seat and 110–130° with respect to the horizontal plane;
- the backrest should preferably be equipped with lumbar support and support between the third and fifth lumbar vertebrae (i.e. a height of 10–18 cm from the seat surface;

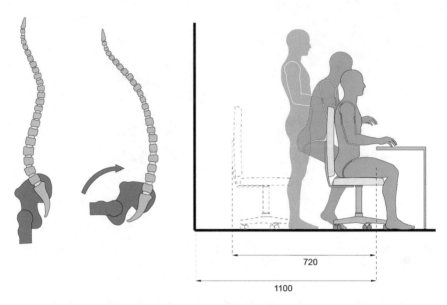

Fig. 15.6 Representation of the shift from an upright position to a seated one. Reworked from Grandjean and Kroemer (1997) and UNI EN ISO 14738:2009. In the seated posture, the vertebral column tends to modify the lordosis of the lumbar area, i.e. the anterior convex of the thoracic curvature, in kyphosis or posterior curvature. Sitting down involves a backward rotation of the pelvis (indicate by the arrow), bringing the sacrum to an upright position and turning the lumbar lordosis into a kyphosis. When moving from an upright position to a seated one: (1) the thigh rises; (2) the upper part of the pelvis rotates backwards; (3) the sacrum rises; the spine in the lumbar area passes from a position of lordosis to an erect or kyphotic position

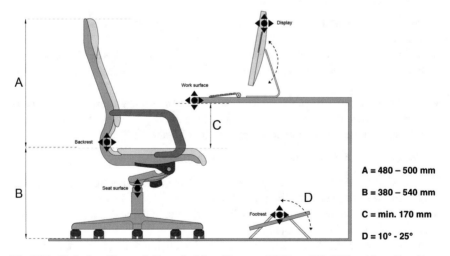

Fig. 15.7 Chairs for office work. Reworked from Kroemer (2017, pp. 328–329) and from Grandjean and Kroemer (1997, pp. 81–82)

- the different positions of the hands and elbows needed to work on the monitor, to write and draw by hand etc. require the adoption of solutions that allow the height of the seat surface and the inclination of the back to be adjusted (Figs. 15.8 and 15.9).

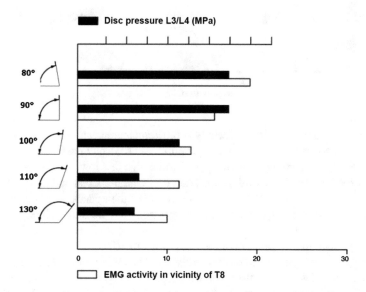

Fig. 15.8 Influence of the various positions on the pressure of the vertebrae discs. Reworked from Grandjean and Kroemer (1997, p. 77). Influence of the various positions on the pressure of the vertebrae discs (1 MPa = 10.2 kgf/cm^2; L3 and L4 = third and fourth lumbar vertebra). The value 0 on the pressure scale corresponds to an angle of 90°

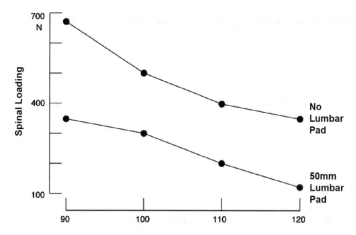

Fig. 15.9 Compression of the spine in relation to the inclination of the backrest. From 90° (vertical) to 120° (reclined), with and without lumbar support. Reworked from Pheasant and Haslegrave (2006, p. 127)

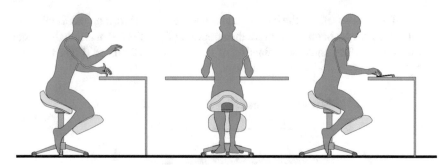

Fig. 15.10 Semi-seated. Taken from Kroemer (2017, p. 327)

The curvature of the spine, especially in the lumbar area, has been of great interest to orthopoedists and seating designers. Their ideas on correct posture and body support have generated countless seating proposals: high and low, hard and cushioned, inclined forward and backward, shaped and flat, saddle-shaped and curved in other ways. The surface of the seat supports the weight of the upper part of the body in a comfortable and safe way, while hard surfaces generate pressure points, which can be avoided with upholstery, cushions or other surface materials that elastically or plastically adapt to the contours of the body (Fig. 15.10).

Some scholars argue that a backrest is not necessary when sitting because, without it, the back muscles stabilise the torso, which, presumably, is a good muscle exercise, although many people find that a backrest is desirable for several reasons listed below:

- a back support can support part of the weight of the upper torso, which reduces the load that the spine must otherwise bear;
- a lumbar pad that slightly protrudes in the lumbar area helps maintain lumbar lordosis, which is considered beneficial;
- a properly formed back support allows the muscles to relax (Figs. 15.11 and 15.12).[13]

15.4   Work Stations for Manual and/or Precision Activities

The sizing of the work station for precision activity and, in particular, the distance between the seat-work surface and the work-floor level also depend on the height at which the arms must be placed and, above all, on the position of the hands and the inclination of the forearms.

In cases where the work surface is used in an upright position, the fatigue resulting from standing for a long time must be taken into account, along with any supports, such as stools or ischial supports with variable height (sit/stand seat), or semi-standing

---

[13] See Kroemer (2017, pp. 326–327).

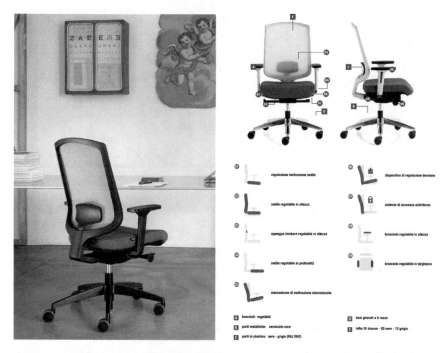

Fig. 15.11   Armchair with variable height and inclination. Vaghi catalogue

Fig. 15.12   Open space work stations: dividing elements. Herman Miller catalogue

RAILS   **PROP STOOL**   **JUMP SEAT**

Fig. 15.13 Positions and supports at the work table Taken from Chengalur et al. (2004, p. 254)

seats, which allow you to alternate between standing and sitting offer numerous advantages, such as[14]:

- they can support more than 60% of body weight;
- they are easily adaptable for the semi-upright position and for the seated position.

  Meanwhile the disadvantages can be summarised as:

- the use of stools or ischial supports can cause localised pressure and a decrease in blood circulation;
- they allow a limited number of positions;
- the legs tend to slip after a certain time (Fig. 15.13).

15.5   Shelves and Containers

The definition of the position and dimensions of shelves, bookcases, wall units, etc. represents one of the most common design problems, one that is more frequent than general furnishings; the equipment and all the elements must be reached and used by hands which can, in this case, apply to both the workplace and domestic settings.

The reference anthropometric parameters are those needed to deter-mine the horizontally and vertically accessible areas, both in the upright and seated position, and relate to stature, the ability to move relative to the arm support, the bending and rotation of the torso, the effort the ability to support the hand and, finally, the limitations and difficulties in movement that derive from the possible use of aids (sticks, hangers and, in particular, wheelchairs).

---

[14]See UNI EN ISO 14738/2009 Safety of machinery—Anthropometric Requirements For The Design Of Workstations At Machinery.

The dimensional constraints concern the height of the floors or the items that must be reached by hand and the depth the work surfaces and the spaces necessary for the person to move in operations and when opening doors, drawers and entrances.

The space needed for movement is obviously greater for people who use assistive devices for movement, such as sticks or crutches; in the case of people who use a wheelchair, the spaces necessary for the rotation and the side approach of the chair and the space necessary for the insertion of the feet for the frontal approach must be considered.[15]

The activities we can refer to are generally related to grasping and storing objects inside the shelf or container, and involve the movement of the arms up or down, the inclination of the torso or the need to kneel or extend the body when standing on tiptoe to reach surfaces that are too low or too high. When the containers are closed and/or contain drawers or pull-out elements, the required movements will involve grabbing, pressing or pulling handles and knobs, turning keys or other closing elements and finally extracting and reinserting drawers and/or extractable shelves, opening and closing doors, etc.

The zones of regular vertical and horizontal accessibility allow us to identify heights and depths that can be effortlessly reached through arm movement and moderate inclination of the torso. The dimensional constraints in this case must refer to the distances that can be reached by the users with height and reduced movement and, in particular, by women belonging to the 5th percentile and people who use wheelchairs (Fig. 15.14).

The maximum area of accessibility—between 60 and 140 cm from the floor—corresponds to the distances reached by the hand: at the top with the sole rotation of the arms and from the sitting position, at the bot-tom with the inclination of the back of about 30° (95th percentile) from the standing position.

In this area, which also corresponds to the distances that can be reached effortlessly from the upright position (referring to woman in the 5th percentile), all the elements on which safety and the accessibility of the environment or object depend must be placed, in particular: handles, switches, plant and equipment controls, alarm buttons, microphones, opening and closing controls for doors, windows, etc. (Fig. 15.15).

All the most frequently used elements and all the elements that can favour the autonomy of people who use a wheelchair and, in general, of people who have difficulty moving, must also be placed in the same area. In the areas immediately above and below—between 140 and 180 cm and between 60 and 40 cm from the floor—shelves and containers can be placed for objects of less frequent use, which can still be used effortlessly by people of average height.

Similar considerations must be made for the horizontally accessible zones, for which the reference parameter is the extension of the arms horizontally. The optimal size of about 50–55 cm refers to a woman in the 5th percentile and is less than the 60–65 cm size that is typically used for cabinets, kitchens and basic containers.

---

[15]See Ministerial Decree 236/1989 "Technical requirements necessary to ensure the accessibility, adaptability and visitability of private buildings and subsidised and facilitated public residential buildings, for the purpose of overcoming and eliminating architectural barriers".

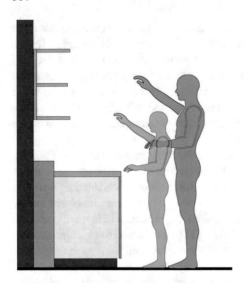

Work station for office work: Size constraints and anthropometric parameters	
Size constraints	**Anthropometric parameters**
1. Width of the work surface	Reachability zones related to the movement of the arms and hands
2. Depth of the work surface	
3. Distance between work surface and seat	Width of the legs
4. Distance between seat and foot support surface	Height of the knees
	Width of the legs
5. Legroom:	Height of the knees
height	Space for movement necessary for the extension of the legs
width	
depth	
6. Width of the seat surface	Width of the hips
	Width of the elbows
7. Height of the armrests	Height of the elbows
8. Distance between seat surface and foot support surface	Length of the legs
9. Width of the back support	Width of the shoulders

Fig. 15.14 Shelves and containers: size constraints and anthropometric parameters

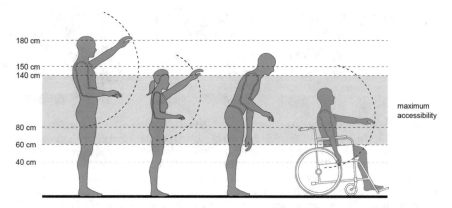

Fig. 15.15 Vertical reachability zones. Elaboration on ISO 7250-2:2017 data. The measures indicated refer, from left to right, to: (1) height of man in the 95th percentile with an inclination of the arm of about 45° (hand at head height); (2) height of woman in the 5th percentile with an inclination of the arm of about 45° (hand at head height); (3) height of man in the 95th percentile with a back inclination of about 30°; (4) height of man in the 50th percentile sitting on a wheelchair

The reference to the gripping distance of the hand from the sitting position is also necessary to guarantee the accessibility of spaces and furnishings to wheelchair users; similar considerations pertain to the placement of the command, opening and adjustment elements on which safety and accessibility depend (see Law 13/89). The reference to the sitting position, however, is very useful in any case where container shelves and work surfaces must, or can be, used from this position.

The layout of the office work station can ensure that many of the storage, consultation, reordering, selection activities, etc. can be used via containers placed near them on the table, which favours using them while seated and taking advantage of, for example, a common chair with wheels, or using stools or ischial supports to avoid having to continually get up or move.

Similar operations can take place in domestic environments and, in particular, in the kitchen, where utensils, crockery and small appliances can be arranged to be used while still seated.

Compliance with the dimensional constraints, defined on the basis of anthropometric parameters, obviously clashes with the need to exploit the available space. In this case, it is also possible to use solutions that allow us to adapt the initial configuration of the furnishing system to the anthropometric characteristics of the users and to the different functions for which they can be used (Figs. 15.16, 15.17, 15.18 and 15.19).

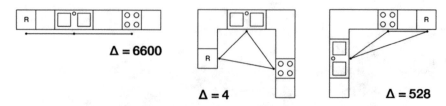

Fig. 15.16 Size references for three domestic kitchen layouts. Taken from Pheasant and Haslegrave (2006, p. 106). The sequence of operations generally proceeds from left to right in a horizontal line (usually: from the sink, the worktop and the hob). For left-handed people, the sequence of operations is generally the opposite. The layout of the kitchen must minimise the need for movements, respect the sequence of horizontal work operations and provide for a circulation space of min. 120 cm in front of the equipped wall. The ideal layout is: (1) in line on one side only; (2) L-shaped on two-contiguous sides; (3) U-shaped on three sides. The figure shows the three standard arrangements, referred to a 30 cm modular grid, which include the refrigerator (R), the hob and the sink. For each arrangement, the work triangle and its width (D) are shown. The adaptability to the different dimensional requirements can be obtained with technical measures already widely used to vary the height of the worktops or internal shelves of fitted wardrobes and walls, for office chairs etc., which allow the standard dimensions to be varied depending on the anthropometric characteristics of the user or the activity he is carrying out

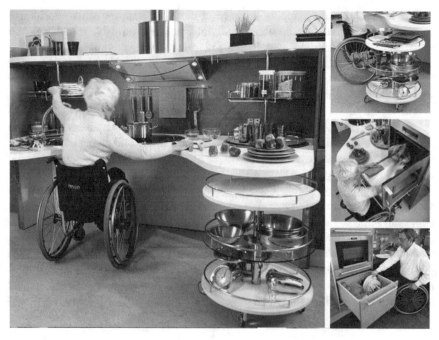

Fig. 15.17 Domestic kitchens: solutions to facilitate maximum accessibility. Snaidero catalogue, skyline kitchen

Fig. 15.18 Domestic kitchens: solutions to facilitate maximum accessibility. Valcucine catalogue, new logic system kitchen

15.6 Handles

The dimensions, shape and position of handles, knobs and, in general, all the elements that are used to open and close doors and containers must be defined according to the size of the hand that characterises the group of reference users, the ability to grip and, in particular, the ability to accurately and easily implement movements that allow precise grip and exercise of power.

In both cases, the extent and precision of the movements varies according to age and, without going into the details of the factors of change linked to the process of growth and ageing, it should generally be taken into account that, in children, the ability to hold it is strongly reduced compared to that of adults, because of the size of the hand and fingers and the lower muscular strength.

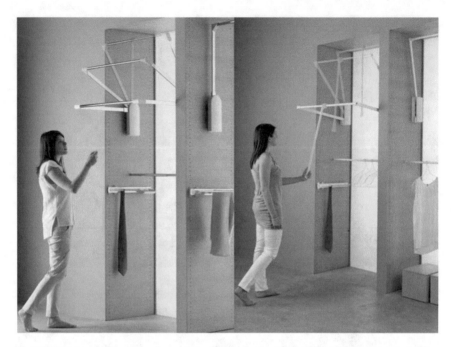

Fig. 15.19 Adaptability of use of containers, shelves and worktops. Servetto catalogue. Left: Electric Servetto serie 3; right: Servetto 2004. Servetto® hangers for wardrobes and walk-in closets, rotating baskets, guides for vertical movement of shelves, are technical features that are already widely used by the current production of home furnishings that allow you to "bring" accessories and furnishings in the maximum accessibility range, ma-king the best use of space availability

In the elderly, the ability to grip can be reduced due to both the progressive decrease in muscle strength and due to joint problems or pathologies that reduce the agility and precision of movements.

The holding capacity is obviously linked to the force that the hand is able to exercise during the exercise of power or a precise grip (Fig. 15.20).

The recommendations for hand grip capacity[16] concern the minimum dimensions of the objects that must be grasped and moved and the extension of the requested movements. For high gripping forces, the following must be ensured:

- dimensions between 50 and 70 mm for objects that must be grasped and moved; the size that allows the maximum gripping force (grip strength) is about 60 mm for men and 55 mm for women;
- neutral positions of the forearm (with the thumb facing upwards) or supine positions (with the palm facing upwards). The grip strength is optimised in the supine position and reduces to 87% with the forearm in the prone position (with the palm facing downwards);

---

[16]Cfr. Adams (2006, pp. 365–375), Chengalur et al. (2004, pp. 343–354).

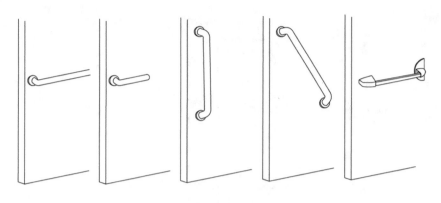

Handles	
Size constraints	**Anthropometric parameters**
1. Height off the ground	Hand grip height (in an upright position and from the wheelchair). The height of the handles must be between 85 and 95 cm (recommended 90 cm)
2. Width or diameter	Size and grip capacity
3. Arrangement with respect to the door or entrance	Approach movements allowed by the wheelchair

Fig. 15.20 Handles: size constraints and anthropometric parameters

- neutral positions of the wrist (parallel to the longitudinal axis of the fore-arm) (Figs. 15.21, 15.22 and 15.23).[17]

15.7   Buttons, Switches, Manual Controls and Touchscreens

The dimensional and morphological aspects concern the support or grip space necessary for the action of the fingers, in particular, grasping, tightening and acting upon knobs and supporting and pressing keys and buttons with one or more fingers.

---

[17]"The strength of wrist flexion (when not assisted by the forearm) varies approximately from 90 N (105 N for men and 75 N for women) with a bend angle of 90° towards the torso, and approximately 45 N (50 N for men and 40 N for women) with an extension angle of 75° in the opposite direction to the body, with an increase of 10 N at an extension of 90°. The extension force remains fairly constant, with an average of approximately 55 N (65 N for men and 50 N for women) and an increase up to 67 N (75 N for men and 60 N for women) in correspondence with a 90° extension" (Adams 2006).

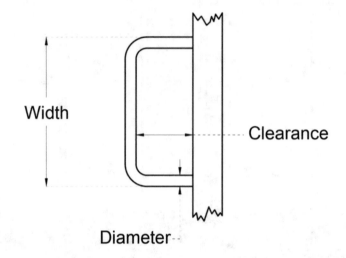

Fig. 15.21 Recommended dimensions for handles on equipment components or on containers

	Measurements	Value (mm)
Clearance	One hand Two hands	64 64
Width	One hand Two hands	120 240
Diameter	One hand	Min 6 13 19 Max 38
Edges	Exposed edges should be rounded to a minimum radius of:	1
Corners	Exposed corners should be rounded to:	13
Grip	Fingers should be able to curl to a minimum of:	120° (as in a hook grip)

The usability of all manual controls, in particular, keys and buttons, depends both on the size and the distance between the keys, that is, on the space for movement offered to the fingers and the possibility of recognising the keys and their position through touch contact.

As for touch screen displays, there seems to be an engineering trend to reduce the production costs of mechanical switches, lights and input devices (Chengalur et al. 2004). In fact, the usability of the touch screen is guaranteed by the management and movement of a single finger, for adjustment, processing, selection, insertion etc. operations. In addition, the touch screen is useful for reducing the workload when the input types are limited and well defined (Fig. 15.24).[18]

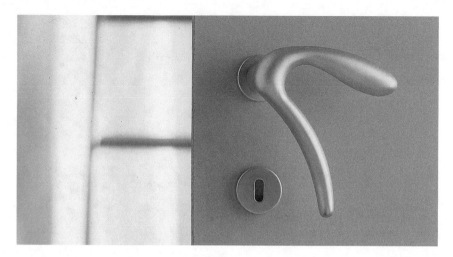

Fig. 15.22   Leonardo handle. Produced by Ghidini Pietro Bosco

Fig. 15.23   Marylin handle and wind handle. Olivari catalogue

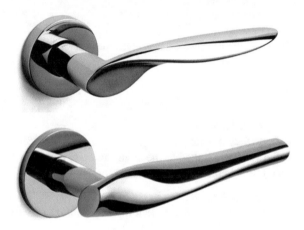

Fig. 15.24   Touch screen control recommendations (dimensions in mm). Reworked from Chengalur et al. (2004, p. 327)

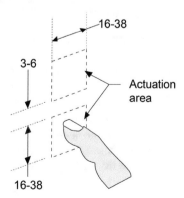

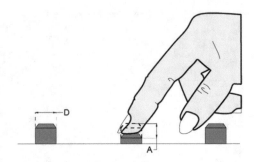

Parameter	Recommended Design Values (in mm)
Diameter D Fingertrip activation Palm or thumb activation Emergency push buttons	10-19 min 19 min 25
Displacement D Finger activation Palm or thumb activated	3-6 3-38

Fig. 15.25 Push buttons and switches: recommended sizes. Reworked from Chengalur et al. (2004, p. 307)

Manual controls must also take into account:

- user characteristics;
- frequency of use;
- the variables linked to the environmental conditions, giving particular consideration to whether the elements are, or can be, used:
  - inside or outside (or in cases, both);
  - in conditions of sufficient lighting and/or in conditions of darkness;
  - with bare hands or while wearing gloves or protective clothing (Figs. 15.25, 15.26 and 15.27).

The ability to recognise the elements by touch allows them to be identified and handled with only the hands (i.e. without looking) and, moreover, the possibility of understanding the outcome of the action just carried out. The lowering of the key, therefore, represents the feedback through which we can assess whether the finger pressure has been successful. Similarly, the click following the rotation of a knob, the resistance given by the lock of the knob and so on constitute the noticeable feedback at the touch of the hand.

The touch screen displays, meanwhile, provide physical or audio feedback in response to the press of a digital button, such as buzzing, vibrating, "clicking", etc.

---

[18] See Chengalur et al. (2004, pp. 325–3).

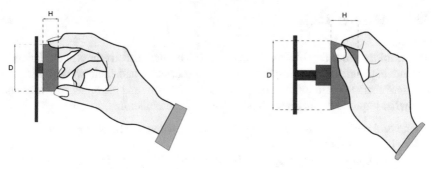

Parameter	Recommended Design Values (in mm)
Fingertrip operation (precision grip) Diameter D Diameter minimum, for very low torque Depth H	10-100 6 12-25
Palm grasp operation (power grip) Diameter D Depth H, minimum	35-75 15

Fig. 15.26 Handles: recommended sizes. Reworked from Chengalur et al. (2004, p. 309)

Recommended separation for various types of control			
Control	Type of use	Distances in mm	
		minimum	desiderable
Push button	one finger	12	51
Toggle switch	one finger	20	51
Crank	one hand	51	100
	two hands	76	127
Lever	one hand	51	100
	two hands	76	127
Knob	one hand	25	51
	two hands	76	127

Fig. 15.27 Distance from adjacent control. Minimum and recommended measures to avoid accidental and involuntary touching of adjacent controls. Reworked from Chengalur et al. (2004, p. 301)

15.8 Manual Tools

The usability and safety of manual tools, as well as their effectiveness and efficiency, depend both on the compatibility of their dimensions and their weight with the physical characteristics of the users, and on the positions and movements of the hand and wrist required by their use. It is essential to remember that, in general[19]:

- a neutral position of the wrist is preferable, i.e. where the hand is aligned with the longitudinal axis of the forearm;
- the maximum gripping force can be exercised with the whole hand (exercise of power with the opposition of the thumb to all the other fingers) while the various types of precision grip (pinch grip) involve a notable decrease in the force that can be exercised;
- movements that require extreme deviations of the wrist or deviations, that should be repeated for significant times are not recommended, even if they are not extreme.

The dimensions of the handle we can discuss are[20]:

- *length: for tools that require an appreciable application of force (such as pliers, screwdrivers, etc.), 130 mm (minimum 100) is recommended, which must be increased by 15 mm when gloves are required;*
- for tools with a double grip (such as scissors, pliers, etc.), the *distance* between the point of application of the force and the gripping point must be between 65 and 90 mm. Users and/or operators who have small hands or short fingers will have difficulty using handles that exceed 100 mm;
- for tools with cylindrical handles (such as screwdrivers etc.), the *diameter* to ensure the best exercise of power is 40 mm. However, a diameter between 3 and 5 cm is recommended; a 12 mm diameter is recommended for precision gripping (a diameter that is acceptable between 8 and 16 mm is recommended);
- for exercise of power for tools such as hand saws, drills etc., an *inclination of the handle* between 0° and 15° with respect to the vertical line is recommended, which allows the position of the wrist closest to the neutral position. In the case of a saw, the length of the hole must not be less than 12 cm and the width must be at least 60 mm (Figs. 15.28 and 15.29).

The space for the support of the fingers must have a minimum size of 100 mm. The space for the insertion and the support of the fingers must have dimensions of 120 mm in length and 60 mm in width to allow the insertion of the hand wearing gloves.

Hand tools must also[21]:

- ensure electrical and thermal insulation;
- be free of sharp parts, sharp edges or parts in relief;

---

[19]See Chengalur et al. (2004, pp. 110–112).

[20]See note 19.

[21]See note 19.

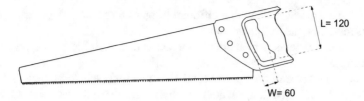

Fig. 15.28 Sizes and inclination of the handle (dimensions in mm). Reworked from Occhipinti and Colombini et al. (1996) and Chengalur et al. (2004, p. 350)

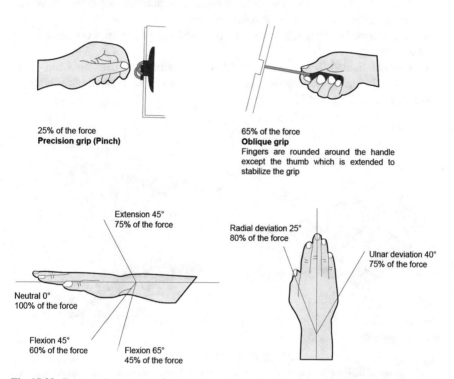

Fig. 15.29 Decrease in grip strength at different wrist and hand postures. Reworked from Colombini et al. (1996, p. 728)

- be coated with materials that are impermeable to oily substances, solvents and chemical agents;
- prevent the accumulation of dirt on their surface;
- facilitate use with an adequate conformation and texture of the handle.

Further recommendations regarding the usability and safety of manual instruments are listed as follows[22]:

---

[22] See note 19.

- equip the electrical tools with a different trigger other than a finger trigger, as it offers more options for the operator;
- equip the instruments, which are used to cut and to exert high forces, with stops and protections;
- the triggering force of the "power tool" (ignition tool) should be high enough to avoid accidental activation, but not so high as to strain the fingers used. Activation forces should be about 1 kgf (10 N);
- the weight of a manual instrument should be less than 2.3 kg if used away from the body or above the shoulders. If the weight is greater, this instrument should be supported by a counterweight or balancing system;
- the recommended weight for instruments used for precision work should not exceed 0.4 kg;
- it is recommended to use materials that absorb vibrations near the hand or on the sides, in turn protecting the operator from the risk of exposure to vibrations (Figs. 15.30, 15.31 and 15.32).

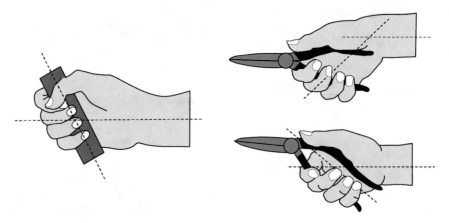

Fig. 15.30 Wrist positions. The optimal position is when the wrist is in a neutral position and the thumb is aligned with the longitudinal axis of the forearm. Reworked from Kroemer (2017, p. 76)

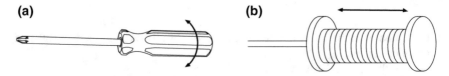

Fig. 15.31 Shape of the handle. Reworked from Eastman Kodak Company (1983, p. 149). The grip can be facilitated by the provision of elements to contrast the movement or slipping of the fingers and the texture of the surface of the handle: (1) for rotation or torsion actions, longitudinal grooves or reliefs (**a**) are recommended; (2) for pushing or pulling in the direction of the tool's longitudinal axis horizontal grooves or ridges (**b**) are recommended

Fig. 15.32 Utensili manuali. Tratto da catalogo Bahco, gamma Ergo

References

Adams SK (2006) Hand grip and pinch strength. Int Encycl Ergon Hum Factors 1:365–376

Chengalur SN et al (2004) Kodak's ergonomic design for people at work. Wiley, New Jersey (1st ed.: Rodgers SH (1983) Ergonomic design for people at work. Wiley, New York)

Colombini D et al (1996) Orientamenti per la riprogettazione del lavoro per compiti ripetitivi. La medicina del lavoro 87(6):728–749

Di Martino V, Corlett EN (1999) Organizzazione del lavoro e ergonomia. FrancoAngeli, Milano

DM—Ministero dei Lavori Pubblici 14 giugno (1989, n. 236) Prescrizioni tecniche necessarie a garantire l'accessibilità, l'adattabilità e la visitabilità degli edifici privati e di edilizia residenziale pubblica sovvenzionata e agevolata, ai fini del superamento e dell'eliminazione delle barriere architettoniche

Eastman Kodak Company (1983) Ergonomics design for human at work. Van Nostrand Reinhold, New York

Grandjean E, Kroemer KHE (1997) Fitting the task to the human: a text book of occupational ergonomics. Taylor & Francis, Londra e Philadelphia (1st ed.: Grandjean E (1963) Fitting the task to the man. Taylor & Francis, Londra)

Kroemer KHE (2017) Fitting the human: introduction to ergonomics/human factors engineering. CRC press, Boca Raton (1st ed.: Grandjean E (1963) Fitting the task to the man. Taylor & Francis, Londra)

ISO 9241-11 (1998) Ergonomic requirements for office work with visual display terminals (VDTs)—Part 11: Guidance on usability

Leplat J (2000) Activity. In: Karwowski W (ed) International encyclopedia of ergonomics and human factors. Taylor & Francis, Londra e Philadelphia

Occhipinti E, Colombini D (1996) Alterazioni muscolo-scheletriche degli arti superiori da sovraccarico biomeccanico: metodi e criteri per l'inquadramento dell'esposizione lavorativa. In: La medicina del lavoro, 87(6)

Pheasant S, Haslegrave CM (2006) Bodyspace: anthropometry, ergonomics and the design of work. CRC press, Boca Raton (1st ed.: Pheasant S (1986) Body-space: anthropometry, ergonomics and design. Taylor & Francis, Londra e Philadelphia)

UNI EN 614-1 (2009) Sicurezza del macchinario – Principi ergonomici di progettazione – Parte 1: Terminologia e principi generali

UNI EN 614-2 (2009) Safety of machinery. Ergonomic design principles. Interactions between the design of machinery and work tasks, p 9

UNI EN ISO 10075-1 (2003) Principi ergonomici relativi al carico di lavoro mentale – Termini generali e definizioni

UNI EN ISO 14738 (2009) Sicurezza del macchinario - Requisiti antropometrici per la progettazione di postazioni di lavoro sul macchinario

UNI EN ISO 7250-1 (2017) Dimensioni del corpo umano da utilizzare la progettazione tecnologica – Parte 1: Definizioni delle dimensioni del corpo umano e dei punti di repere anatomico

Chapter 16
Interface Design: Intervention Criteria and Good Practices

16.1 Introduction

Interactive products can traditionally be used by means of a physical element, often comprising an interface. The interface is the physical and cognitive intermediary we use to manage and control our digital devices, or the information contained on a web platform.

The Design of the interface that allows the interaction between people and devices is a crucial feature of Interaction design. This is often defined as the User Interface (UI) and comprises all of the elements of system that come into contact with people on a physical, perceptive and conceptual level.

The design of user interfaces generally refers to two categories of devices: input devices (for inserting information) and output devices (which return user information).

As stated by Saffer (2010, p. 122), interfaces are the tip of the iceberg. The visible part is really only the tip; the part below the surface, which cannot be seen, is transmitted to the user, however. In practice, the interface displays the invisible functions of a product and, as a result, the user can access and benefit from these functions. In the field of Services design, the interface typically consists of physical spaces, products and people.

The complexity of the design intervention lies in the designer's ability to handle a series of challenges. Firstly, the interface design assumes that the flow of interactions will be analysed, taking into account aspects such as: the perceptive dimension, the nature of the people-interfaces and people-interface-context interactions, the accessibility of the contents, reliability, appeal, etc.

In summary, the interface of an interactive system is comprised of all of the parts of the system that people come into contact with physically, perceptively and conceptually (Benyon 2010, p. 46):

This chapter was authored by Alessia Brischetto.

F. Tosi, *Design for Ergonomics*, Springer Series in Design and Innovation 2, https://doi.org/10.1007/978-3-030-33562-5_16

- the *physical interaction* occurs when the user physically interacts with a device; for example, by pushing the buttons or levers on the device or through a gestural interaction, the device responds by providing aural or visual feedback or opens another window or function.
- the *perceptive aspects* are related to the display of information on a screen, which we can see or perceive through sound or vibrations.
- the *conceptual aspects* are engaged when we interact with a device and try to understand what it does and how we should use it. The device supplies messages or other inputs that guide us through a given activity.
- *input and output*: the interface must generally supply some stimuli so that users can carry out a given activity, such as starting the system or entering data into it. This type of mechanism is defined with the term input. The interface must also supply some stimuli/mechanisms so that the system can communicate what is happening to the user by means of feedback (defined with the term output). This content may take the form of information, images, videos or animations.

When we talk about user interfaces, we are talking about designing interactive systems. For this reason, disciplines like Human-Computer Interaction (HCI) and Interaction Design are integral in design sectors like the User Interface (UI), Graphical User Interface (GUI) and the User Experience (UX).

When designing an interface, one must consider the entire human-computer interaction, as well as the human-human interaction created by using digital systems. Digital systems and network infrastructures are, by definition, interactive systems composed of a series of interconnected devices: some can be attached, some can be incorporated into products or placed in a given environment and some can be transported, like *personal devices*.

Given the complexity of the usage context and the pervasive nature of new technologies—which allow widespread connectivity through personal devices—designers need to consider the entire usage context. The design of the user interface requires the adoption of design strategies aimed at satisfying aspects like the versatility and adaptability of various user profiles and different usage contexts.

In this chapter, we will tackle the strategic design themes and aspects used to develop human-level interfaces. In particular, the theoretical aspects and intervention criteria for designing user interfaces will be illustrated and explained through case studies.

16.2   Interface Design: Design Intervention Strategy

People interact with a computer or devise via an interface, by means of instruments like a keyboard, a mouse, a touch screen, microphone, *Near Field Communication* (NFC) systems or environmental devices, such as sensors for example. Today, when we talk about interfaces, based on the different contexts related to Interaction design, we are referring to websites (including Web 2.0), *Computer Supported Cooperative Work* (CSCW) and *mobile computing*. In this chapter, we will look at tangible interfaces and the physical, cognitive and emotional aspects that must be considered during the design phase.

As we saw in Chap. 3, in Garrett's UX diagram (Fig. 3.5), the user interface is the visible surface of the system and concerns the aesthetics and the immediate sensory experiences of the user on a superficial level (when interacting with the product or interface). During this phase, we must examine all of the physical, cognitive and sensory aspects that allow the user to interact with the system effectively. It is implicitly understood that anything preceding the surface is fundamentally important for the system's development (strategy, purposes, information architecture, etc.). As discussed in Chap. 6, the development of a system or product requires the definition of an intervention strategy, the involvement of the users in each phase of product development and the validation of ideas through specific methodologies.

In Interface design, as in other fields of Design, we must follow an intervention strategy that allows us to bring together the various skills involved in the development of the system. An interface is connected to a software system and the development of its design thus requires collaboration between diverse professionals. In order to ensure the effective utility and performance of the entire system (software, hardware, interface). A good website or mobile application can be appealing from an aesthetic perspective, but if their functions or contents are difficult to figure out, aesthetics will not be enough. Conversely, a system that is well-designed from a software perspective but that is not very usable or appealing will be seen as poor in performance terms.

For these reasons, referring back to Garrett's model (2011), the purpose and the structural level must be defined based on the identified user needs. The structural level allows us to define the product's preliminary functions; subsequently, the skeletal level allows us to define the shape and the interaction methods (feedback, navigation systems, etc.) of the system's function. The surface level, therefore, will be interdependent on all of the elements that are identified in the earlier phases and will, in turn, be validated through specific evaluation methods that require the users' involvement.

16.3 The Standards on Accessibility of Interactive Systems

The design development of the "surface" (or user interface) of the system requires us to consider aspects like usability and accessibility. To support the design phase, designers have a series of standards-based tools.

Based on the intervention environment, the standard indications can vary. In terms of developing web pages, the sector standards generally refer to the Web Accessibility Initiative (WAI). The WAI published the first Web Content Accessibility Guidelines (WCAG) version 1.0 in May 1999; these provide indications for developers on how to design accessible web pages for people with motor and sensory disabilities.

The WAI deals with Web accessibility in a broad sense, that is, not only contents but also the tools used to create web pages, browsers and, more generally, technologies for accessing the web. The Web Content Accessibility Guidelines (WCAG) are particularly important for content accessibility.

The WCAG were published as recommendations by the World Wide Web Consortium (W3C) on 5 May 1999. An accessory document was published as a W3C note on 6 November 2000. Moreover, they help to make web content usable on all devices, including mobile devices (PDA—Personal Digital Assistant—and smartphones). The WCAGs are recognised as the de facto standard and are the basis for legislation and evaluation methods in many countries.

In 2008, the WCAG work group published the WCAG 2.0, which are based on different requirements than WCAG 1.0 (which instead refer to HTML and CSS programming languages), as they deal with accessibility issues in greater detail. These indications are still under discussion, as experts and researchers from around the world believe that there is still much to be done before we have a shared standard. The Interaction Design guru Alan Cooper[1] responded to this problem with his famous axiom:

"Respect the standard unless there is a better alternative." The standards and tools we have are useful for the definition of the basic requirements of a system, but as Saffer (2010) states: *"We should feel free to propose a new system for cutting and gluing, but it must be absolutely superior to what users are used to, thus creating a new standard."* In terms of interactive systems, the ISO offers a new series of standards;

The following are among the most pertinent:

- ISO 9241-210:2010, Ergonomics of human-system interaction—Part 210: Human-Centred Design for interactive systems;
- ISO 9241-110:2006, Ergonomics of human-system interaction—Part 110: Dialogue principles;
- ISO 14915-2:2003, Software ergonomics for multimedia user interfaces—Part 2: Multimedia navigation and control.

In addition to technical standards, the ISO supplies other documents: ISO Technical Specification (ISO/TS), ISO Public Available Specification (ISO/PA ISO Technical Specification (ISO/TS), ISO Public Available Specification (ISO/PAS) and ISO Technical Report (ISO/TR). These are seen as less relevant in the scientific community but must be considered to be a valid starting point when one begins to work in the design sector.

In the next paragraph, we will look at ISO 9241 in detail, in particular, the dialogue principles, or the fundamental features to consider when studying the interaction between a user and an interactive system.

---

[1] See www.cooper.com.

16.3.1 Evaluating Interaction Quality: The Seven ISO 9241 Dialogue Principles

ISO 9241 is an International Organization for Standardization standard regarding Ergonomics and Human-Computer Interaction. It was originally called "Ergonomic requirements for office work with visual display terminals (VDTs)" and essentially dealt with ergonomic aspects of the video terminals used in office work. With time, its goals have expanded, and it now deals with the general usability issues of interactive systems. It comprises many documents; the most interesting for UI-design purposes is Part 110 (Dialogue principles). This describes seven "dialogue principles", or seven features that each dialogue between a user and an interactive system must consider. In this document, the term "dialogue" is used to mean "human-system interaction", that is, the sequence of user actions (inputs) and system responses (outputs) needed to achieve a goal.

User actions include data entry, navigation and system control. Let's look at the seven dialogue principles of ISO 9241 in detail:

1. **Suitability for the task**

A dialogue is suitable for an activity when it supports the user in completing the activity effectively and efficiently. In a dialogue that is suitable for the activity, the user can focus on said activity rather than the technology chosen to perform it. The dialogue should:

- present the user with information for completing the activity;
- avoid presenting the user with unnecessary information for completing the activity;
- supply the user with information for completing the activity, if necessary;
- propose input and output formats that are appropriate for the activity (default values to be selected, where possible);
- reduce the steps in the dialogue to those that are strictly necessary for completion;
- be consistent between the source document format and the interface.

2. **Self-description**

A dialogue is self-descriptive when, at any time, the user is fully aware of where they are, what stage they are at, what actions they can perform and how to perform them. The dialogue should:

- guide the user through completion of the activities with information;
- minimise the need for manuals or other external information to what is absolutely needed;
- inform the user about changes in the dialogue status;
- supply information about the input, if the system requires it;
- visually design the interface so the dialogue is clear to the user;
- supply the user with acceptable units and values for data entry;
- make the dialogue easy to understand through appropriate terminology.

3. Conformity with user expectations

A dialogue conforms with users' expectations if it corresponds to predicted contextual needs and commonly accepted conventions. The dialogue should:

- use vocabulary that is familiar to users when performing the activity;
- supply appropriate feedback about input and the user's actions, if appropriate;
- reflect data structures and organisation that the user sees as natural;
- follow linguistic and cultural conventions in the appropriate formats;
- supply feedback based on users' needs;
- use understandable units of measurement;
- be consistent in the behaviour and aspect of the dialogue;
- design the input positions based on users' expectations;
- formulate feedback or messages to the user in an objective and constructive manner.

4. Suitability for learning

A dialogue is suitable for learning when it supports and guides the user in learning how to use the system. The dialogue should:

- provide rules and useful concepts for learning purposes;
- supply appropriate support to help the user become familiar with the dialogue;
- give feedback or explanations to help the user understand the system they interact with conceptually;
- supply sufficient feedback about activity results for the user to know if they have completed it successfully;
- allow the user to explore the steps without negative repercussions;
- allow the user to initially complete the activity by entering only the essential information, while the system provides non-essential information for default settings.

5. Controllability

A dialogue can be controlled when the user is able to initiate and control and direction and rhythm of the interaction until the goal has been achieved. The dialogue should:

- adapt to the needs and features of the user and not be dictated by the system's function;
- give the user full control over dialogue;
- allow the user to restart from the last completed step, where possible;
- allow for the deletion of the last phase;
- allow amounts to be entered ("If different amounts of data may be relevant to a task then the user should be able to control the amount presented");
- enable movement between available input/output devices;
- allow users to modify the default values, if appropriate for the activity;
- grant the user access to data that has been created or modified.

6.   Tolerance to errors

A dialogue is tolerant to errors if, despite clear input errors, the desired result can be obtained via an absent or minimal corrective action on the user's part. Tolerance to errors is achieved using damage control, error correction or error management for errors that have been flagged. The interactive system should:

- help the user to detect and avoid input errors;
- prevent any user input that could cause undefined statuses or faults in the interactive system;
- supply the user with an explanation to help correct the error, when flagged;
- provide active support for solving errors (for example, positioning the cursor where the correction is needed);
- in cases where the interactive system is able to automatically correct errors, it should advise the user to correct or ignore them;
- allow the user to postpone the correction of an error, unless the correction is necessary for the dialogue to continue;
- provide the user with further information on the error and the required correction;
- allow the validation/verification of information before the interactive system processes the input;
- minimise the steps required to correct errors;
- supply explanations and confirmation requests before performing potentially damaging actions stemming from a user action.

7.   Suitability for individualisation

A dialogue is capable of individualisation when users can modify the interaction and the presentation of information to satisfy their individual skills and needs. The interactive system should:

- consider the characteristics of the user;
- allow the user to choose between alternative displays for individual needs;
- adapt the amount of information/explanations (for example, error message details, help information) according to the user's level of knowledge;
- enable input/output speed settings to satisfy individual needs;
- enable a choice between various dialogue techniques;
- enable the user to select the interaction levels and methods that best satisfy their needs;
- enable the user to select the way in which their input/output information is displayed (type and format);
- allow, where appropriate, users to add or reorganise, in a specific manner, the dialogue elements or supports in order to support their personal preferences when performing tasks;
- provide interactions levels and methods based on the users' varying needs and characteristics.

The principles of ISO 9241-110 are important from both a conceptual and design point of view. In practice, they define a quality model that can be used to evaluate the usability of a system or to compare the usability of two similar systems (Polillo 2010, p. 136).

This is why it is necessary to analyse each system and score it based on how much each individual principle is applied. In this way, it is possible to display the parameters measured on a star diagram, as seen in Fig. 16.1 (in which a scale from 0—minimum score to 4—maximum score—is used), which allows us to see the quality of the system.

In the diagram shown in Fig. 16.1, we can see the profile of a system that has received a maximum level in terms of self-description and suitability for learning, but a low level in terms of the suitability for individualisation parameter.

As noted in the ISO 9241-210:2010 standard (see Chaps. 5 and 6), once the need to develop a system, product or service has been identified, and the reference problem has been identified as a result, there are four essential steps that should be followed in order to integrate the usability requirements into the development process for a product/system: understand and specify the usage context (a); define the user and organisational requirements (b); produce design solutions (c); verify the design solutions (d).

The evaluation of the quality of the interaction, by applying the dialogue principles, can be applied in phase (a), by analysing similar systems to the one we are designing, and in phase (d), that is, where we verify our design solutions. This type of evaluation can be conducted using methods such as Task Analysis and the Hierarchical Task Analysis. The information is collected through field observations, such as Contextual Inquiry, interviews and questionnaires, Thinking Aloud, Ethnography and many other methods (as seen in Chap. 6). When designing an interface, another strategic aspect pertains to the usage experience and, as we saw in Chaps. 3 and 6, there are a range of instruments to evaluate the emotional aspects of the iteration; these include technological tools that allow us to analyse the level and the emotional nature of the users' involvement, or to measure and detect the frequency of errors or aspects linked to parameters like attention and cognitive load.

16.4   Design Tools: Wireframes

Now we will look at the design development of the interface. The designer must be able to organise the collected information and produce prototypes that can be tested again with the users' involvement.

These include not only wireframes (we have blueprints and service plans in the Service design sector), which allow us to define the layout of the interface and the primary functions of the system, but also the paper, interactive and physical prototypes and the evaluation of the latter.

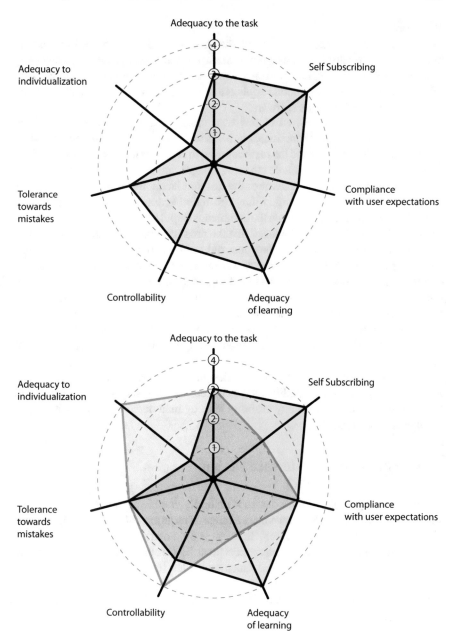

Fig. 16.1 Quality profile of a system based on the seven dialogue principles of ISO 9241 (above) and compared to due quality profiles (below)

Wireframes

Wireframes are documents that display the structure and hierarchy of information. Wireframes generally have three main areas: the wireframe itself, accompanying notes and information on the wireframe. This content organisation technique can be used to analyse web pages, mobile applications or specific functions, such as volume control for a video or music software.

The interface is synthesised in this way, casting aside the elements of Visual design. The structural and functional elements of the system are favoured, including the contents, the functions and the methods used to access or manipulate the functions and contents. The content includes the test, images, icons, animations, etc.

Too many details are unnecessary: the test and icons, for example, can be summarised or identified with specific fillings (different textures, patterns, colours, etc.). For example, the functions include the commands (input boxes, icons, levers, buttons) for a certain function, or the feedback that the product supplies for these commands.

The annotations displayed in the upper right of Fig. 16.2 immediately communicate the technical aspects. They are useful for communicating with the developers and technicians working on the system design and, moreover, allow us to note the correlation between the on-screen functions and the types of interactions needed from the user to access tasks and sub-tasks in the system.

The wireframe technique can be developed in parallel to the Task Analysis and the Hierarchical Task Analysis. The graphic design portion is typically carried out afterwards.

This technique allows us to define the design constraints for the parameters such as accessibility and system quality (as seen in Sect. 16.3.1), to verify compliance with legal constraints and to plan any usability tests by manufacturing paper, interactive or physical prototypes.

16.5   The Basics of Interface Design

A fundamental part of Interface design is naturally graphic design. Interfaces, as we saw in the previous sections, must meet a range of requirements and, if designed correctly, allow a human-system interaction that is fluid and rewarding. The functional and cognitive elements of the interaction must be carefully designed and comply with a range of compositional and functional rules (Galitz 2007). These include commands with buttons and navigation tools, labels, the position of elements on the screen, etc. (the functions are inserted based on the type and size of the digital support employed).

Layout and Visual Flow[2]

At the core of Screen design is the good practice of defining a layout, that is, the visual and compositional structure, based on the hierarchy of information and available inputs.

---

[2]There are programs for the creation of wireframes and the layout of an interface, including: Promoted: Justinmind, Wireframe. cc., Moqups., UXPin., Fluid UI., Balsamiq Mockups., Axure RP., Pidoco, UX Adobe.

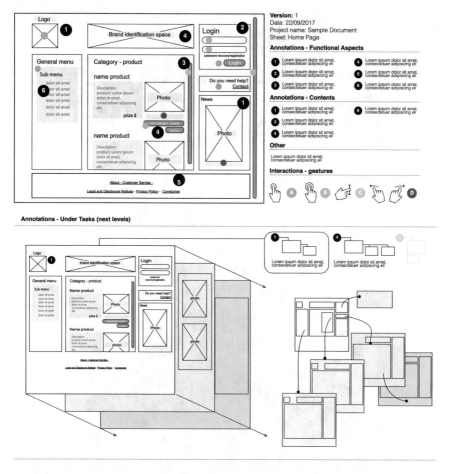

Fig. 16.2 Wireframe of a web site, navigation on a touch screen. For example, tablet—horizontal view. Wireframes are structural and functional elements of an interface. Navigation interactions can be touch-operated or can be carried out using external devices (like a mouse or a PAD). For this reason, it is good practice to develop all of the navigation and display methods simultaneously (different sizes and orientations)

The layout must typically be organised using a grid system, which helps the designer to organise the contents of the interface within a consistent screen. On an operational level, as seen in Fig. 16.3, we divide the screen into basic rational areas, which include the gutters (the white spaces that separate rows and columns).

It is important not to fill the screen with too many functions. This creates not only clutter but what is known as visual noise. The eye is encouraged to move from one element to another without taking in a clear and linear flow and it would be difficult to decode the hierarchy of information and efficiently locate the system's contents. Gestalt's psychology maintains that the closer objects are to each other, the more likely the human mind is to assume they are related (see Chap. 13).

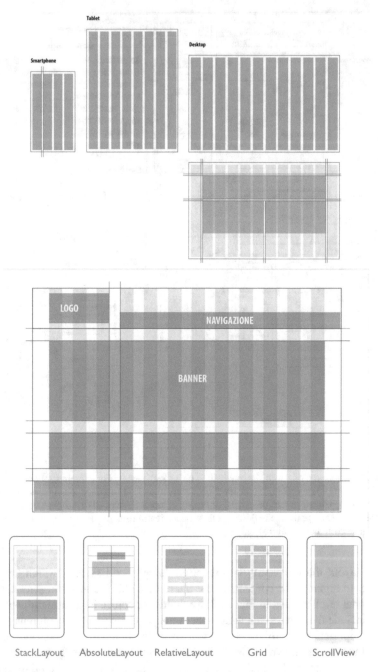

Fig. 16.3 Examples of grid layouts for smartphones, tablets and desktops. Above is an example of a grid system used for the main screen of many applications and web pages. The upper panel is used for the menus and/or icons; the lower area is used for general contents or the work area. Below are examples of grids for smartphone layout

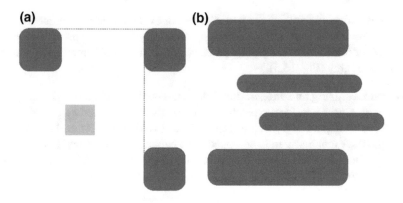

Fig. 16.4 Rules of visual flow. Reworked from Saffer (2010, p. 127)

The visual flow is a crucial aspect, as the organisation of the layout and the white spaces is useful for supplying the correct hints to the user about where to look; the same is true for colours and animations, which must be carefully chosen and not exaggerated with saturated colours and animations and redundant and chaotic pop-ups. In terms of visual flow, we will now try to understand what errors we must avoid by means of simple rules.

The first rule pertains to the alignment (Fig. 16.4): aligned objects appear in relation to each other (A). The same is true for objects positioned beneath other objects. The latter will appear to be less important than those positioned above them (B).

As we saw in Chap. 6, there are many software tools for measuring the cognitive load and the user experience, including technologies like Eye Tracking and Face Radar (see Sect. 3.2 regarding the User Experience).

These tools can support us in understanding the level of user attention and what they look at most closely. A less expensive and easy-to-use technique, which is particularly useful in the early stages of design development, is the "squint" test. In practice, squinting in front of the screen allows us to blur the details and perceive which elements of the interface are more important. This test is particularly useful when defining the interface layout.

The layouts can also be tested by manufacturing paper prototypes which, when positioned in the screen, allow us to cheaply and quickly evaluate how easy it is to understand the commands and the functions for the users.

Colour

The choice of colours is a fundamental aspect in Interface design. Colour represents one of the most important channels for cognitive decoding, both in terms of aesthetics and, in particular, the decoding of information. The colour and size of the texts is equally important, as is, moreover, the contrast of colours. These are all essential elements for facilitating reading. In terms of contrast, we must generally avoid the effect of chromostereopsis, which occurs when two colours placed side-by-side (or superimposed on each other) seem to vibrate. For example, blue text or a button on a red background. We must also note that approximately 15% of the population suffers from colour blindness; these people perceive green and red differently, for example, as shades of grey (Fig. 16.5).

Physical Gesturing and Interaction Interaction with an interactive device, particularly with a user interface, typically occurs by means of a hardware device or by directly using fingers on a touch screen (referred to as Gesture Based-User Interface in English). A touch screen is equipped with haptics, a technology that can detect touch, by implementing tactile feedback on a hardware surface. There are mainly due implementations of this technology: *resistive touch and capacitive touch.*

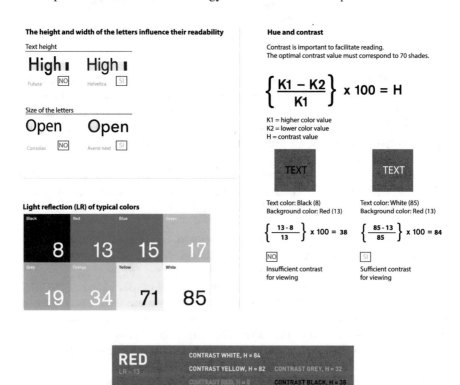

Fig. 16.5 Indications for text height, tone and contrast and light reflection. Reworked from Arthur and Passini 1992, p. 17)

Resistive touch uses pressure to locate the focal point of the touch. Capacitive touch uses electrical signals and is adapted to multi-touch. Interaction with the Gesture Based-User Interface (mobile devices like smartphones, tablets, etc.) is based on simple gestures using one, two or more fingers, which the user drags and creates symbols on the touch screen; the interface translates this to a command (Angelini et al. 2015). For example, a crossed sign created by a finger can be translated as an action to delete something in an application. Gesture based-user interfaces are very popular in apps and interactive game (Fig. 16.6).[3]

There are also multi-touch screens (typically applied to media larger than 17 in.). This technology can simultaneously recognise two or more contact points on the surface. It tracks multiple points, allowing the interface to recognise gestures. This permits advanced functions, such as skimming, pinch-zooming and more. Multi-touch uses capacitive touch screen technology for the input.

16.6 Assistive Technologies

In terms of accessibility for those with visual impairments, a touch screen device uses simple tactile interactions, audio, tactile feedback (vibration), accelerometer and other technologies based on sensors.

Due to the absence of a raised tactile surface, it is difficult to use the *Braille* system on a touch screen *interface*. To this day, the shared practices used to facilitate access and use for people with visual impairments use assistive technologies as primary resources, applying them to supports, hardware devices and software.

Assistive technologies (AT) refer to the entire collection of resources and services that help to provide or expand the functional capacity of people with disabilities and, as a result, promote independent living. Several expressions are used as synonyms for assistive technologies,[4] such as "technical aids", "technical support" and "adaptive technologies".

The rise of information technology and new media has created important new opportunities for the disabled and people who have perceptive or functional difficulties but has also created new problems in terms of accessibility.

Traditionally, it has always worked to adapt technologies designed for the general market (and therefore based on the features and skills of the conventional "average user") to be usable by people with disabilities, via the use of rehabilitative and assistive technologies. This is done via the software modifications, the use of special devices and the combined use of special program and interactive devices. Production policies like this, however, have many limitations in terms of their availability and

---

[3]For further information, consult Basic gestures for most touch commands, developed by Craig Villamor, Dan Willis, Luke Wroblewski (2010). Website: www.static.lukew.com.

[4]Assistive technologies are considered to be fundamental support tools for developing autonomy and encouraging social participation for the disabled, as they allow people to relate with the environment around them.

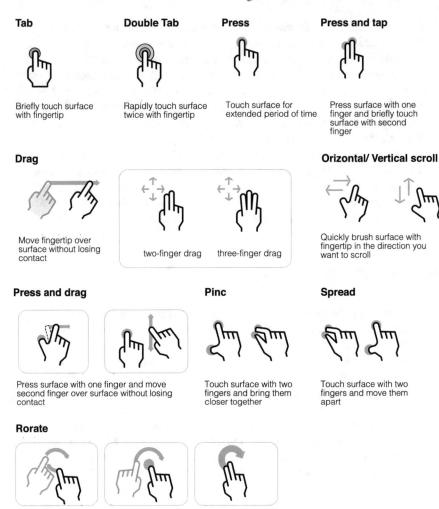

Fig. 16.6 Gesture based-user interface. Example of the most-used tactile commands on smartphones and tablets

cost: adaptations are normally available when a technology is older and require continuous updates, with significant costs and social impacts.

Like other inclusive design sectors, research and experiments have progressively focused on the need to overcome the concept of adaptation, that is, designing for the average user and then subsequently adapting the products to the needs of people with disabilities. Conversely, the goal must be designing with consideration for the abilities and preferences of all potential users (Universal Design and Inclusive Design; see Chap. 9).

Taking interactions with IT systems into consideration, it is not possible that a single interface can be used in each application (from controlling a household appliance to a complex multi-media interaction online) and by every user. Therefore, while it is necessary to attempt to unify the individual application environments, it is also necessary, at the same time, to create complex and scalable interfaces that allow us to customise the interactions, based on users' skills and preferences. From a conceptual perceptive, the potential avenues of operation are[5]:

- the simplification of interactions, based on the concept that the complexity of the systems must be increased to a level the user cannot see, bringing interactions back to known metaphors;
- the introduction, as part of the increased complexity, of a sufficient intelligence to make the systems adaptable, either automatically or semi-automatically, to the users' characteristics or the usage context, and capable of adjusting to the methods with which they are used.

The use of these concepts, which are not based on the analyses of average users' needs but on those of all potential users, make it possible to produce a technology that can be directly used by the majority of people. In support of these considerations, we find the ICF Classification (International Classification of Functioning, Disability and Health),[6] which evaluates disability based on the personal and social dimension as follows:

1. Functioning and Disability, the terms described above, which, in turn, comprise:

    (a) corporeal functions and structures;
    (b) activity and participation;

2. contextual factors, which include:

    (a) environmental factors;
    (b) personal factors.

These two dimensions cannot be understood separately. On the contrary, the specificity of the ICF lies in considering the individual health conditions and contextual factors as interdependent.

As we saw in Chap. 9 (see Sect. 9.4 and Fig. 9.2), the ICF classification allows us to describe the health conditions of the individual and to outline the user profile for people who will use the technology or aid, identifying the features using descriptors (Fig. 16.7).

The classification also allows us to define the individual's level of autonomy and independence in terms of a given technological aid and the implications in terms of accessibility. Each component is placed at a different level, which proceeds in a branched manner, each branch of which is associated with an abbreviation.

---

[5]Cit. Europa Minerva Platform (www.minervaeurope.org).

[6]An operational support for the definition of the dimension of the disability, which can be traced back to the ICF classification. To see the entire ICF classifications: www.openicf.it, www.icf-italia.org, www.who.int/classifications/icf/en/ (see Chap. 9).

TECHNOLOGIES AND THE ICF: SUMMARY TABLE	
Corporeal functions and structures	- User profile; - Analysis of the type of disability and the level impairment for the choice of the suitable technology; - Identification of the functions that can be improved.
Activity and participation;	- Factors that can be modified on a performance and skill level;
Environmental factors;	- Classification of technologies (ICT and TA) - Nature of the technology as a facilitator or barrier;
Personal factors;	- Impact of personal factors on the use of the technologies.

Fig. 16.7 Technologies and the ICF: summary table Reworked from Guglielman (2014)

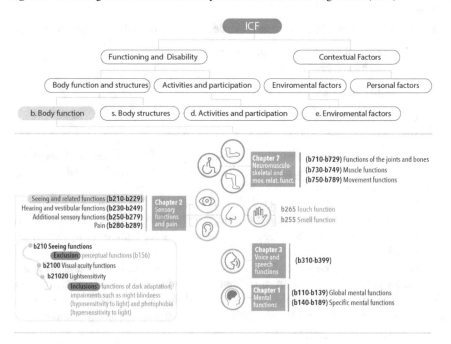

Fig. 16.8 General diagram of International Classifications World Health Organization and ICF core set for disability evaluation

With these classification levels, we can identify the aspects related to the pathology and their level of severity: in particular, the qualifiers use numeric codes to describe the extension and severity of the pathology and the disability stemming from it (e.g. b167.3: (b) refers to corporeal functions; (b1) mental structures; (b16) thinking; (b167) mental functions and language; (3) qualifier: serious problem (see Fig. 16.8).

On an operational level, the classification allows us to define the disability level in terms of the real and potential usage context, thus providing the cognitive elements we need to identify the range of users and to plan the intervention strategy to be adopted (Brischetto 2005). Based on the evaluation goals, this classification can be followed by the selection of the standards tools for the ergonomic sector and, in particular, those aimed at Human-Centred and Inclusive Design.

References

Angelini L, Lalanne D, Hoven E, Khaled O, Mugellini E (2015) Move, hold and touch: a framework for tangible gesture interactive systems. Machines 3(3):173–207

Arthur P, Passini R (1992) Wayfinding: people, signs, and architecture

Benyon D, Turner P, Turner S (2010) Design interactive systems. Pearson Custom Publishing

Brischetto A (2005) L'approccio universal design. In: Tosi F, Rinaldi A (eds) Il design per l'home care: l'approccio human-centred design nel progetto dei dispositivi medici. Didapress, Firenze, pp 104–122

Galitz WO (2007) The essential guide to user interface design: an introduction to GUI design principles and techniques. Wiley, New York

Garrett JJ (2011) The elements of user experience—centered design for the web and beyond. New Riders, Berkeley

Guglielman E (2014) E-learning accessibile: Progettare percorsi inclusivi con l'Universal Design. Learning Community, Roma

ISO 9241-210:2010 (2010) Ergonomics of human-system interaction—part 210: human-centred design for interactive systems

ISO/IEC 25010:2011 (2011) Systems and software engineering—systems and software quality requirements and evaluation (SQuaRE)—system and software quality models

Polillo R (2010) Facile da usare: una moderna introduzione all'ingegneria dell'usabilità. Apogeo, Milano

Saffer D (2010) Designing for interaction: creating innovative applications and devices. New Riders, Berkeley

Villamor C, Willis D, Wroblewski L (2010) Touch gesture reference guide

W3C (1999) Web content accessibility guidelines (WCAG) 1.0. www.w3.org/TR/WCAG10

W3C (2008) Web content accessibility guidelines (WCAG) 2.0. www.w3.org/TR/WCAG20

World Health Organization (2001) International classification of functioning, disability and health: ICF. World Health Organization. www.who.int/classifications/icf/en/

Appendix A
Ergonomics and Design—Design for Ergonomics: The Design Process

A.1 Introduction

As has already been mentioned in this book's introduction, the relationship between Ergonomics/Human-Centred Design (E/HCD) and Design is today seen in the complete integration of the theoretical and methodological contents of E/HCD into the culture and practice of design and, on an operational level, their application into the design process and the creation of products, environments, services and, more generally, systems (hereafter referred to as products[1] for the purpose of brevity), and in the ergonomic quality evaluation methods that are specifically aimed at the design sector.[2]

The methodological and operational approach used by the Laboratory for Ergonomics and Design in the Architecture Department of the University of Florence is presented below, while the diagrams showcases some case studies of interventions developed by the Laboratory in recent years.

The contents and methods from Ergonomics and Design are handled in a strictly applicative way in this appendix. Refer to the previous chapters for the theoretical assumptions and greater detail about each of the evaluation methods.

The conception, development and manufacturing process for the product is generally described as a sequence of varying phases which, starting from the client's request (public institution, business, association or private committee)—which is described in the so-called design brief—develops through a preliminary analysis phase using a reference framework, followed by the planning and development phase of the design and the production phase. The design and manufacturing phases are subsequently followed by placing the product on the market, purchasing it, using it and, finally, disposing of it (Fig. A.1).

---

[1] For the definition of the terms interaction, system and product used in this book, see Footnotes 2, 3 and 4 in Sect. 1.1 in Chap. 1. For the definition of the term product, in particular, refer to Sect. 1.1.

[2] For detail about this subject, in particular, the methods for evaluating the ergonomic quality of products, refer to Chaps. 5 and 6.

© Springer Nature Switzerland AG 2020
F. Tosi, *Design for Ergonomics*, Springer Series in Design and Innovation 2,
https://doi.org/10.1007/978-3-030-33562-5

Fig. A.1 Design process

Many professional figures come into play throughout this process, all of whom are essential for the development of the product and its optimal manufacturing. Figures like the designer, or, more simply, the design team, play a decisive role, but it is certainly not a unique or self-sufficient one. The designer is required to work with other skilled professionals, both within and outside the company, who are involved in the design process (managers for marketing, production coordination, production line programming, planning of sales and purchases, company development strategy, etc.) and to understand the results of studies and the information produced.

The designer/ergonomist will also work closely with other professionals in the field of ergonomics and on the basis of the results of their studies (for example, assessments of postural loads and repetitive movements, workplace stress in the field of workplace health and psychology; evaluations of the operators' attention spans, the structure of the dialogue interface of information and communications systems and, generally, the interaction between users and ICT technologies in cognitive ergonomic terms, etc.).

Ergonomics & Design Design for Ergonomics

where and how the design sets in: the levels of intervention

basic interventions

- adaptations of individual components/services
- redesign of physical systems (interior design, layout/internal paths, etc.) interventions to adapt/redesign the organizational system
- interventions to adapt/redesign the organizational system
- design of individual system elements (physical and/or organizational)
- global intervention on the single product
- intervention in the entire process of design, construction, use, maintenance, disposal
- global intervention on the product/service system
- on the product line
- on the company line
- ....

optimal interventions

Fig. A.2 Levels of intervention

Last, but certainly not least, the scope of the design action, and the role required by the designer or design team, naturally depends on the requirements and the availability of the company.

The required intervention may be limited to evaluating what already exists and proposing new ideas, or may refer to the comprehensive development of design proposals, from a single product to an entire range (Figs. A.2 and A.3).

The role of Design—and the designer—therefore is part of the process of designing and manufacturing products. It develops in parallel with the complex nature of all of the other factors involved. There are also many theories and interpretations about the role and definition of design capacity, which can certainly not be defined as simply "creative capacity" and yet is closely tied to it.

As Trabucco (2015, p. 11) says: "Today, 'design' indicates both the profession that bestows aesthetic value and originality to a physical or virtual artefact and the artefact itself. In fact, it is frequently said that a certain item is 'designer'. Both product and profession are characterised by expressive research, technological innovation, formal heterodoxy and contemporaneity." And, "the intentional nature of the design is the condition for the innovative work, while the satisfaction of a need poses a question about the necessity of the endeavour. It is a design activity, one which may deal with a physical or digital item, a strategy or a new business. Knowledge, technical skills, applications and, sometimes, ethical contents change, but combining all of these constitutes the design activity aimed at manufacturing an essentially aesthetic

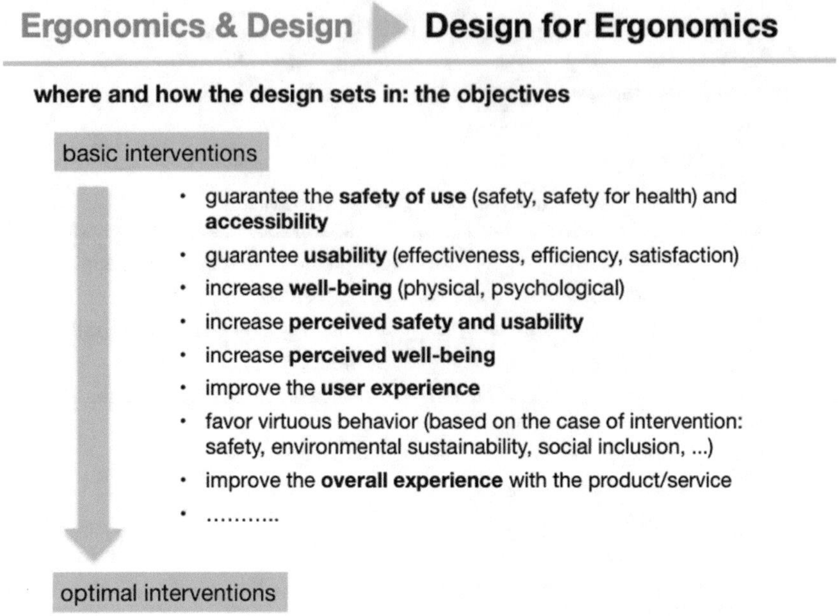

Ergonomics & Design ▶ Design for Ergonomics

where and how the design sets in: the objectives

basic interventions

- guarantee the **safety of use** (safety, safety for health) and **accessibility**
- guarantee **usability** (effectiveness, efficiency, satisfaction)
- increase **well-being** (physical, psychological)
- increase **perceived safety and usability**
- increase **perceived well-being**
- improve the **user experience**
- favor virtuous behavior (based on the case of intervention: safety, environmental sustainability, social inclusion, …)
- improve the **overall experience** with the product/service
- …………

optimal interventions

Fig. A.3 Intervention goals

item that did not exist before: something useful that will be produced, encountered, understood, judged and purchased based on its ability to respond to a given need, yet which will find its value in the evocative quality that it is able to express" (Trabucco 2015, p. 75).

Many methods have been developed to stimulate, guide, channel and develop the creative capacity, in order to generate new ideas and make it possible to manufacture them. As we have seen in previous chapters, the progressive approach between the Human-Centred Design methods and Design Thinking methods (see Chaps. 6, 7 and 8) moves in this direction, aimed at focusing the overall training and development process for products on the real needs of people, identifying conscious and as yet unexpressed needs and desires and adding the creative, inventive and innovative capacity of not only the designer, but of everyone involved in the development and use of the product, to the design process.

The innovative value of the design research process described below lies in the integration of the iterative process from HCD and the knowledge of "human factors" from Ergonomics and Human Factors into design and the training and development process of designs in this field.

From a strictly methodological point of view, the iterative process from HCD is applied through evaluation methods that are specifically aimed at the field of design, in particular, at evaluating and designing usability and the User Experience.

A.2   The Design Process

The preliminary phase of each design is awareness—that is, analysis and evaluation—of the reference framework within which the design shall be developed.

The ability to develop new ideas and translate them into innovative and producible intervention solutions is based on awareness and interpretation of all the factors in play and their varying levels of complexity: the client's requests, the required objectives, the productive materials and technologies that are available and/or that can be used, and, in terms of the product to be manufactured, the characteristics of similar products that are already on the market (so-called comparative analysis or bench-marking), the potential solutions that have already been developed or that are available on the market or in other sectors. Finally—and naturally—the availability of economic, temporal and human resources, etc.

As we have seen in the previous chapters, the HCD approach is based on a cyclical process, in which the methods for evaluating usability and global quality allow us to identify people's needs and to then verify the requirements during the initials phases of the project.

Once the reference framework has been defined and, naturally, based on the design brief, the planning and development phase starts from the awareness of the usage context—the basic reference point of the Human-Centred Design process. This is formed by analysing and evaluating the contextual variables, that is, all of the factors that contribute to defining the interaction between the individual and the product, that is, by definition: the users, the activities carried out and their goals, the physical, organisational and technological reference environment and, of course, the products being examined.

One must then respond to the basic questions: *what (must be designed)? who? for what activities and for what purposes? when? how? where?* (Fig. A.4)

At the same time, the analysis of tasks (Task Analysis—TA), a basic method in Ergonomics and this design process, is applied to similar products to those being designed during this initial phase. This allows us to highlight the critical factors, potential spaces for innovation and the intervention alternatives. The usage scenarios can also be used during this initial phase to detail how and where existing products are used by different users, in different contexts, etc.

Both the TA and the scenarios will then be redeveloped for the finished design, in order to demonstrate its validity and capacity for innovation (Fig. A.5).

At the same time, competing products (made either by the company or a competitor) on the market will also be examined.

The investigative phase is conduction during the initial phase for previously existing products, through both task analysis (Task Analysis—TA) developed for each user category, and direct observation and the administration of questionnaires and interviews. These methods allow us to highlight the critical factors, potential spaces for innovation and intervention alternatives.

The usage scenarios can also be developed during this initial phase. They are constructed for the type of product being designed and aim to highlight the way in

Define the context variables - Operational scheme

WHAT IS THE PRODUCT

- primary and secondary functions
- primary and secondary goals
- types of users to whom it is addressed
- types of employment (domestic, professional, etc.)

WHO IS THE USER

- age
- sex
- nationality
- profession
- budget

- characteristics and abilities (physical, sensory, cognitive)
- generic or professional user
- competence of use
- level of experience
- ....

WHAT THE PRODUCT IS USED FOR

- types of activities carried out/planned
- level of psycho-physical commitment required
- ....

WHEN THE PRODUCT IS USED

- in which period of the year, day, etc.
- with what duration (for how long)
- how often (occasional, continuous temporary, etc.)
- ....

WHERE THE PRODUCT IS USED

- the environment of user/product interaction (physical, social, organizational, technological)

HOW THE PRODUCT IS USED

- methods of use by users
- possible habits of use

Fig. A.4 Evaluating and interpreting the usage context variables

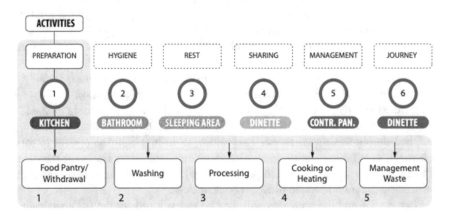

Fig. A.5 Example of Task Analysis (TA) applied to the preparation activity, as it pertains to the kitchen

which people use that product (for example, during the course of a day, at special times, etc.) and the possible new methods that can be proposed. The scenarios can also be very useful for illustrating the proposals and the innovative capacity of initial concepts.

Evaluation methods based on the users' involvement, in particular, samples of users that are indicative of the design's target market, are of particular interest during this phase.

The goal is to reveal conscious or unexpressed needs and to collect information about them in such a way that they can be compared and verified for the case at hand and for subsequent interventions. The most suitable methods for this include direct observation and Thinking Aloud, which consists in using the product and describing aloud what its strengths and weaknesses are, difficulties in using it and aspects that could be improved. The most significant aspect, other than the wealth of information that can be gathered using this method, is the discrepancy between the user's behaviour and what they describe "aloud" (the user says: "this step is really easy", while they perform it with great difficulty or in an incorrect manner).

In terms of evaluating and interpreting people's needs, and the subsequent identification of the design requirements, a scale of priorities can be indicated, ranging from the basic requirements of safety, functionality and usability to the emotional and aesthetic dimension of the interaction between people and the products/systems and the User Experience, which includes "all of the emotions, beliefs, preferences, user perceptions, physical and psychological responses, behaviours and realisations that occur before, during and after use".[3] Though the list can obviously be integrated and modified for each intervention case, examples include:

- The *physical usability* of parts and components, which involves the anthropometric-dimensional study of the so-called physical inter-face (buttons, levers, manoeuvrable parts, screens, etc.) and its accessibility, and the verification of the movements and efforts required.
- The *manageability* of devices, furnishings, accessories and components and command and adjustment elements. For example, these aspects pertain to the phases for assembly and disassembly ad opening, closing and adjusting parts and components.
- *The visibility and legibility of information*, which involves strictly perceptive elements (ability to distinguish between different products, able to see and recognise parts and components, size of the characters, use of colours, etc.) and cognitive elements linked to the correct interpretation of information.
- *Usability of adjustment, programming and command interfaces*, that is, the effectiveness and comprehensibility of the information provided by the dialogue interfaces, in particular, digital interfaces (appropriateness of the information, correct hierarchy of selection and decision-making procedures, comprehensibility of dialogue/selection/decision-making options, presence of signals and warnings, etc.).
- *Aesthetic and emotional value*, and all of the aspects related to the User Experience, and, therefore, evaluations of appreciation, repulsion, indifference, etc. These can be attributed to formal solutions, choices relating to materials (surface treatments,

---

[3] See EN ISO 9241/210:2010 standard, Ergonomics of human-system interaction. Human-Centred Design for interactive systems—terms and definitions. See Chap. 5.

visual appeal, etc.) and, in general, responsiveness to personal tastes and/or the widespread taste of a social group.

The real development of the design will start from the definition of the initial concepts, and their subsequent evaluation and selection on the basis of their responsiveness to design requirements (and the company's requests). The development of the final design starts from the selection or synthesis of the initial concepts, and the renderings and physical models that allow for the design (and its compliance with requirements) to be verified.

The goal is to define the design, and to put it into production, only when all of the aspects have been verified, so that they do not have to repeat steps and incur the economic and temporal costs that this entails.

Intervention Cases

Claudia Becchimanzi and Ester Iacono

This section summarizes 5 research projects carried out by the Ergonomics and Design Laboratory of the Department of Architecture of the University of Florence.

F. Tosi, *Design for Ergonomics*, Springer Series in Design and Innovation 2,
https://doi.org/10.1007/978-3-030-33562-5

KITCHEN ECOLOGY AND ERGONOMICS | DESIGN RESEARCH
Technological innovation and using kitchen environments and accessories
SUMMARY SHEET

GENERAL DATA ABOUT DESIGN	**Partners:** • Effeti Industrie | Italy **Research institution involved (University of Florence)** • TAED (Department of Architectural Technology and Design) **Financing:** Tuscany (PORCReO 2011–12). **Research programme:**Kitchen ecology and ergonomics: technological innovation and using the kitchen environment and accessories to develop the company competitively. **Scientific director:** Francesca Tosi **Development & coordination director:** Alessandra Rinaldi **Tutor:** Simone Lucii with Alessia Brischetto, Linda Carlocchia, Irene Bruni, Antonio Carretta, Valentina Frosini, Meri Seto, Daniele Busciantella Ricci and Giovanni Tallini
AIMS OF THE RESEARCH PRO-GRAMME	The objective of this design research was to develop solutions that were radically inno-vative from a technological and user point of view, aimed at reducing the environmental impact of the kitchen system. Starting from the redefinition of usability and technologies linked to kitchen furnishings and accessories, the specific goal was to define new usage scenarios for the domestic setting, which refocus the production on not only the aesthe-tic-functional innovation and economic development, but, more importantly, respect for mankind and the environment.

APPLIED METHODOLOGY	The research work was developed through a variety of activities, which, starting with analyses and evaluations of critical issues, contributed to the definition of the final con-cept. During the investigation phase, the needs and expectations of potential users were identified, by applying methods from Human-Centred Design:
	    • Task Analysis;
	    • Direct observation;
	    • Interviews/questionnaires
	    • Thinking Aloud.
	The evaluations gave particular consideration to the following aspects:
	    •dimensional and functional;
	    •cognitive, pertaining to management of automation;
	    •perceptive, pertaining to the interpretation of living space;
	    •skill- or limitation-related, or pertaining to the habits and lifestyles of
	    users.
	The data collected in this way has been processed and summarised in a design proposal that aligns with the work of all the other research groups involved.

Contribution of the Ergonomics Laboratory to Design

The analysis of users' needs and expectations through the Human-Centred Design approach allowed us to identify the areas for typological and technological innovation in the kitchen system and its innovative components, performance and services. Introducing these elements in a mature and consolidated setting, such as the domestic kitchen, allows us to interpret and encourage innovation of cooking- and eating-related lifestyles and habits, in particular as they pertain to the containment of consumption, the correct use and consumption of energy sources and waste reduction.

The guidelines identified within the design research have been tested and verified via the "Well Living in the kitchen: technological innovation and use of the kitchen environment" workshop. This involved some young designers experimenting with an innovative design formula known as Creativity @Fabbrica. In fact, the workshop took place within the Effeti company, where young creatives, led by researchers and senior professors from the University, collaborated directly with production experts and sector technicians. The result was a series of projects based on the Human-Centred Design approach, which led to an expansion of the company's product range, not only in terms of the combinations of fittings, accessories and colours, but, more importantly,l in terms of housing and user types. The workshop produced designs in four areas that were identified as highly innovative for the sector: the smart table, the wall-furniture system, the wearable accessory and the smart floor.

RESULTS ACHIEVED

For more information:

www.dida.unifi.it/vp-353-laboratorio-ergonomia-design.html

BIKE INTERMODAL | DESIGN RESEARCH – *WP ERGONOMICS*
Multi-modal integration of cycling mobility through product and process innovations in the design of the bicycle.SUMMARY SHEET

GENERAL DATA ABOUT DESIGN	**Leading player:** Trilix, Italy **Partners:** • UNIFI | Department of Architecture, Italy • Trilix | Italy • Ticona | Germany • Maxon Motors | Switzerland • ATAF | Italy • LPP Ljubjana | Slovenia • Urban Technologies | Italy **Research institution involved (University of Florence)** • TAED (Department of Architectural Technology and Design) • Department of public health. **Financing:**Project is funded by the EU– FP7 – SST "Bike Intermodal" (2010–2014) **Project director:** Alessandro Belli, Tecnologie Urbane | Italy **Scientific director for WP Ergonomics:** Francesca Tosi **Working group** Alessandra Rinaldi (scientific coordinator) Alessia Brischetto (collaborator) With: Grazia Tucci and Valentina Bonora - University of Florence – Dip.DICEA, Geomatic Laboratory for GECO Conservation Vincenzo Cupelli, Giulio Arcangeli and Marco Petranelli – University of Florence (Department of Public Health)
AIMS OF THE RESEARCH PROGRAMME	The design research, which was funded by the European Union as part of the 7th Framework Programme, aimed to design a super-compact and lightweight bicycle with electrically-assisted pedals. This would be an efficient transportation method for urban mobility. The work started from the idea that the synergy between the bicycle as a means of urban transport and public (train, tram, subway, bus) and/or private (taxi, car) transport systems can increase the increase bicycles' share of total urban and suburban travel. This can be a push factor in encourage users not to use private cars in urban centres.
APPLIED METHODOLOGY	The research work was developed through two different surveying and analysis activities on a sample of select users (expert city bikers), in order to define the **requirements, both** functional and dimensional, of the design. The former allowed us to survey the anthropometric data of the users, according to five different height groupings, and their bicycles (saddle/handlebar/pedal relationship), by applying the following methods: • Task Analysis; • Direct observation; • Interviews/questionnaires; • Thinking Aloud; • 3D photogrammetry. The second activity was the biomechanical analysis of postures and movements while pedalling and was developed by the Occupational Medicine research group (UNIFI).

The developed design led to the design of an ultra-light vehicle, with a frame that uses tensioning cables, inspired by the idea of tensile structures. The choice of frame materials aimed for lightness, favouring magnesium, a light alloy that is already widely used in for motorbikes, over more widely used metallic materials, such as aluminium or steel. The general architecture of the bicycle, and the range of postural configurations indicated by ergonomic studies, have been considered, guaranteeing maximum pedal efficiency, protection from stressful effects and problems or damage to the spinal column, neck and arms, even on short rides, the effectiveness and ease of controlling the vehicle, and the subsequent increase in feelings of well-being. The posture adopted is an upright one, between 5° and 15° in terms of torso inclination, which is the most comfortable for a relaxed ride and for short journeys in urban settings.

RESULTS ACHIEVED

Contribution of the Ergonomics & Design Lab (LED)

WP Ergonomics brought the acceptable range of potential postural positions to the design, ensuring optimal pedal efficiency, protection against stress and damage to the spinal column, neck and arms, and the efficiency and ease of control of the vehicle. Moreover, the usability, which refers to the comfort and pleasure, both real and perceived, of the vehicle when pedalling and riding has also been evaluated and optimised, also in relation to the product's usage context.

For more information:

www.dida.unifi.it/vp-353-laboratorio-ergonomia-design.html

www.bike-intermodal.eu

WORK STATIONS – BRUNELLO CUCINELLI | RESEARCH PROJECT
Ergonomic evaluation and first phase in the design of the work stations for the Brunello Cucinelli facility in Perugia.

SUMMARY SHEET

GENERAL DATA ABOUT DESIGN	**Partners:** • Perugia Check up s.r.l.	Italy • University of Florence	DIDA Department of Architecture, Italy **Research institution involved (University of Florence)** • DIDA Department of Architecture, University of Florence **Financing:** Agreement between the DIDA Department of Architecture and Perugia Check up s.r.l. Period: 2014–2015 **Scientific director:** Francesca Tosi **Scientific coordinator:** Alessandra Rinaldi with: Alessia Brischetto Daniele Busciantella Ricci Mattia Pistolesi

AIMS OF THE RESEARCH PROGRAMME

The general objective of the design project aims to improve the quality of the working environment and optimising/redesigning the operators' work stations in the company's prototype laboratory. The research was based on the scientific and methodological approach of Ergonomics and Design and its theoretical and operational tools. The operational aim was the development of intervention solutions to improve the ergonomic quality of three work stations that are currently in use in the company. The interventions were based on the assessment of the operators's needs and expectations, as they pertain to the work activities carried out. The analysed positions were: flat machine and mending.

APPLIED METHODOLOGY

The research work was developed through a variety of activities, which, starting with analyses and evaluations of critical issues, contributed to the definition of the final concept. During the investigation phase, which was developed at the LED Ergonomics and Design Laboratory, the needs and expectations of potential users were identified, by applying methods from Human-Centred Design:

- Task Analysis;
- Direct observation;
- Interviews;
- Thinking Aloud.

The evaluations gave particular consideration to the following aspects:

- dimensional and functional;
- cognitive, pertaining to management of automation;
- perceptive, pertaining to the interpretation of work space;
- pertaining to the abilities and limitations of operators and specific skills.

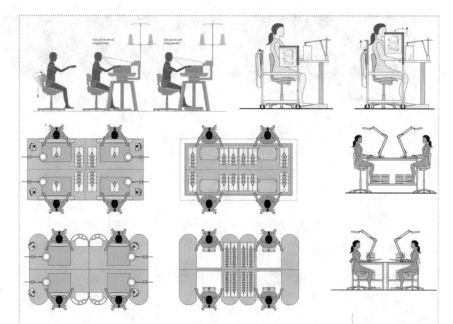

Each evaluation phase has been carried out in conjunction with the working group from the University of Bologna, Department of Medical and Surgical Sciences (F. Violante, S. Mattioli, R. Bonfiglioli).

RESULTS ACHIEVED

The design followed the following phases:

Phase 1: Analysis and evaluation using Ergonomics for Design methods (for the 3 identified work stations)

A) Study of the work stations used in the prototype lab and the posture of the relevant operators (stitching machine, mending machine and flat sewing machine), using the chief methods of Ergonomics and Design.

B) Analysis/evaluation of collected data.

C) Conclusions and drafting of summary documentation.

Phase 2: Adaptation of the analysed work stations (for each work station analysed: flat mending machine)

D) Identification of potential intervention solutions for partial adaptation of each work station analysed, through minimal interventions.

E) Choice of the feasible solution (with the company's approval) for each work station. Development of operational indications via designs and graphic diagrams.

For more information:

www.dida.unifi.it/vp-353-laboratorio-ergonomia-design.html

TRIACA | DESIGN SEARCH
Technological solutions for Reducing the Environmental Impact of the camper during the usage phase
SUMMARY SHEET

GENERAL DATA ABOUT DESIGN	**Leading player:** Trigano Spa

Partners:

- Espansi Tecnici Srl
- Dielectrick Srl

Research institution involved (University of Florence)

- DIDA Department of Architecture, University of Florence
- Consorzio Polo Tecnologico Magona
- Department of Social, Political & Cognitive Sciences – DISPOC (University of Siena)
- CUBIT – Consortium Ubiquitos Technologies

Financing: Tuscany (POR CReO 2007–13).

"Design" working group – DIDA Department, University of Florence
Scientific director: Francesca Tosi

Giuseppe Lotti (scientific director for sustainable design)
Vincenzo Legnante
Alessia Brischetto
Irene Bruni
Daniela Ciampoli
Stefano Follesa
Marco Mancini
Marco Marseglia

AIMS OF THE RESEARCH PROGRAMME

The research was aimed at creating an innovative camper, in terms of its environmental sustainability, energy efficiency and ergonomic characteristics. These objectives were developed through the evaluation of both the production and usage phase and ended up in the proposal of design solutions that were based on the use of sustainable materials, technologies for containing consumption, distributive, functional and formal innovation of the space and internal furnishings, all of which aim to improve the user experience of the overall vehicle.

APPLIED METHODOLOGY

The research work was developed through a variety of activities, which, starting with analyses and evaluations of critical issues, contributed to the definition of the final concept. During the investigation phase, the needs and expectations of potential users were identified, by applying methods from Human-Centred Design:

- Task Analysis;
- Direct observation;
- Interviews/questionnaires;
- Thinking Aloud.

The evaluations gave particular consideration to the following aspects:

- dimensional and functional;
- cognitive, pertaining to management of automation;
- perceptive, pertaining to the interpretation of living space;
- pertaining to the abilities and limitations of potential users, as well as users' habits and lifestyles.

The data collected in this way has been processed and summarised in a design proposal that aligns with the work of all the other research groups involved. The developed solutions have been verified using virtual simulations.

"Concept 230" is an innovative caravan, which focuses particularly on the interiors and energy saving, along with the advantages in terms of reduced environmental impact through the product's life cycle. The final proposals were developed for interventions that were deemed to be "possible areas for innovation". In particular:

Living area: an open space with optimal "functional flexibility".

The study of the space has led to a restructuring of the vehicles interiors and their related equipment, which aims to improve living conditions and the ability of the interior to be transformed for the well-being and comfort of the users. The camper's internal volume has been designed for exceptional usability and to be accessible and adapted for optimal use.

Ergonomics and optimal usability in small environments: toilets.

Toilet spaces have been evaluated from the perspective of physical size and usability, in order to ensure optimal freedom of movement and the functionality and ease of use of the internal components. Particular attention was focused on taps (with air mixing), which allows us to separately control the overhead shower, basin and free-standing shower head.

Design serves efficiency, to create a trendy and highly functional kitchen.

The kitchen section has been designed with a special focus on the innovation of the materials for the entrances (in honeycombed aluminium, coated with MDF) and the kitchen surfaces (made from PRAL, an anti-bacterial material that is suitable for food usage). The components have been designed to ensure optimal functionality and versatility (removable electrical induction hob, sinks that has a cover that can be used as a drain or chopping board).

Technological innovation of the structure: floor, walls and roof.

In terms of the camper's structure, a resin with basalt fibre has been used; this material is fully recyclable and safe to touch and use during the processing phase, which ensures a lighter weight and similar mechanical characteristics and performance to fibreglass. The inside floor is covered in Ecomalta.

Technical innovation and roof transformability.

The roof above the bed can be lifted, and allows the dinette to be extended during the day and the bed at night. The lifting roof is made of a dual fabric, ensuring thermal insulation and waterproofing. Furthermore, the innovative touchscreen control unit consumption to be monitored around the clock: this optimises available resources and adapts/updates remaining autonomy, thanks to configurable settings based on the number of users or the duration of use. The interface allows the interior lighting to be regulated by selecting established scenarios (relaxation, lounge, sleep and other settings) and managing resources (fuel, electricity). A similar system was developed for water management, designed as a logbook which, based on known information (number of passengers, days spent on the camper), returns a forecast on the amount of daily water available, in relation to the duration of the journey.

For more information:

www.dida.unifi.it/vp-353-laboratorio-ergonomia-design.html

RESULTS
ACHIEVED

SMART RUNNING (2nd EDITION) | WORKSHOP
Definition of scenarios and concepts for products and systems for outdoor physical activities.

GENERAL DATA ABOUT DESIGN	**Partners:** • Technogym Study and Research Centre | Italy **Research institution involved (University of Florence)** • DIDA Department of Architecture, University of Florence **Financing:** Workshop in conjunction with the Technogym Study and Research Centre (Giuseppe Fedele and Francesco Calcaterra, engineers), 2016. **Scientific director:** Francesca Tosi **Coordinator:** Alessandra Rinaldi **Tutors:** Alessia Brischetto, Daniele Busciantella Ricci, Mattia Pistolesi **Participants:** Rosario Lo Turco, Angelo Iannotta, Wang Haoyu, Stefano Gabbatore, Cristina De Alfieri, Wu Fang Yuan, Niccolò Mazzoni, Alessio Tanzini, Matilde Muscatello, Le Diem Huong, Vu Tuong Quyen, Gianluigi Cantanamessa, Mariana Motta, Ester Nerini, Carolina Oliveira, Pu Ling Xing, Valentina Zamorano, Eleonora Didio, Rachele Nencioni, Wang Xiang, Ji Lu, Liu Xu, Martino Nicola, Monica Moggia, Soheil Ehsani, Maria Borrego, Armando Moreno
AIMS OF THE RESEARCH PROGRAMME	The "Smart Running" workshop, which was organised in collaboration with the Technogym company, was aimed at defining the scenarios and concepts for the products and systems for outdoor physical activities. This involved 25 students from the Master's programme in Design, who worked intensively with a group of Technogym designers and engineers for four days across two weeks. The design proposals were then evaluated and awarded by the company.
APPLIED METHODOLOGY	The workshop was developed through a variety of activities, which, starting with analyses and evaluations of critical issues, contributed to the definition of the final concept. During the investigation phase, which was developed at the LED Ergonomics and Design Laboratory, the needs and expectations of potential users were identified, by applying methods from Human-Centred Design: • Task Analysis; • Direct observation; • Interviews/questionnaires • Thinking Aloud. The evaluations gave particular consideration to the following aspects: •dimensional and functional; •cognitive, pertaining to management of automation; •perceptive, pertaining to the interpretation of open space for sports; •pertaining to the abilities and limitations, habits and lifestyles of potential users.

The Workshop is based on a shared research and innovation objective, which aims to define new design solutions for product lines and/or systems. The results led to concepts that were developed by selected groups of students, under the guidance of instructors.

The Workshop was structured across 2 weeks of full-time design work and developed starting from basic design research (market orientations, existing production, technological opportunities and innovative materials, processing systems, etc.) carried out by teachers and researchers who oversaw the Workshop and conducted in close collaboration with the company.

The designs consist of a series of products designed for outdoor activities, some of which require the use of machinery and equipment, while others allow us to perform physical activities by simply using our body weight.

The chief characteristics are simplicity and easy of use, with ground signs being added to certain designs to more accurately explain their use.

The proposed forms, which are adjustable and highly sculpted, allow us to combine the utility of an outdoor gym with the appeal of sleek and evocative shapes. In certain cases, the designs include the integration of technologies, such as bluetooth and/or solar panels, that let us charge our electrical devices.

RESULTS ACHIEVED

Suggested Readings

Bandini BL (2008) Ergonomia olistica, il progetto per la variabilità umana. FrancoAngeli, Milano

Branzi A (1999) Introduzione al design italiano, una modernità incompleta. Baldini & Castoldi, Milano

Branzi A (2008) Il design Italiano 1964–1990. Electa, Milano

Bürdek BE (2008) Design: Storia, teoria e pratica del design del prodotto. Gangemi, Roma

Chengalur SN et al (2003) Kodak's ergonomic design for people at work. Wiley, New York (1st ed.: Rodgers SH (1983) Ergonomic design for people at work. Wiley, New York)

Di Bucchianico G, Kercher P (2016) Advances in design for inclusion. Springer, Svizzera

Grandjean E (1986) Il lavoro a misura d'uomo: trattato di ergonomia, Edizioni Comunità, Milano (1st ed.: Grandjean E (1979) Physiologische Arbeits-gestaltung: Leitfaden der Ergonomie, Ott, Thun)

Green SG, Jordan PW (2002) Pleasure with products, beyond the usability. CRC Press, London and New York

Hall ET (1996) La dimensione nascosta. Bompiani, Milano (1st ed.: Hall ET (1966) The hidden dimension, Doubleday, Garden City)

Jordan PW (1998) An introduction to usability. CRC Press, London

Jordan PW et al (1996) Usability evaluation in industry. Taylor & Francis, London and New York

Karwowski W (2006) International encyclopedia of ergonomics and human factors. CRC Press, London and Philadelphia (1st ed.: Karwowski W (2001) International encyclopedia of ergonomics and human factors. Taylor & Francis, London and New York)

Kroemer KHE (2017) Fitting the human: introduction to ergonomics/human factors engineering. CRC Press, Boca Raton (1st ed.: Grandjean E (1963) Fitting the task to the man. Taylor & Francis, London)

Lucibello S, La Rocca F (2015) Innovazione e utopia nel design italiano. Rdesign-press, Roma

Maldonado T (1989) Disegno industriale un riesame. Feltrinelli, Milano

Mantovani G (2000) Ergonomia, lavoro, sicurezza e nuove tecnologie. Il Mulino, Bologna

Manzini E (1990) Artefatti. Domus Academy, Milano

Manzini E (2015) Design, when everybody designs: an introduction to design for social innovation. MIT Press, Cambridge

Norman DA (1998) The invisible computer. MIT Press, Cambridge

Norman DA (2004) Emotional design. Basic Books, Cambridge, 2004

Norman DA (2010) Living with complexity. MIT Press, Cambridge

Norman DA (2013) The design of everyday things. Basic Books, New York

Panero J, Zelnik M (1983) Spazi a misura d'uomo. Be-Ma, Milano (1st ed.: Panero J, Zelnik M (1979) Human dimension and interior-space. Whitney Library of Design, New York)

© Springer Nature Switzerland AG 2020   369

F. Tosi, *Design for Ergonomics*, Springer Series in Design and Innovation 2,
https://doi.org/10.1007/978-3-030-33562-5

Pheasant S, Haslegrave CM (2006) Bodyspace: anthropometry, ergonomics and the design of work. CRC Press, Boca Raton (1st ed.: Pheasant S (1986) Body-space: anthropometry, ergonomics and design. Taylor & Francis, London and Philadelphia)

Preece J et al (2004) Interaction design. Apogeo, Milano (1st ed.: Preece J et al (2002) Interaction design, beyond human-computer interaction. Wiley, New York)

Rubin J, Chisnell D (2011) Handbook of usability testing: how to plan, design, and conduct effective tests. Wiley, Indianapolis (1st ed.: Rubin J (1994) Handbook of usability testing: how to plan, design, and conduct effective tests. Wiley, New York)

Shorrock S, Williams C (2017) Human factors and ergonomics in practice: improving system performance and human well-being in the real world. CRC press, Boca Raton

Spadolini MB (2013) Design for better life, longevità: scenari e strategie. FrancoAngeli, Milano

Stanton NA et al (2014) Guide to methodology in ergonomics: designing for human use. CRC Press, Boca Raton (1st ed.: Stanton NA, Young MS (1999) A guide to methodology in ergonomics: designing for human use. Taylor & Francis, London and New York)

Steffan IT (ed) (2014) Design for All —The project for everyone. Methods, tools, applications. Maggioli, Rimini

Tosi F (2005) Ergonomia progetto prodotto. FrancoAngeli, Milano

Tosi F (2006) Ergonomia e progetto. FrancoAngeli, Milano

Tosi F (2016) La professione dell'Ergonomo nella progettazione dello ambiente, dei prodotti e dell'organizzazione. FrancoAngeli, Milano

Tosi F, Rinaldi A (2015) Il Design per l'Home Care. L'approccio Human-Centred Design nel progetto dei dispositivi medici. Didapress, Firenze

Trabucco F (2015) Design. Bollati Boringhieri, Torino

Verganti R (2009) Design driven innovation. Etas, Milano

Wilson JR (1995) A framework and a contest for ergonomics methodology. In: Wilson JR, Corlett EN (eds) Evaluation of human work, Taylor & Francis, London

Wilson JR, Corlett EN (1995) Evaluation of human work, Taylor & Francis, Londra (1st ed.: Wilson JR, Corlett EN (1990) Evaluation of human work. Taylor & Francis, London and New York)

Wilson JR, Sharples S (2015) Evaluation of human work. CRC Press, Boca Raton (1st ed.: Wilson JR, Corlett EN (1990) Evaluation of human work. Taylor & Francis, London and New York)

Printed in the United States
By Bookmasters